3rd Padjadjaran International Physics Symposium (PIPS 2017)

Journal of Physics: Conference Series
Volume 1080

Bandung, Indonesia
14 - 15 November 2017

Editors:

Risdiana
Yukio Furukawa
Shukri Sulaiman
Andrivo Rusydi
Husin Alatas

ISBN: 978-1-5108-7202-8
ISSN: 1742-6588

Printed from e-media with permission by:

Curran Associates, Inc.
57 Morehouse Lane
Red Hook, NY 12571

proceedings.com

Some format issues inherent in the e-media version may also appear in this print version.

Printed by Curran Associates, Inc. (2018)

For permission requests, please contact the Institute of Physics at the address below.

Institute of Physics
Dirac House, Temple Back
Bristol BS1 6BE UK

Phone: 44 1 17 929 7481
Fax: 44 1 17 920 0979

techtracking@iop.org

Additional copies of this publication are available from:

Curran Associates, Inc.
57 Morehouse Lane
Red Hook, NY 12571 USA
Phone: 845-758-0400
Fax: 845-758-2633
Email: curran@proceedings.com
Web: www.proceedings.com

Table of contents

Volume 1080

3rd Padjadjaran International Physics Symposium

14–15 November 2017, Bandung, Indonesia

Accepted papers received: 2 August 2018
Published online: 17 September 2018

Preface

Papers

IOP Conf. Series: Journal of Physics: Conf. Series **1080** (2018) 011001 doi:10.1088/1742-6596/1080/1/011001

Preface

The present volume of *Journal of Physics: Conference Series* contains selected papers of the 3rd Padjadjaran International Physics Symposium (PIPS 2017) held on November 14-15, 2017 in Bandung, Indonesia. PIPS 2017 is organized by the Department of Physics, Faculty of Mathematics and Natural Sciences, Universitas Padjadjaran. PIPS 2017 aimed to improve communication among researchers in the field of physics and related fields for their contribution to the role of physics in sustainable development.

About 90 researchers from six countries have participated and shared experiences on the development and progress of physics and its applications. After the peer-reviewed process, 51 papers covering a wide range of theoretical physics, material sciences, computational, instrumentation, nuclear physics, earth sciences and energy are accepted for publication in this volume.

Editors,
Risdiana (Universitas Padjadjaran, Indonesia)
Yukio Furukawa (Waseda University, Japan)
Shukri Sulaiman (Universiti Sains Malaysia, Malaysia)
Andrivo Rusydi (National University of Singapore, Singapore)
Husin Alatas (Institut Pertanian Bogor, Indonesia)

IOP Conf. Series: Journal of Physics: Conf. Series **1080** (2018) 011001 doi:10.1088/1742-6596/1080/1/011001

Committee

Advisory Board

Okihiro Sugihara	(Utsumoniya University, Japan)
Shukri Sulaiman	(Universiti Sains Malaysia, Malaysia)
Yukio Furukawa	(Waseda University, Japan)
Poki Chen	(National Taiwan University of Science and Technology, Taiwan)
I Made Joni	(Universitas Padjadjaran, Indonesia)
Muhamad Mat Salleh	(Universiti Kebangsaan Malaysia, Malaysia)
Rustam Siregar	(Universitas Padjadjaran, Indonesia)
Jae-Sung Kim	(Sook-Myung Women's University, South Korea)
Husin Alatas	(Institut Pertanian Bogor, Indonesia)
Camellia Panatarani	(Universitas Padjadjaran, Indonesia)
Andrivo Rusydi	(National University of Singapore, Singapore)
Risdiana	(Universitas Padjadjaran, Indonesia)

Organizing Committee

Lusi Safriani (Chairman)
Ayi Bahtiar (Co-Chairman)
Sri Suryaningsih (Secretary)

Annisa Aprilia
Togar Saragi
Sahrul Hidayat
Setianto
Norman Syakir

IOP Conf. Series: Journal of Physics: Conf. Series **1080** (2018) 011001 doi:10.1088/1742-6596/1080/1/011001

IOP Conf. Series: Journal of Physics: Conf. Series **1080** (2018) 011002 doi:10.1088/1742-6596/1080/1/011002

Peer review statement

All papers published in this volume of *Journal of Physics: Conference Series* have been peer reviewed through processes administered by the proceedings Editors. Reviews were conducted by expert referees to the professional and scientific standards expected of a proceedings journal published by IOP Publishing.

IOP Conf. Series: Journal of Physics: Conf. Series **1080** (2018) 012001 doi:10.1088/1742-6596/1080/1/012001

Improved charge carrier transfer in squaraine-capped ZnO nanoparticles layer for electron transport of hybrid solar cells

A Bahtiar[*], **S H Tusaddiah and L Safriani**

Department of Physics, Faculty of Mathematics and Natural Sciences, Universitas Padjadjaran, Jl. Raya Bandung-Sumedang km 21 Jatinangor, Sumedang, West Java, Indonesia 45363

[*]ayi.bahtiar@phys.unpad.ac.id

Abstract. Blend of conjugated polymer poly(3-hexylthiophene) or P3HT and Zinc Oxide nanoparticles (ZnO-NPs) has been intensively used as active material for high performance hybrid solar cells. However, agglomeration of ZnO-NPs hinders efficient charge carrier or charge transfer both from P3HT to ZnO-NPs and its transport within ZnO-NPs which lead to low performance of solar cells. Capping of ZnO-NPs is currently applied to avoid this agglomeration effect. In this study, we used small molecule squaraine (SQ) to cap ZnO-NPs. SQ was chosen due to its lowest unoccupied molecular orbital (LUMO) level lower than conduction band of ZnO, therefore, SQ can assist charge transfer to ZnO-NPs. Moreover, SQ molecule also enhances light absorption due to its high light absorption in the near infrared region and large spectral overlap between the P3HT emission and SQ absorption. Here we report our current study on charge carrier transport in the blend of P3HT and SQ-capped ZnO-NPs by using photoluminescence spectroscopy with UV and blue excitation wavelengths. ZnO-NPs were synthesized by using sol-gel method. The capping process was employed by mixing small amount of SQ solution into the ZnO-NPs gel before its blend with P3HT. Small molecule SQ was successfully capped ZnO-NPs as observed from its TEM image. ZnO-NPs solution has absorption wavelength less than 380 nm, therefore, no emission spectrum is observed when it is excited with 473 nm excitation wavelength. However, a clear emission spectrum is observed from SQ-capped ZnO-NPs solution, which shows that charge transfer was occurred from SQ to ZnO-NPs. Similar behavior was also observed in the blend of P3HT and SQ-capped ZnO-NPs. Our study shows that small molecule squaraine can improve charge carrier transfer from ZnO-NPs and therefore the performance of solar cells might be increased.

1. Introduction

Hybrid organic-inorganic solar cells have been intensively studied to obtain high performance and low cost solar cells [1-4]. Combination of superior optoelectronic properties of inorganic materials and low-cost preparation methods of organic semiconducting molecules make hybrid organic-inorganic materials as promising materials for optoelectronic devices, such as diode and solar cells. Blend of conjugated polymers poly(3-hexylthiophene) or P3HT and zinc oxide nanoparticles (ZnO-NPs) has been recently developed for high performance solar cells. Combination of high hole mobility and versatile processing of P3HT and high electron mobility of ZnO-NPs is expected to produce high performance solar cells. Zinc oxide has also low exciton binding and its conduction band is lower than LUMO of P3HT which facilitates easier for charge transfer from excited electrons in P3HT into ZnO-NPs. Moreover ZnO-NPs can be synthesized at lower temperature which is suitable for mixing with conjugated polymers. Although several efforts have been done, the Power Conversion Efficiency

(PCE) of hybrid solar cells is still below 2%, far below reported fully organic BHJ solar cells, like P3HT:fullerene indene-C60-bisadduct (ICBA) blend solar cells above 7% [5]. Oosterhout *et al.* [6-7] reported 2% PCE hybrid solar cells using control morphology of blend P3HT:ZnO films. The main problem of this low PCE is caused by the difficulty of morphology control of hybrid layer and the understanding of charge carrier motion or transport both in P3HT or in ZnO, which is still not fully revealed yet [8]. The low PCE of P3HT:ZnO nanoparticles is mainly caused by the difficulty of morphology control of hybrid layer, especially in controlling aggregation of ZnO nanoparticles. Agglomeration or aggregation of ZnO-NPs forming large size particles will decrease its surface area and hinders charge transport from P3HT to ZnO, thus lowers the performance of solar cells. Capping ZnO nanoparticles with small molecules and polymers is therefore commonly used to avoid the agglomeration in ZnO-NPs.

Recently, several groups have used small semiconducting molecules or surfactants such as naphthalene [9], pyridine [10] and fullerene derivative PCBA [11] to overcome ZnO agglomeration and have successfully increased the electric current and PCE of hybrid solar cells. Li et. al. studied theoretically hybrid P3HT:ZnO nanoparticles capped with monolayer of fullerene derivative PCBA [11]. They found that the external quantum efficiency (EQE) of hybrid P3HT:ZnO solar cells can be tripled due to induced interfacial dipole and shifted-up the LUMO level of P3HT relative to the conduction band edge of ZnO. Therefore, the capping molecules should possess high electron mobility and suitable LUMO level both with ZnO and P3HT to facilitate more charge transfer, thus finally to increase the PCE of solar cells.

In this study, we used small molecules squaraine (SQ) to cap ZnO nanoparticles. The intense absorption and emission property of squaraine, which is associated with the strong charge-transfer interaction, is ideally suited for applications related to the photosensitization phenomenon. Squarine absorption bands with the extinction coefficient of ~ 10^5 L $mol^{-1}cm^{-1}$ generally become rather broad in the solid state near the IR region and photostable as a consequence of the strong charge transfer or intermolecular interactions and a tendency to aggregation. Recently, it has been reported a near infrared light-emitting field-effect transistor based on squraine compounds with relative balance electron and hole mobilities in the order of 10^{-4} cm^2/Vs [12]. Moreover, the hole mobility increases up to 10^{-3} cm^2/Vs as the devices are thermally annealed. The PCE of SQ:fullerene solar cells have reached more than 6% [13]. The highest occupied molecular orbital (HOMO) and LUMO levels of SQ are -5.3 eV and -3.6 eV, respectively [14] which are lower than that of P3HT (-5.1 eV and -3.1 eV) [15]. The most important parameter for capping is the LUMO of SQ, which is higher than the Conduction Band (CB) of ZnO at -4.2 eV. Therefore, capping ZnO-NPs using squaraine is expected not only to avoid agglomeration of ZnO-NPs, but also to improve charge transfer from P3HT to ZnO-NPs, thus the PCE of solar cell can be increased.

2. Materials and Experimental Methods

2.1. Materials and Samples Preparation

Conjugated polymer P3HT, small molecule SQ, methanol and chlorobenzene were purchased from Sigma Aldrich. Chemical structure of P3HT and SQ are shown in figure 1. Zinc acetate dehydrate and sodium hydroxide were purchased from Merck. All materials are used as received without any further purification. ZnO nanoparticle was synthesized using sol-gel method, as described in our previous work [16].

The ZnO-NPs solution was prepared by dissolving ZnO nanoparticles gel in methanol. Squaraine was dissolved in methanol. Capping of ZnO-NPs by SQ was prepared by mixing ZnO-NPs solution and few drop of SQ solution and then was ultrasonicated for 15 minutes to obtain the homogeneous solution. Meanwhile the blend of P3HT:ZnO-NPs, P3HT:SQ and P3HT:ZnO-NPs:SQ were dissolved in chlorobenzene.

IOP Conf. Series: Journal of Physics: Conf. Series **1080** (2018) 012001 doi:10.1088/1742-6596/1080/1/012001

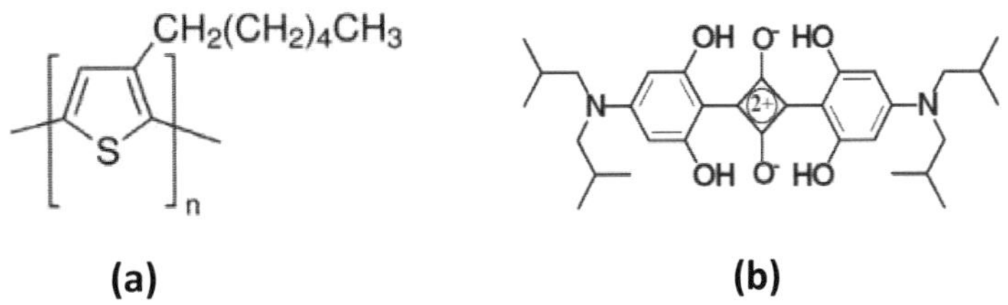

Figure 1. Chemical structure of a). Poly(3-hexylthiophene) or P3HT, and b). Squaraine or SQ.

2.2. Characterization
The particle sizes of ZnO-NPs and SQ-capped ZnO-NPs were measured by use of Transmission Electron Microscopy (TEM JEOL JEM 1400). The optical absorption of each solution and their blend were measured by use of T70+UV-Vis spectrometer. In order to study charge carrier transfer from SQ into ZnO-NPs, the photoluminescence (PL) of uncapped ZnO-NPs solution and SQ-capped ZnO-NPs solution were performed by fluorometer with excitation wavelength of 325 nm (UV-light) and 473 nm (blue light). We also measured the PL spectra of blend solution of P3HT:ZnO-NPs, P3HT:SQ and P3HT:SQ:ZnO-NPs.

3. Results and Discussion
Figure 2 shows TEM images of uncapped-ZnO-NPs and SQ-capped ZnO-NPs diluted solution in methanol. The average diameter of ZnO-NPs is 40 nm. Agglomeration is still can clearly observed even in a much diluted solution. Capping molecule SQ covers the surface of ZnO-NPs as shown with bright images of smaller particles than ZnO-NPs. This results show that capping process was successfully carried out, although agglomeration of ZnO-NPs is still preserved. The capping mechanism of SQ molecules into ZnO-NPs is still unclear, whether due to chemically bonding of SQ molecules to hydroxyl groups of ZnO-NPs or mechanically bonding. The different size of ZnO-NP between uncapped-ZnO-NP and SQ-capped-ZnO-NPs is caused by agglomeration during synthesis.

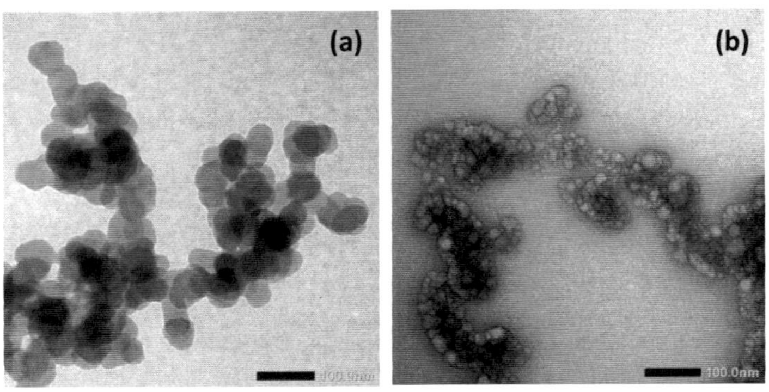

Figure 2. TEM images of (a) ZnO nanoparticles (ZnO-NPs) and (b) Squaraine-capped ZnO nanoparticles (SQ capped ZnO-NPs). Dark image is ZnO-NPs and bright image is for squaraine.

UV-Vis absorption spectra of ZnO-NPs, SQ and SQ-capped ZnO-NPs solutions are shown in figure 3. ZnO-NPs absorb light at wavelengths shorter than 370 nm, with wavelength cut-off at 369 nm, while SQ solution absorbs at longer wavelengths ranging from 620 nm to 680 nm. We calculated the optical bandgap of ZnO nanoparticle from wavelength cut-off and obtained 3.38 eV. This value is higher than that of its bulk (3.37 eV), which indicates that the synthesized ZnO forms nanoparticles as confirmed by TEM image of ZnO-NPs with average diameter of 40 nm. Absorption spectra of SQ-capped ZnO-NPs clearly consist of the optical absorption of SQ molecule and ZnO-NPs. This shows that SQ can cover or cap the ZnO-NPs as also confirmed by TEM image displayed in figure 2.

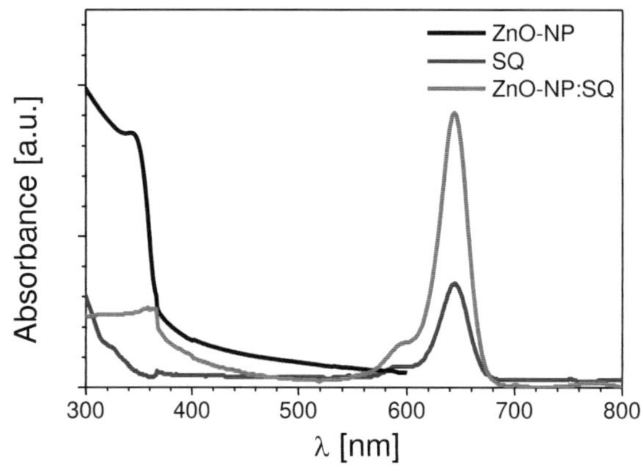

Figure 3. UV-Vis spectra of ZnO-NPs, SQ and SQ-capped ZnO-NPs solutions in methanol.

PL spectra of ZnO-NPs, SQ and SQ-capped ZnO-NPs are displayed in figure 4(a). For comparison, their absorption spectra are also displayed in the figure. The excitation wavelength is 325 nm (UV-light). Two PL peaks are observed in ZnO-NPs solution related to the intrinsic defect at 3.22 eV (Zinc vacancy or UV emission) and at 2.32 eV related to the interstitial zinc defect [17]. The defect at 1.62 eV which is caused by oxygen vacancy [17] is not observed due to overlap with the second generation of excitation wavelength (650 nm) as shown by huge peak intensity in PL spectra. The proses of absorption and photoluminescence in ZnO-NPs are energetically illustrated in figure 4(b).

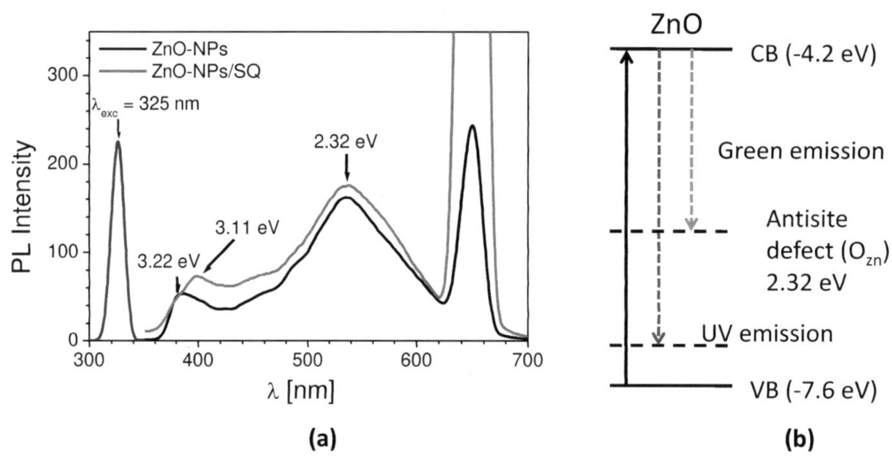

Figure 4. Photoluminescence spectra of (a) ZnO-NPs, SQ-capped ZnO-NPs and (b) Schematic process of absorption and emission in ZnO-NPs. The excitation wavelength is 325 nm.

In order to study the effect of SQ to the charge transfer, we measured PL spectra of uncapped and SQ-capped ZnO-NPs. Their PL spectra are shown in figure 5(a) No emission of ZnO-NPs is observed when excited by 473 nm. This can be easily understood because the excitation energy is smaller than the energy gap of ZnO-NPs, therefore the required energy for charge to be excited into conduction band of ZnO is not sufficient. Meanwhile, the SQ molecules exhibit PL emission peak at 700 nm, which shows that the energy gap is 1.7 eV. This energy gap is similar to the difference between HUMO and LUMO level of SQ (1.7 eV) as mention above.

Photoluminescence spectrum of SQ-capped ZnO-NPs solution as shown in figure 5(a), however, shows a significant PL peak at wavelength around 600 nm. By using the excitation wavelength at 473 nm, the emission of excited electron is not possible to take place in ZnO-NPs due to insufficient energy. Therefore, this PL peak is clearly caused by charge transfer from SQ molecules to ZnO-NPs and then subsequently emitted in ZnO-NPs. This result shows that small molecule SQ assists more charge transfer to ZnO nanoparticle through Forster Resonance Energy Transfer (FRET) process [18]. The charge transfer processes in SQ-capped ZnO-NPs can be schematically illustrated in figure 5(b).

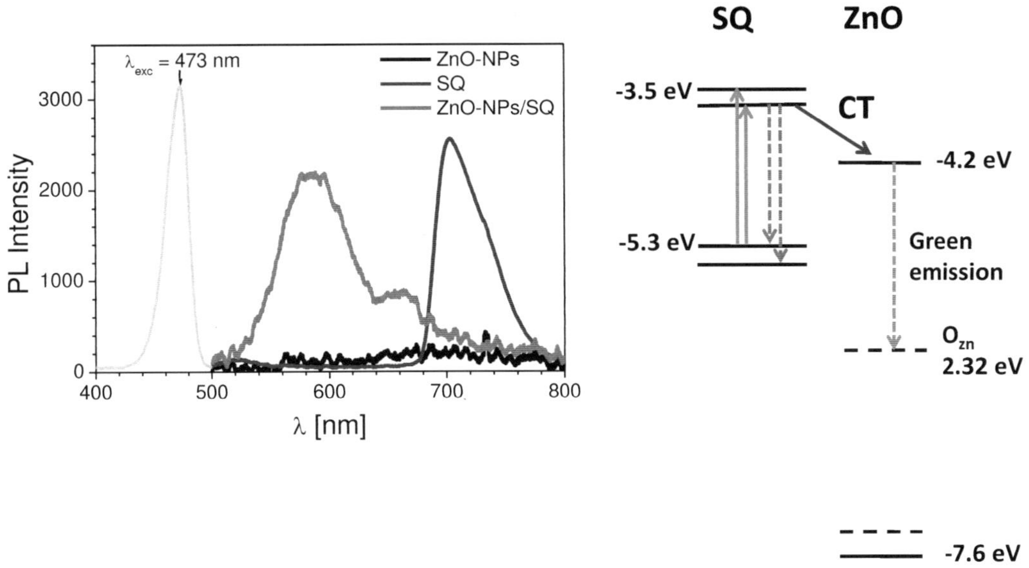

(a) (b)

Figure 5. Photoluminescence spectra of (a) ZnO-NPs, SQ-capped ZnO-NPs and (b) Schematic process of absorption and emission in SQ-capped ZnO-NPs. The excitation wavelength is 473 nm.

Figure 6 shows the PL spectra of blend solution of P3HT, P3HT:ZnO-NPs, P3HT:SQ and ternary P3HT:SQ-capped ZnO-NPs. The PL of P3HT shows broad spectrum with a strong peak at 584 nm, corresponds to energy gap of 2.12 eV, which is close to the difference value of HUMO (-5.1 eV) and LUMO level (-3.1 eV) of P3HT [15]. The broad spectrum of PL indicates the broad of energy bands or energy levels both in HUMO and LUMO levels. A shoulder PL peak at 626 nm is attributed to aggregation of P3HT chains [19-21]. Blend hybrid P3HT:ZnO-NPs solution has similar PL spectrum with P3HT only, which shows no observed charge transfer occurs from P3HT into ZnO-NPs. Other study showed that the charge transfer from P3HT chains to ZnO-NPs can be observed in PL spectra of P3HT:ZnO-NPs blend solution by varying the ZnO-NPs concentration. The decreased of PL spectra with the increased of ZnO-NPs concentration indicates the charge transfer is occurred from P3HT to ZnO-NPs [2].

The PL spectrum of P3HT:SQ solution shows two distinguished peaks at 567 nm and at 696 nm. The first peak is assigned to emission in P3HT and the second strong peak is attributed to the emission peak in SQ molecules. We cannot clearly observe the charge transfer from P3HT chains to the SQ molecules. In the solution of P3HT:SQ-capped ZnO-NPs, three PL peaks are observed. The first peak at 567 nm is due to emission in P3HT, the second peak is caused by the emission in ZnO-NPs. The clear distinguished peak at 675 nm is clearly attributed to the emission in SQ molecules. Again, however, it is very difficult to distinguish charge transfer from P3HT to ZnO-NPs and to SQ molecules or from SQ molecules to ZnO-NPs due to strong overlap of their PL emission spectra. Additional time resolved photoluminescence spectroscopy is therefore highly required to clarify the effect of SQ into charge transfer in the blend of P3HT:SQ-capped ZnO-NPs. However, from figure 5, it is clearly confirmed that SQ molecules facilitate more charge transfer from SQ to ZnO-NPs.

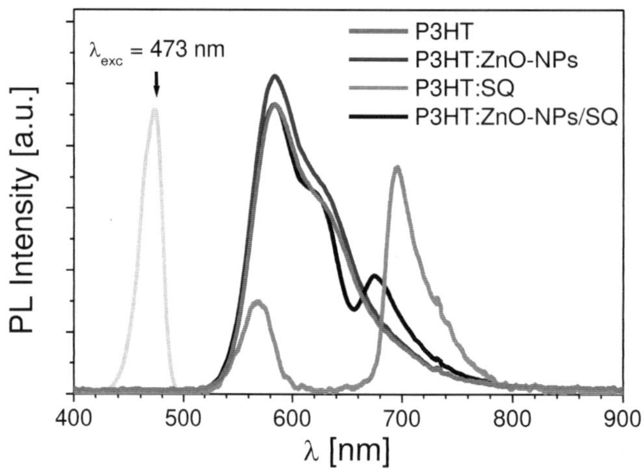

Figure 6. Photoluminescence spectra of solutions of P3HT, blend P3HT:ZnO-NPs, P3HT:SQ and ternary P3HT:SQ-capped ZnO-NPs (P3HT:ZnO-NPs/SQ). The excitation wavelength is 473 nm.

4. Conclusion

We have successfully synthesized ZnO-NPs with average diameter of 40 nm and capped ZnO-NPs by small molecule squaraine. The SQ molecules facilitate charge transfer from SQ into ZnO-NPs as confirmed by the appearance of PL peak even though if it was excited with photon energy smaller than bandgap of ZnO-NPs. We believe that by capping ZnO-NPs with SQ molecules, the performance of hybrid P3HT:ZnO-NPs solar cells can be improved due to the enhanced charge transfer from SQ to ZnO-NPs.

Acknowledgement

This work is financially supported by Directorate General for Higher Education (DIKTI), Ministry of Research, Technology and Higher Education Indonesia under Research Grant for International Collaboration and Scientific Publication 2015, Contract No. 393/UN6.R/PL/2015 February, 16, 2015.

References

[1] Beek W J E, Wienk M M and Janssen R A J 2004 *Adv. Mater.* **16** 1009
[2] Beek W J E, Wienk M M, Kemerink M, Yang X and Jannsen R A J 2005 *J. Phys. Chem. B* **109** 9505
[3] Beek W J E, Wienk M M and Janssen R A J 2006 *Adv. Funct. Mater.* **16** 1112
[4] Moet D J D, Koster L J A, de Boer B and Blom P W M 2007 *Chem. Mater.* **19** 5856
[5] Guo X, Cui C, Zhang M, Huo L, Huang Y, Hou J and Li Y 2012 *Energy Environ. Sci.* **5** 7943
[6] Oosterhout S D, Wienk M M, van Bavel S S, Thiedmann R, Koster L J A, Gilot J, Loos J, Schmidt V and Janssen R A J 2009 *Nat. Mater.* **8** 828
[7] Oosterhout S D *et al.* 2011 *Adv. Energy Mater.* **1** 90
[8] Bhat S V, Govindaraj A, Rao C N R 2011 *Sol. Energy Mater Sol. Cells* **95** 2318
[9] Chen J Y, Hsu F C, Sung Y M and Chen Y F 2012 *J. Mat. Chem.* **22** 15726
[10] Said A J *et al.* 2010 *J. Phys. Chem.* **114** 11273
[11] Li Z, Zhang X and Lu G 2012 *J. Phys. Chem. C* **116** 9845
[12] Anthopoulos T D, Tanase C, Setayesh S, Meijer E J, Hummelen J C, Blom P W M and de Leeuw D M 2004 *Adv. Mater.* **16** 2174
[13] Chen G, Sasabe H, Wang Z Q, Wang X F, Hong Z R, Yang Y and Kido J 2012 *Adv. Mater.* **24** 2768
[14] Chen G, Yokoyama D, Sasabe H, Hong Z, Yang Y and Kido J 2012 *Appl. Phys. Lett.* **101** 083904

[15] Scharber M C, Muhlbacher D, Koppe M, Denk P, Waldauf C, Heeger A J, Brabec C J 2006 *Adv Mater.* **18** 789
[16] Yuliah Y and Bahtiar A 2013 *AIP Conf. Proc.* **1554** 139
[17] Lin B, Fu Z, Jia Y 2001 *Appl. Phys. Lett.* **79** 943
[18] Huang J S, Goh T, Li X, Sfeir M Y, Bielinski E A, Tomasulo S, Lee M L, Hazari N and Taylor A D 2013 *Nat. Photon.* **7** 479
[19] Brown P J, Thomas D S, Koehler A, Wilson J S, Kim J S, Ramsdale C M, Sirringhaus H and Friend R H 2013 *Phys. Rev. B* **67** 064203
[20] Niles E T, Roehling J D, Yamagata H, Wise A, Spano F C, Moule A J and Grey J K 2012 *J. Phys. Chem. Lett.* **3** 259
[21] Wang L and Beljone D 2016 *Cryst. Eng. Comm.* **18** 7297

IOP Conf. Series: Journal of Physics: Conf. Series **1080** (2018) 012002 doi:10.1088/1742-6596/1080/1/012002

The effect of thickness on microwave absorbing properties of barium ferrite powder

E Handoko[1*], I Sugihartono[1], S Budi[2], M Randa[3], Z Jalil[4] and M Alaydrus[5]

[1]Department of Physics, Universitas Negeri Jakarta, Jl. Rawamangun Muka 13220, Jakarta, Indonesia
[2]Department of Chemistry, Universitas Negeri Jakarta, Jl. Rawamangun Muka 13220, Jakarta, Indonesia
[3]Balitbang Kementerian Pertahanan Republik Indonesia, Jakarta Indonesia
[4]Department of Physics, Syiah Kuala University, Banda Aceh, Indonesia
[5]Department of Electrical, Engineering Universitas Mercu Buana, Jakarta, Indonesia

[*]erfan@unj.ac.id

Abstract. We have investigated the effect of thickness on microwave absorbing properties of barium ferrite ($BaFe_{12}O_{19}$) powder prepared by a solid-state reaction method with a sintering temperature of 1100°C for 5 hours. The sample was analyzed using XRD, SEM, PERMAGRAPH, and vector network analyzer (VNA). The microwave absorbing properties were measured at the frequency in the range of 8.2 – 12.4 GHz. Permeability, permittivity and reflection loss (RL) were calculated at a given thickness based on transmit line theory. The results show the optimal RL of −23.07 dB can be obtained at 10.72 GHz with a thickness of 3.0 mm, bandwidth with RL less than −5 dB and −10 dB can reach 1.2 GHz (10.4−11.6 GHz) and 0.8 GHz (10.6−11.4 GHz), respectively.

1. Introduction

Recently, the microwave absorbing materials have been used in military purposes such as radar-absorbing material (RAM), stealth technology, telecommunication for wireless system and electromagnetic interference (EMI) absorbers [1-2]. The two important things required as microwave absorbing materials are impedance and ability to reduce the incident electromagnetic (EM) wave. The value of the intrinsic impedance of material should be close to the intrinsic impedance of free space. Thus, it is necessary to make the value of magnetic permeability similar to the electrical permittivity [3]. Moreover, the reflection loss (RL) with different sample thicknesses can be calculated to evaluate EM absorption abilities of an absorber [4]. In EM wave absorber technology, barium ferrite ($BaFe_{12}O_{19}$) has large values of permeability, high resistivity, good magnetic and dielectric properties at microwave frequencies that are suitable and as one of the potential candidates for microwave absorbing materials [5]. The exploration of barium ferrite as a microwave absorber material has stimulated a lot of inventions. However, barium ferrite ($BaFe_{12}O_{19}$) composition does not work quite well in the giga-hertz frequency range which is known as snoek's limit [6-8].

In this paper, the effect of layer thickness on microwave absorbing properties of barium ferrite ($BaFe_{12}O_{19}$) was investigated. The thickness variations to reflection loss (RL) in the frequency range from 8.2 to 12.4 GHz from $BaFe_{12}O_{19}$ was calculated using the transmit line theory [4-9]. Furthermore, structure and magnetic properties investigation of the $BaFe_{12}O_{19}$ are also discussed.

2. Experimental

2.1. Materials and synthesis

Barium ferrite was synthesized through a solid-state method using $BaCO_3$ and Fe_2O_3 as precursors. The mixtures of $BaCO_3$ and Fe_2O_3 were pulverized by high energy ball mill for 1 hour and pelletized under a pressure of 5 MP to increase the hardness of barium hexaferrite. Furthermore, the mixture was sintered in a furnace at 1100°C for 5 hours. Finally, sintered barium ferrite was crushed for 1 hour to obtain magnetic powders.

2.2. Characterization

Crystal structure identification of barium ferrite ($BaFe_{12}O_{19}$) powder was characterized by PANalytical X-ray diffractometer (XRD) using CuKα radiation (λ =1.5418 Å) in the range of 2θ = 20°–70°. scanning electron microscopy (SEM) SEC type SNE4500 M examination was used for analyzing the morphological structure of magnetic powders. Hysteresis loop for magnetic properties was recorded using PERMAGRAPH techniques type Electromagnet EP 3 from magnet physic Dr.Steingroever GmbH and measured at room temperature with an applied field up to 1 tesla. Keysight PNA-L N5232A vector network analyzer (VNA) was used to analyze microwave absorbing properties for X-band (8.2–12.4 GHz) frequencies. The magnetic powders of barium ferrite were mixed with epoxy resin to be converted into a microwave absorbing composite. The mixing weight ratio of ferrite to epoxy resin was 4:1 and placed in a sample holder with a dimension of width 22.86 mm and height 10.16 mm. The reflection (S_{11}) and complex transmission (S_{21}) coefficients were measured with VNA. To obtain relative complex permeability ($\mu = \mu' - j\mu''$) and permittivity ($\varepsilon = \varepsilon' - j\varepsilon''$) of barium ferrite ($BaFe_{12}O_{19}$), coefficients were used in the Nicolson-Ross-Weir (NRW) algorithm with a thickness of 5 mm [10]. Reflection loss (RL) was calculated using the transmit line theory by the following equations:

$$\mathrm{RL\ (dB)} = 20\ \mathrm{Log}\left[\frac{(Z-1)}{(Z+1)}\right] \tag{1}$$

$$Z = \sqrt{\frac{\mu}{\varepsilon}}\tanh\left[\left(\frac{-j2\pi f d}{c}\right)\sqrt{\mu.\varepsilon}\right] \tag{2}$$

According to these equations, relative complex permeability ($\mu = \mu' - j\mu''$) and complex permittivity ($\varepsilon = \varepsilon' - j\varepsilon''$), f is frequency and d is the thickness of barium ferrite ($BaFe_{12}O_{19}$), and c is the light velocity.

3. Results and discussion

3.1. Structure identification

Figure 1 shows the XRD pattern of barium ferrite ($BaFe_{12}O_{19}$) powder sintered at 1100°C for 5 hours. Base on XRD pattern, a barium ferrite has successfully formed a single phase with polycrystalline structure. The calculation result using High Score Plus software shows that lattice constant value is a = b = 5.8882 Å and c = 23.1988 Å. The results are not much different from the indicated literature that is a = b = 5.8920Å and c = 23.1830 Å (International Centre for Diffraction Data, ICDD no. 00-084-0757). The result proves that the formation temperature of barium ferrite is about 100°C lower than as indicated in the literature [11].

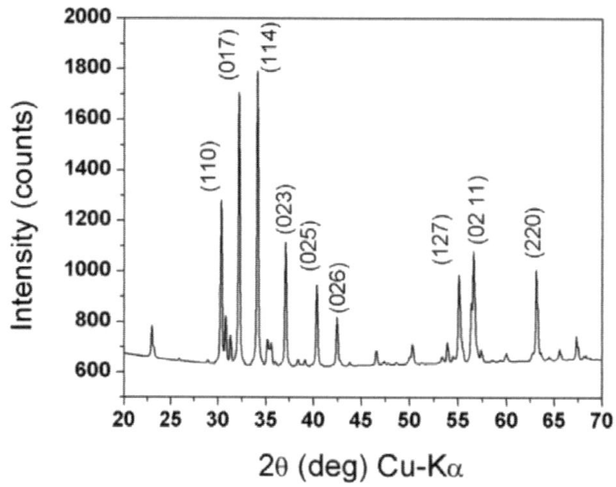

Figure 1. XRD patterns of barium ferrite ($BaFe_{12}O_{19}$).

3.2. Morphology

SEM micrograph of the barium ferrite is shows in figure 2. After milling, the powder of sintered sample is relatively the same in size. The average of powder size is less than about 10 μm. Therefore, the powder of barium ferrite is suitable to be mixed with epoxy resin for preparation a microwave absorbing material.

Figure 2. The SEM of morphological structure of barium ferrite ($BaFe_{12}O_{19}$).

3.3. Magnetic and dielectric propertiess

Figure 3 shows hysteresis loop of barium ferrite ($BaFe_{12}O_{19}$) at room temperature. The values of residual magnetization (M_r) at 0.213 T, is coercive field (H_c) 235.5 kA/m and saturation magnetization (M_s) 0.371 T. The results of magnetic properties indicate a good correlation between the magnetic properties and structural. A high H_c for sample (about 235.5 kA/m) is caused by strong uni-axial anisotropy along the c-axis. The result confirms that the barium ferrite powder is a permanent magnet.

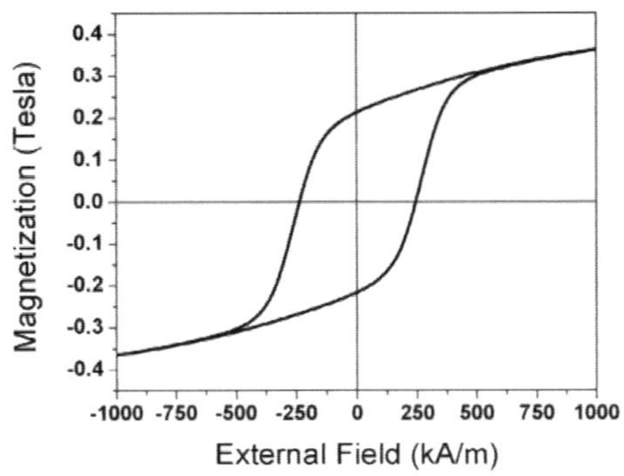

Figure 3. Room temperature hysteresis loop of barium ferrite ($BaFe_{12}O_{19}$).

Figure 4(a) shows the real (μ') and imaginary (μ") parts of relative complex permeability while figure 4(b) shows the real (ε') and imaginary (ε") parts of permittivity at the frequency range of 8.2 – 12.4 GHz of barium ferrite. The relative complex permeability (μ = μ′ − jμ″) and permittivity (ε = ε′ − jε″) determine the microwave absorbing properties of the absorbing materials. The real (μ') and imaginary (μ") permeability symbolize the storage and the attenuation abilities of magnetic energy, respectively. While the real (ε′) stands for the storage ability and imaginary (ε″) permittivity represents dissipation ability of electric energy [4, 12-14]. According to the relative complex permeability and permittivity of the sample, a microwave absorbing properties can be estimated. As shown in Fig. 4a, the value of μ' and μ" are in the range of 0.85–1.02 and 0.01–0.12, respectively. The values of μ' and μ" with increasing frequency decreases in the range of about 8.75–9.25 GHz and 9–9.75 GHz, respectively. On the other hand, the value of μ' and μ" increases in the range of about 9.25–10.25 GHz and 8.2–9 GHz. The changes of the magnetic tangent loss of barium ferrite indicate that the sample has a magnetic loss to electromagnetic waves in some degree. Figure 4(b) shows the value of ε' and ε" in range 4.15–4.42 and 0.08–0.42, respectively. According to the results, it suggests that barium ferrite has good absorption properties due to the magnetic loss and the dielectric loss [15].

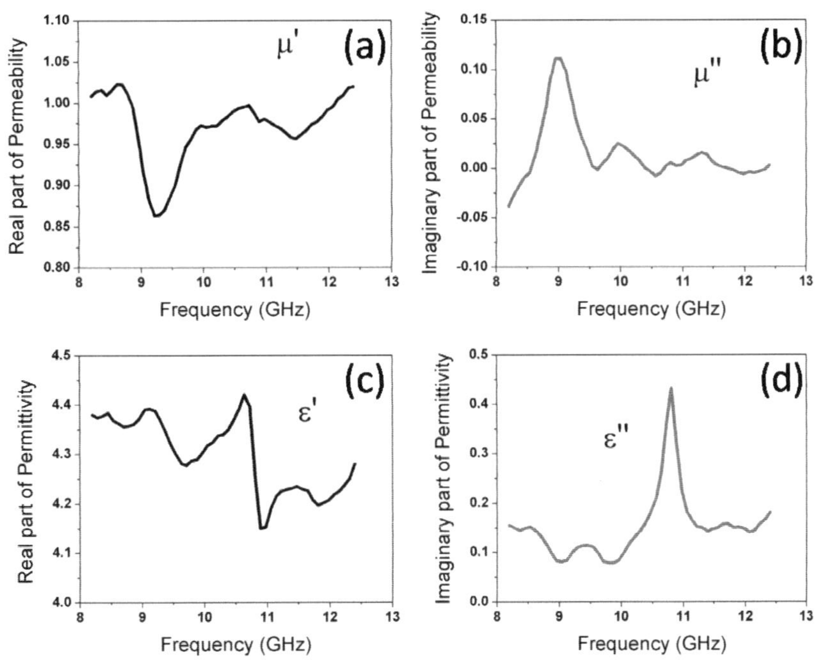

Figure 4. (a) The Real (μ'), (b) imaginary (μ") parts of relative complex permeability, (c) the real (ε') and (d) imaginary (ε") parts of permittivity of barium ferrite ($BaFe_{12}O_{19}$).

3.4. Microwave absorbing propertiess

The reflection loss (RL) values of barium ferrite ($BaFe_{12}O_{19}$) were calculated using the transmit line theory by the following equations1 and 2 based on the real (μ') and imaginary (μ") parts of relative complex permeability and the real (ε') and imaginary (ε") parts of permittivity are shown in figure 4. The figure 5(a) shows RL values of barium ferrite ($BaFe_{12}O_{19}$) at different thickness 0.2 mm, 0.5 mm, 1.0 mm, 2.0 mm and 3.0 mm. It is found that the minimum RL value in barium ferrite ($BaFe_{12}O_{19}$) is influenced by thickness of barium ferrite ($BaFe_{12}O_{19}$) as shown in table 1.

The microwave absorbing properties of barium ferrite($BaFe_{12}O_{19}$) were exhibited in figure 5(a). The optimal RL of −23.07 dB can be obtained at 10.72 GHz with thin thickness of 3.0 mm, bandwidth with RL less than −5 dB and −10 dB can reach 1.2 GHz (10.4−11.6 GHz) and 0.8 GHz (10.6−11.4 GHz), respectively. Table 1 shows the electromagnetic absorption properties of barium ferrite ($BaFe_{12}O_{19}$). It can be found that the thickness of sample is useful for the enhancement of the electromagnetic absorption. The enhanced absorption properties were attributed to the two-dimensional contour (see figure 5(b)) and more interfacial polarization relaxation. The results indicate that a thicker absorber layer has a wider frequency bandwidth.

Table 1. Electromagnetic Absorption Properties of barium ferrite ($BaFe_{12}O_{19}$).

sample thickness (mm)	minimum RL (EMI) value (dB)	frequency range (GHz) (RL < -5 dB)	effective bandwidth (GHz) (<-5 dB)	frequency range (GHz) (RL < -10 dB)	effective bandwidth (GHz) (<-10 dB)
0.2	-0.99	-	-	-	-
0.5	-3.15	-	-	-	-
1.0	-3.28	-	-	-	-
1.0	-9.19	10.6-10.9	0.3	-	-
2.0	-3.99	-	-	-	-
2.0	-19.08	8.6-9.5	0.9	8.7-9.2	0.5
2.0	-8.28	12.3-12.4	0.1	-	-
3.0	-12.81	8.5-9.4	0.9	8.9-9.3	0.4
3.0	-23.07	10.4-11.6	0.8	10.6-11.4	0.8
3.0	-14.82	10.4-11.6	1.2	10.6-11.4	0.8

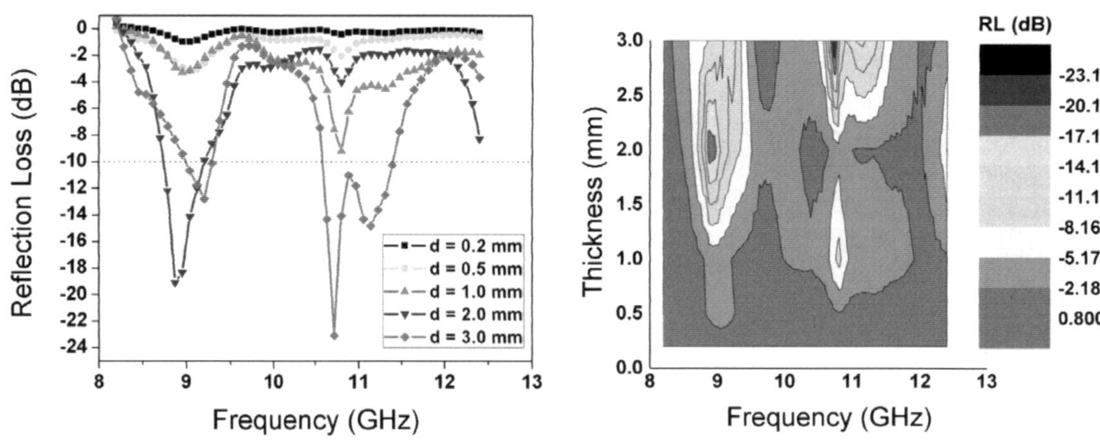

Figure 5. (a) Reflection loss and (b) two-dimensional contour of the simulated RL values barium ferrite ($BaFe_{12}O_{19}$) different thicknesses.

4. Conclusions

Barium ferrite ($BaFe_{12}O_{19}$) have been prepared using $BaCO_3$ and Fe_2O_3 as precursors by solid-state method. XRD characterization shows that barium ferrite phase has successfully formed as a single phase of polycrystalline structure after sintering at 1100°C for 5 hours. SEM measurement shows morphological structure of the powders with relatively same in size less than 10 μm. Magnetic properties show that the sample has a high H_c and as permanent magnet barium ferrite. According to VNA measurement, the real (μ') and imaginary (μ") parts of permeability and the real (ε') and imaginary (ε") parts of permittivity of the magnetic powders of barium ferrite vary at the frequency range of 8.2–12.4 GHz. The optimal RL of −23.07 dB can be obtained at 10.72 GHz with thickness of 3.0 mm.

Acknowledgments

Supported by Kementerian Riset, Teknologi dan Pendidikan Tinggi Republik Indonesia under grant INSINAS 2018 with No. RD-2016-0146 gratefully acknowledged.

References

[1] Oikonomou A, Giannakopoulou T and Litsardakis G 2007 *Journal of Magnetism and Magnetic Materials* **316** e827
[2] Gairola S P ,Verma V, Singh A, Purohit L P and Kotnala R K 2010 *Solid State Communications* **150** 147
[3] Ni Q Q, Melvin G J H and Natsuki T 2015 *Ceram. Int.* **41** 9885
[4] Zhao B, Fan B, Shao G, Zhao W and Zhang R 2015 *ACS Appl. Mater. Interfaces* **7** 18815
[5] Sözeri H, Mehmedi Z, Kavas H and Baykal A 2015 *Ceramics International* **41** 9602
[6] Ghasemi A, Hossienpour A, Morisako A, Saatchi A and Salehi M 2006 *Journal of Magnetism and Magnetic Materials* **302** 429
[7] Shams M H, Salehi S M A and Ghasemi A. 2008 *Materials Letters* **62** 1731
[8] Itoh M, Nishiyama K, Shogano F, Murota T, Yamamoto K, Sasada M and ichi Machida K 2008 *Journal of Alloys and Compounds* **451** 507
[9] Liu X, Wu N, Cui C, Li Y, Zhou P and Bi N 2015 *Mater. Materials Letters* **149** 12
[10] Handoko E, Mangasi A M and Iwan S 2016 *IEEE Radar, Antenna, Microwave, Electronics, and Telecommunications* p 28
[11] Tan G and Chen X. 2013 *Journal of Magnetism and Magnetic Materials* **327** 87
[12] Zhao B, J. Liu, X. Guo, W. Zhao, L. Liang, C. Ma, and R. Zhang 2017 *Physical Chemistry Chemical Physics* **19** 9128
[13] Handoko E *et al.* 2018 *Mater. Res. Express* in press https://doi.org/10.1088/2053-1591/aac4d7
[14] Zhao B, Zhao W, Shao G, Fan B and Zhang R 2015 *ACS Appl. Mater. Interfaces* **7** 12951
[15] Handoko E, Sugihartono I and Mangasi A M 2017 *IEEE Broadband Communication, Wireless Sensors and Powering (BCWSP)* p 1

IOP Conf. Series: Journal of Physics: Conf. Series **1080** (2018) 012003 doi:10.1088/1742-6596/1080/1/012003

Controlled synthesis of lead-free perovskite Cs_2SnI_6 as hole transport layer in dye sensitized solar cells

Y S Handayani, P Wulandari* and R Hidayat

Physics of Magnetism and Photonics Research Division, Faculty of Mathematics and Natural Sciences, Institut Teknologi Bandung, Jl. Ganesha 10 Bandung 40132, West Java, Indonesia

*wulan@fi.itb.ac.id

Abstract. Hybrid organic-inorganic perovskite material is undergoing rapid expansion over the past few years due to their application in solar cells. In particular, tin (Sn)-based perovskite has been considered to replace lead (Pb)-based perovskite because of toxic issues and it can cause adverse environmental impact. Cesium tin halide based perovskite (Cs_2SnI_6) material appears promising for large scale module fabrication and also oxygen and moisture stable because of high oxidation state of tin. In this study, we successfully synthesized Cs_2SnI_6 perovskite by mixing process of cesium carbonate (Cs_2CO_3) in hydriodic acid (HI) solvent and stannic iodide (SnI_4) in ethanol solvent to form precipitate and all the processes was done in air and room temperature. Cubic crystal structure of perovskite was confirmed by X-ray Diffraction (XRD) measurement in comparison to the crystal structure of Pb-based perovskite. Scanning Electron Microscope (SEM) revealed morphology of Cs_2SnI_6 with the size of 0.4–5 mm. We controlled the solutions of Cs_2SnI_6 by mixing two kinds of solvent (gamma-butyrolactone and acetonitrile) with different volume ratio and the Cs_2SnI_6 thin films were confirmed by used of UV-vis spectrometer. The absorbance spectra give a characteristic two humps feature due to the specific band structure of Cs_2SnI_6 and it indicates the direct band gap about 2.8 eV.

1. Introduction

Over the past few years, metal halide perovskites with superior advantages in optical and electrical properties have been considered as a promising material for applications in photovoltaic devices. Development of metal halide perovskites as solar cell absorber has been increased significantly and demonstrated high power conversion efficiencies over 20%. Three spiro-OMeTAD derivatives were synthesized and employed as hole transporting material (HTM) in the structure of solar cell of $TiO_2/CH_3NH_3PbI_3$/HTM/Au. Highly improved performance of power conversion efficiency (PCE) increase to 16.7% from 15% for conventional p-OMe substituents [1]. Zhou et.al. controlled formation of perovskite layer, suppressed carrier recombination in the absorber, facilitated carrier injection into the carrier transport layer and maintained good carrier extraction at the electrodes. Cell PCE is typically boosted to 16.5% on average with the highest efficiency of ~19.3% in a planar geometry without antireflective coating [2].

In general, metal halide perovskites have a crystal structure with ABX_3 formula. The A, B and X are the organic or inorganic cation ($CH_3NH_3^+$, Cs^+), the metal cation (Pb^{2+}, Sn^{2+}, or Cu^{2+}), and the halide anion (Cl^-, Br^-, or I^-), respectively. The A and B cations have very different ionic radii and the X anions bond to both cations. The A cations coordinate with 12 anions X forming cuboctahedral geometry (AX_{12} polyhedron) and the B cations coordinate with six anions X forming octahedral geometry (BX_6 configuration) [3]. Development of metal halide perovskite have to face the problem regarding to instability and toxicity issues especially for perovskite material based on lead (Pb). Peedikakkandy *et al.* reported synthesis and stability analysis of cesium tin iodide ($CsSnI_3$) prepared through solid state and solution route methods for its application as HTM in dye sensitized solar cell (DSSC) [4]. Lead-free hybrid tin halide perovskites ($MASnI_{0.5}Br_{2.5}$) were prepared and investigated by a rapid screening technique and it exhibits enhanced performance and stability in liquid-junction PEC solar cells, with PCE of 1.51% (an increase of 20.8%) in comparison to $MASnI_3$ perovskites [5]. Wang *et al.* developed a facile solution-phase hot injection approach for shape-controlled synthesis of lead-free and stable single crystalline perovskite derivative Cs_2SnI_6 nanocrystals [6]. Cs_2SnI_6 perovskite as a hole transporter has been demonstrated in solid-state DSSC with a mesoporous TiO_2 film, where doping of this perovskite with additives helps to reduce the internal device resistance and improving cell efficiency about 4.7% [7].

In this research work, we attempt to synthesize a lead-free metal halide perovskite based on Sn (Cs_2SnI_6) by used of simple chemical reaction in solution. We also control the homogenous formation this type of perovskite by used of organic solvents engineering for application in dye-sensitized solar cells (DSSC). It has been known that Cs_2SnI_6 exhibits air and moisture stability because it contains Sn^{4+} ion which is air stable and also has high hole effective masses. Based on this fact, Cs_2SnI_6 can be applied as potential hole transport layer (HTL) in dye sensitized solar cell (DSSC).

2. Experiments

2.1. *Synthesis of Cs_2SnI_6*

Hydriodic acid (HI), cesium carbonate (Cs_2CO_3), and stannic iodide (SnI_4) were purchased from Sigma-Aldrich and used without any further purification. The acidic cesium iodide (CsI) solution was prepared by solved 325 mg of Cs_2CO_3 into 2 mL of HI in a beaker glass and stirred at room temperature to form homogenous solution. A 313 mg of SnI_4 was dissolved in 1 mL of warm ethanol and stirred to form homogenous orange solution. Next, the SnI_4 solution was then added into the acidic CsI solution and stirred vigorously for 10 min to yield black powder precipitate. The final solution was purified for 3 times by used of ethanol (C_2H_5OH, Merck). The top solution was taken out and the black precipitate was re-dispersed and then dried at room temperature. Our synthesized result of black powder Cs_2SnI_6 is quite stable at room temperature.

2.2. *Preparation of Cs_2SnI_6 Thin Film*

In our case, Cs_2SnI_6 perovskite was prepared on thin film structure of glass substrate/TiO_2 compact layer/TiO_2 mesoporous because Cs_2SnI_6 is known has very poor wettability to attach directly on glass substrate. This film structure was also done to facilitate in fabrication of DSSC by used this type of perovskite as hole transporter. Glass substrate was cleaned in ultrasonic bath by de-ionized (DI) water and continued by used of mixed iso-propanol and acetone and finalized by ethanol. Glass substrate was dried by N_2 gas before used. Next, TiO_2 compact layer was spin-coated on the top of the glass substrate at 2000 rpm for 30 s and annealed at 500°C for 10 min. Layer of TiO_2 mesoporous was screen-printed on the top of TiO_2 compact layer and annealed to 500°C. In this study, we controlled the volume ratio of solvent for Cs_2SnI_6 perovskite in order to get homogenous solution and thin film. Cs_2SnI_6 solution was prepared by dissolved 0.1 mg of powder-synthesized Cs_2SnI_6 in 1 mL mixed of gamma-butyrolactone (GBL) and acetonitrile (CH_3CN) solvent (GBL: CH_3CN) in volume ratio of 1:0; 0:1; 1:1; 1:5; 5:1; 1:10, and 10:1. Subsequently, Cs_2SnI_6 solutions were deposited over the TiO_2 mesoporous by drop-casting in all area of thin film and then annealed at 100°C for 10 min.

2.3. *Characterizations*

Optical properties of Cs_2SnI_6 in thin films were characterized by an ultraviolet-visible (UV-vis) spectrometer at room temperature by using an Agilent Technologies 8453 UV-visible spectrometer. The absorbance spectrum was then used to determine band gap of the Cs_2SnI_6 using Tauc-Plot calculation. Scanning Electron Microscope (SEM) of SU3500 model was used to get morphology images of Cs_2SnI_6 powder without additional chemical and mechanical treatment. Measurement of X-ray Diffraction (XRD) was done by Philips Analytical PW 1710 diffractometer.

3. Results and discussions

In this research work, Cs_2SnI_6 perovskite has been successfully synthesized by a simple chemical reaction in solution by mixed an acidic solution of C_SI and a stannic iodide (SnI_4) solution and revealed black precipitate powder of Cs_2SnI_6. Figure 1 shows SEM images of the powder-synthesized Cs_2SnI_6 measured at different magnification. The grain size of Cs_2SnI_6 powder is not homogenous in the range of 0.5 – 4 μm. Further study regarding to the homogenous size and shape of our synthesized Cs_2SnI_6 is still under our investigation.

(a) (b)

Figure 1. SEM Images of the powder-synthesized Cs_2SnI_6 at room temperature under (a) 5000 and (b) 20000 times of magnification.

X-ray diffraction (XRD) pattern of the powder-synthesized Cs_2SnI_6 in room temperature is shown in figure 2. Appearance peaks at 2θ equal to 26.34°, 30.6°, 43.96°, 52.00°, 54.60°, and 63.95° correspond to (222), (400), (440), (622), (444), and (800) planes, respectively. It shows the typical crystal structure of Cs_2SnI_6 perovskite which agrees well with the reference result from Lee *et al.* [7] and $CH_3NH_3PbI_3$ perovskite in our previous work. Rietveld refinement using Match identifies that the powder-synthesized Cs_2SnI_6 has a cubic structure in Fm-3m (225) space group. The lattice parameter is given by 11.6276 Å.

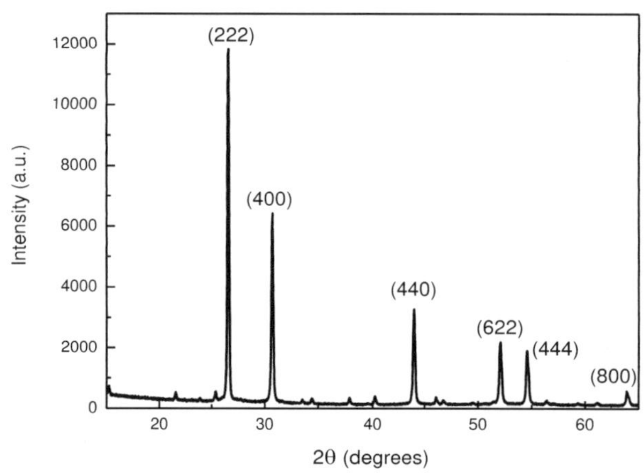

Figure 2. XRD pattern of powder Cs_2SnI_6 perovskite.

In our investigation, Cs_2SnI_6 has good solubility in mixed of GBL and CH_3CN as shown in figure 3(a). Thin films of Cs_2SnI_6 in GBL:CH_3CN solvents with volume ratio of 5:1 and 10:1 show good solubility and homogenous surface coverage compare to the other films with typical colour of bright yellow. Absorbance spectra of Cs_2SnI_6 in figure 3(b) exhibit characteristic two hump feature at 308 nm and 360 nm, which is due to the specific band structures of Cs_2SnI_6. In case of thin film with solvent ratio of 1:10, it is found that the absorbance intensity drastically enhance among to the other thin film samples eventhough the homogeneity of the thin film is less than thin film in solvent ratio of 5:1 and 10:1. In contrast, the film resulted from solvent ratio of 0:1 does not show the characteristic of Cs_2SnI_6 absorbance spectra, which indicated the CH_3CN affect the formation of the perovskite as shown in thin film with inhomogeneous black color (figure 3(a)).

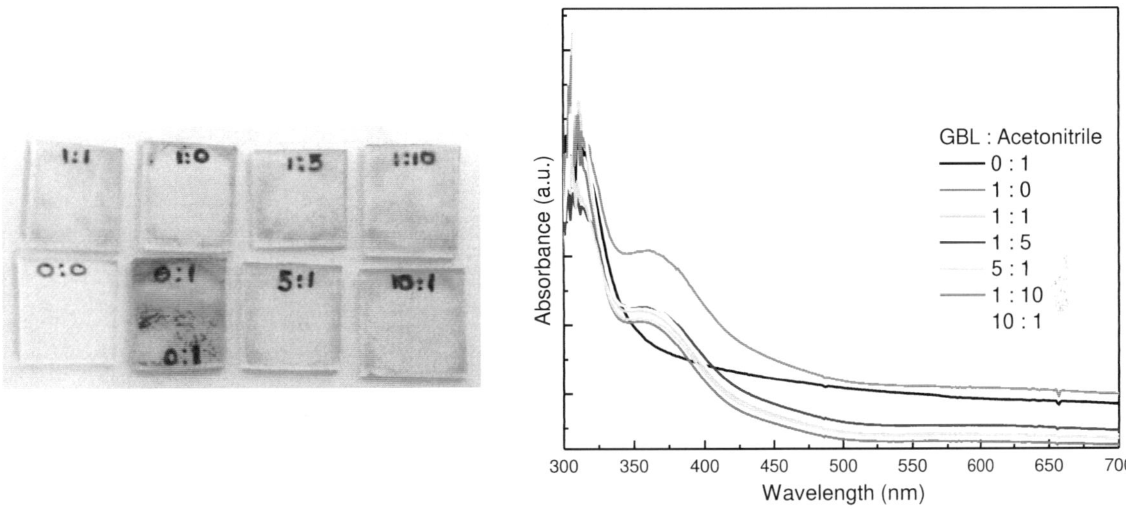

(a) (b)

Figure 3. (a) Top view picture and (b) absorption spectra of Cs_2SnI_6 thin films in various ratio of GBL:CH_3CN solvent.

The band gap of our synthesized-Cs_2SnI_6 thin film was determined from the corresponding UV-visible spectra. Optical band gap calculated from the Tauc plot of the absorbance data was estimated to be 2.8 eV by assuming that Cs_2SnI_6 had a direct band gap. In previous research, Lee *et al.* reported that Cs_2SnI_6 had a direct band gap and gave a value of 1.3 eV [7]. In contrast, Kaltzoglou *et al.* determined the band gap by UV-visible diffuse reflectance spectra and they found band gap of 3.9 eV and 2.7 eV for Cs_2SnI_6 and Cs_2SnBr_6, respectively and it showed indirect band gap [8].

4. Conclusions

We successfully synthesized lead-free metal halide perovskite of Cs_2SnI_6 by used of simple chemical reaction in solution. X-ray Diffraction (XRD) measurement shows the characteristic pattern of Cs_2SnI_6 in a good agreement to the literature. Based on the UV-visible spectra, mixed of gamma-butyrolactone (GBL) and acetonitrile (CH_3CN) is a good solvent for Cs_2SnI_6 in order to form homogenous thin film for solar cell applications. Absorbance spectra of Cs_2SnI_6 thin films give a characteristic two hump feature due to the specific of the band structure of Cs_2SnI_6. The Tauc plot indicates that the direct band gap of Cs_2SnI_6 is resulted in 2.8 eV.

References

[1] Jeon N J, Lee H G, Kim Y C, Seo J, Noh J H, Lee J and Seok S I L 2014 *J Am. Chem. Soc.* **136** 7837

[2] Zhou H P, Chen Q, Li G, Luo S, Song T, Duan H S, Hong Z, You J B, Liu Y S and Yang Y 2014 *Science* **345** 542

IOP Conf. Series: Journal of Physics: Conf. Series **1080** (2018) 012003 doi:10.1088/1742-6596/1080/1/012003

[3] Bhalla A S, Guo R and Roy R 2000 *Mat. Res. Innovat* **4** 3
[4] Peedikakkandy L and Bhargava P 2015 *Materials Science in Semiconductor Processing* **33** 103
[5] Hsu H Y, Ji L, Du M, Zhao J, Yu E T and Bard A J 2016 *Electrochimica Acta* **220** 205
[6] Wang A, Yan X, Zhang M, Sun S, Yang M, Shen W, Pan X, Wang P and Deng Z 2016 *Chem. Mater.* **28** 8132
[7] Lee B, Stoumpos C C, Zhou N, Hao F, Malliakas C, Yeh C Y, Marks T J, Kanatzidis M G and Chang R P H 2014 *J Am. Chem. Soc.* **136** 15379
[8] Kaltzzoglou A *et al.* 2016 *J Phys. Chem. C.* **120** 11777

IOP Conf. Series: Journal of Physics: Conf. Series **1080** (2018) 012004 doi:10.1088/1742-6596/1080/1/012004

Effect of light exposure on the μSR asymmetry of graphene oxide film

Fitrilawati[1*], N Syakir[1], V Marcelina[1], B Adiperdana[1], L Safriani[1], A Aprilia[1], Risdiana[1], R Siregar[1], J Angel[2] and I Watanabe[2]

[1]Department of Physics, Faculty of Mathematics and Natural Sciences, Universitas Padjadjaran, Jl. Raya Bandung-Sumedang km 21 Jatinangor, Sumedang, West Java, Indonesia 45363
[2]RIKEN Nishina Center, 2-1 Hirosawa, Wako, Saitama 351-0198, Japan

*fitrilawati@phys.unpad.ac.id

Abstract. Graphene has received a lot of attention due to its promising properties for electronic devices application such as energy storage device. Due to the amount needed for application, graphene oxide is widely used as a precursor to obtain graphene-like material. The GO film is thermally reduced or photoreduced to obtain graphene-like that known as reduced graphene oxide (rGO). We intend to measure microscopic intrinsic charge carrier motion of GO film by means of muon spin rotation relaxation (μSR). We prepared thick film of GO by stacking a lot of GO sheets using 4 mg/ml GO-water dispersion. The rGO material was obtained by thermally reduced GO film at 200°C for 2 hours under argon flowing. We measured zero field (ZF) μSR measurements of GO and rGO from 10 K and 300 K. During the μSR measurement we exposed the light into the samples. We discuss the effect of light exposure on the μSR asymmetry and compare to μSR asymmetry of rGO sample.

1. Introduction

Graphene oxide (GO) is two dimensional materials that was prepared from graphite using oxidation process and followed by exfoliation. Schematic illustration of the structural model of a GO sheet is shown in figure 1. The GO sheets are decorated with oxygen-containing functional groups such as carboxyl, hydroxyl and epoxy groups [1]. Due to oxygen containing group, the conjugation on GO chains are interrupted and therefore GO is nearly an insulator. The reduction process for removing the oxygen-containing groups from GO that will heal the defects of the conjugated π-electron of GO need to be applied in order to obtain graphene like material.

Thermal reduction is amongst the technique that commonly applied to remove the contained oxygen. A schematic view of the conversion of GO to reduced graphene oxide (RGO) that shown in figure 1, illustrates a recovery of sp**2**-binding sites of RGO. The restoration of sp^2-carbon sites and the recovery of the conjugated π-electron system is indicated by the strongly increasing UV-Vis absorption spectra of RGO as previously reported [2]. Our previous result also shows that change of GO to rGO accompany by the change of structures [3].

We used muon spin rotation relaxation (μSR) spectroscopy to study the microscopic intrinsic charge carrier motion of GO with varied oxygen content. The study of microscopic intrinsic charge carrier motion by μSR spectroscopy has been performed for several organic and hybrid materials [4-

7]. The first muon spectroscopy investigation of graphene was reported in 2011, showing evident muon spin precession that originated from muon–hydrogen nuclear dipolar interactions [8, 9]. In graphene oxide (GO), light illumination has important role on the healing of conjugated bonds by removing oxygen functional group from the molecule, such that the conjugation length increases and GO becomes more conductive [1]. We are interested to probe how the charge transport properties changes related to the change of structure due to light exposure. We performed zero field (ZF) μSR measurements of GO and RGO sample under light illumination. μSR measurement of Si and Ge under light illumination has been reported previously [10].

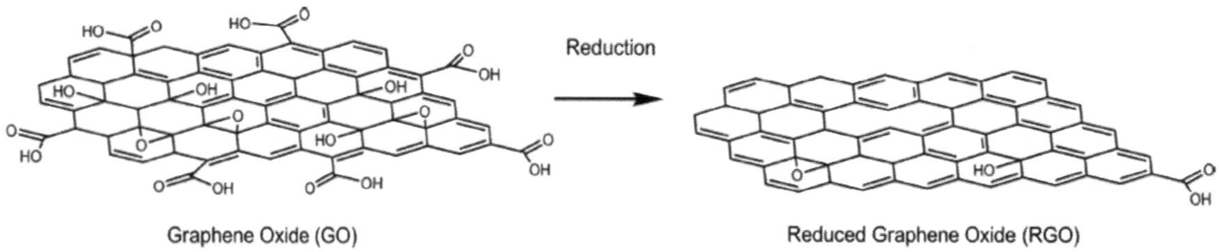

Figure 1. Schematic illustration of the reduction of graphene oxide (GO) to reduced graphene oxide (RGO) [1].

2. Experiments

The free standing GO papers were prepared in petri dish from 4 mg/ml GO dispersed in water, and following by evaporation of water in atmospheric and room until it is almost dried, as previously described [3]. It was further dried at 50°C the vacuum oven overnight and then detaches from the petri dish. In order to obtain graphene-like film that known as rGO, the free-standing GO film then thermally reduced by heating the samples for two hours at 250°C inside the Pyrex tube that flow with Argon gas. We applied Scanning Electron Microscope (SEM) for morphology inspection.

We also exposed thin film of GO with UV light using a home-made UV light source with intensity of 0.12 mW cm^{-2} for varied time exposure and measured its absorbance in the range of 200 nm to 800 nm using UV-Visible spectrometer (PG T70). Sample for the measurement (photoreduction) was very thin and it was deposited in fused silica substrate using spin coating technique.

We have performed zero field (ZF) μSR at varied temperature for both GO and RGO samples under light irradiation at the RIKEN-RAL Muon Facility at the Rutherford-Appleton Laboratory in the UK. The samples were stacked and wrapped in a silver foil and mounted on a silver plate in the cryostat. Measurements were done at temperature varied from 10 K to 300 K using a pulsed positive surface muon beam. In this measurement, the light were exposed to the sample during the μSR measurements.

3. Results

The prepared GO paper were brown-black with a thickness about 30 microns to 50 microns. After thermal reduction for 2 hours at 250°C, the GO paper become darker with a metallic-like. For μSR measurements, about tenth of samples were then stacked in order to get an appropriate mass. Figure 2 shows surface image of GO and rGO samples that measured using SEM. Figure 2(a) shows that GO paper seems quite dense. The surface of rGO film (figure 2(b)) is look similar compare to the GO paper (figure 2(b)), but the rGO paper has more surface roughness in comparison to surface of the GO paper.

While GO sample using for photoreduction experiment was quite thin and transparent. As prepared, it has a brown-look, but after exposed with a UV light with intensity of 0.12 mW cm^{-2}, the sample becomes darker compare to condition before photoreduction process. Quantitatively the change was observed by using UV-Vis spectra that measured at varied time exposure as shown in figure 2. We found that there is an obvious increment of absorption in the range of 230 nm to 800 nm, but the lambda max is remained in 230 nm. The shoulder at 300 nm seems disappeared after exposing

IOP Conf. Series: Journal of Physics: Conf. Series **1080** (2018) 012004 doi:10.1088/1742-6596/1080/1/012004

GO sample for 7 hours' irradiation. The intensity of light used in photoreduction is quite low, but it is still show a change of absorption spectra after light irradiation. It will take long time to observe significant change of absorption spectra of GO. This absorption data has confirmed the effect of light irradiation on GO sample.

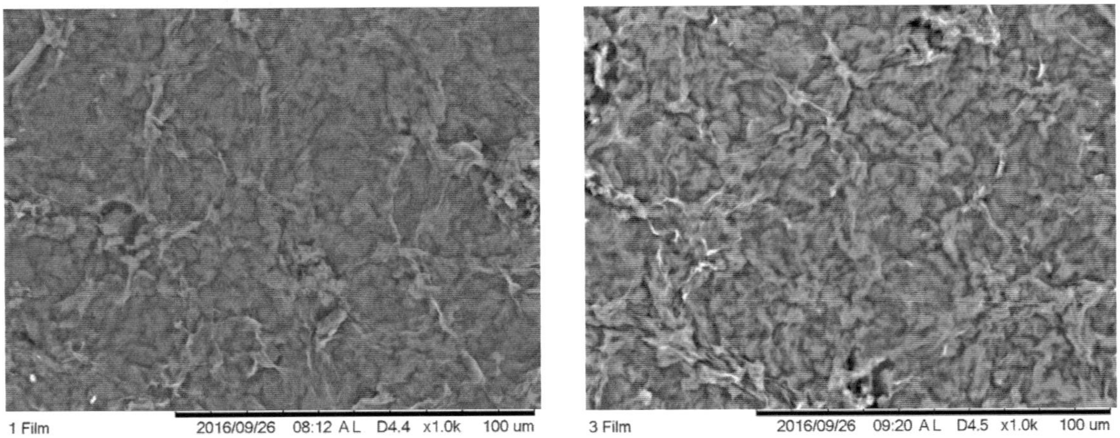

(a) (b)

Figure 2. Comparison SEM images of (a) surface of GO paper and (b) rGO paper.

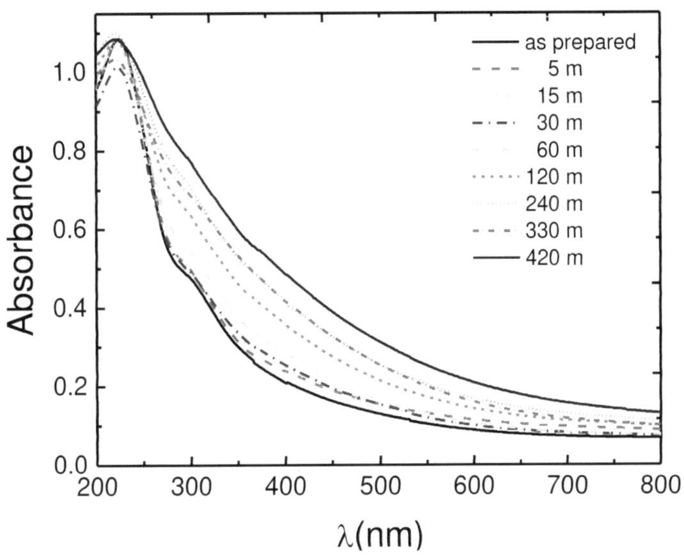

Figure 3. UV-Vis spectra of GO film before and after light irradiation using UV light with intensity of 0.12 mW cm^{-2} at varied time exposure.

The raw data of μSR measurement were shown as an asymmetry parameter as function of time. The asymmetry parameter $A(t)$ at a time t is defined in equation (1).

$$A(t) = \frac{[F(t) - \propto B(t)]}{[F(t) + \propto B(t)]} \quad (1)$$

with $F(t)$ is total muon events counted by the Forward counters, $B(t)$ is total muon events counted by the backward counters, α is the calibration factor reflecting the relative counting efficiencies between the forward and backward counter. The asymmetry of GO and RGO samples at zero field measuring from 10 K to 300 K were shown in figure 4(a) and figure 4(b), respectively. At high temperature of 300 K, the asymmetry of GO and RGO shows a Gaussian-like depolarization. An exponential-like

IOP Conf. Series: Journal of Physics: Conf. Series **1080** (2018) 012004 doi:10.1088/1742-6596/1080/1/012004

depolarization of spectra is observed at low temperatures below 100 K. The degree of depolarization is significant different between asymmetry spectra of GO and RGO. The degree of depolarization in GO samples is larger than that of rGO samples. This phenomenon is probably due to the different intrinsic properties of GO and rGO. When muon implanted to the samples, a muonium, created from a positive muon and an electron, is formed and reacted with other charges in GO or rGO. Since the mobility of GO lower than that of rGO, muonium can feel the intrinsic charge larger in GO than in rGO resulting the large degree of depolarization in GO.

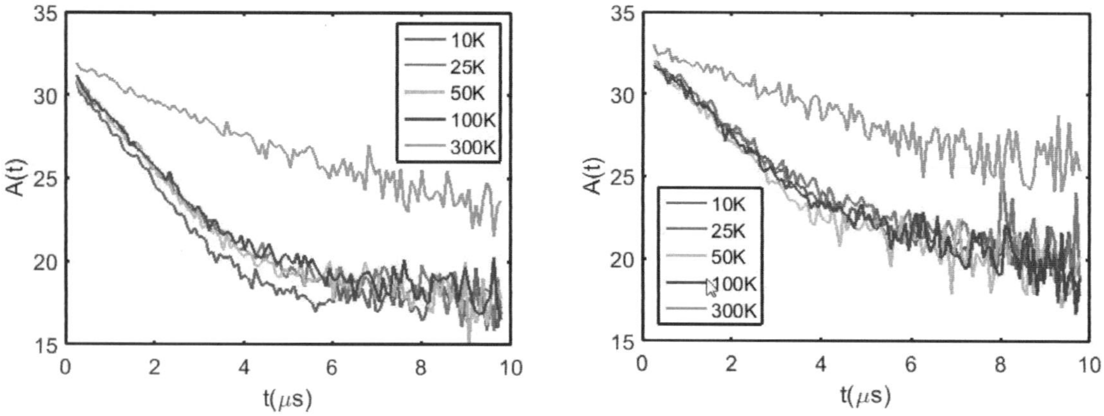

(a) (b)

Figure 4. Asymmetry of (a) GO sample and (b) rGO sample at zero field at varied temperature from 10 K to 300 K.

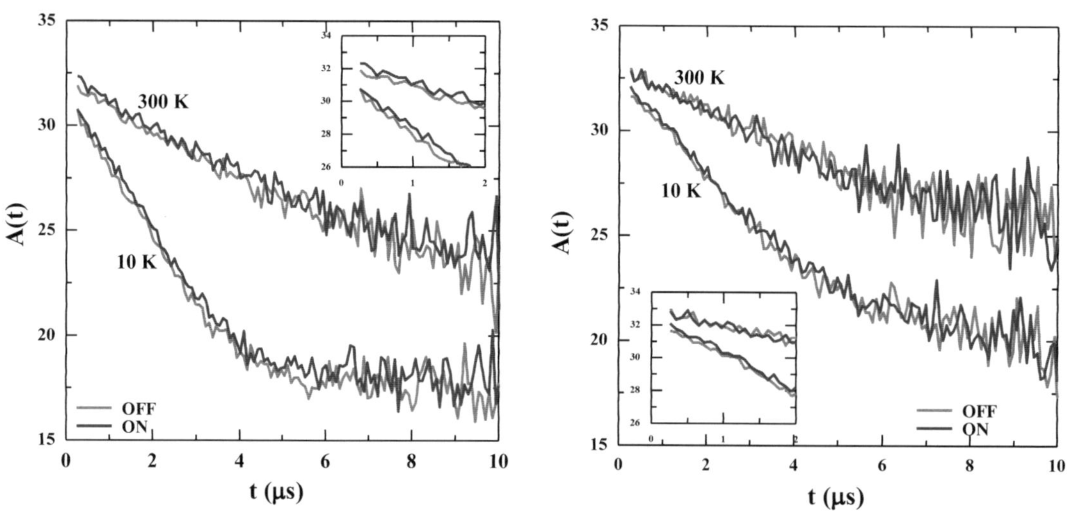

(a) (b)

Figure 5. Asymmetry of (a) GO sample and (b) rGO sample at temperature of 10 K and 300 K in condition of light illumination off (red line) and on (blue line).

Figure 5 shows asymmetry data of GO and rGO before and after light illumination at zero field either at 10 K and 300 K. We can see a small shifted of asymmetry of GO sample after light illumination both at 10 K and 300 K (figure 5(a)). This result shows that both at 10 K and 300 K, μSR asymmetry of GO are affected by light illumination.

The shifted due to light illumination is only small. It is may related to intensity of used white light that not high enough. As compare to change of absorption spectra in figure 3 which the effect only small due to low intensity of light. In order to magnify the effect of light exposure it is necessary to use higher light intensity.

In figure 5(b) we also can see a very small shifted of asymmetry of rGO sample at 10 K after light illumination, but it is not the case at 300 K. This shows that the light illumination seems not change μSR asymmetry of rGO sample at 300 K. There is no distinguishable shifted of Asymmetry rGO at 300 K due to light exposure. More detail result needs more analysis by fitting the asymmetry data.

4. Conclusion

We have obtained asymmetry data of GO and rGO sample using μSR method at zero field measured at varied temperature from 10 K to 300 K. Both GO and rGO samples show that at low temperature the asymmetry consists of fast component and slow component, but at room temperature the asymmetry seems only consist of fast component. At 10 K the GO sample has shorter fast component in compare to rGO sample. Light illumination has a distinguish effect on GO asymmetry, but not on rGO sample.

Acknowledgments

This works was partially supported by Penelitian Unggulan Perguruan Tinggi (PUPT) contract no. 718/UN6.3.1/PL/2017 April 17th 2017 and Academic Leadership Grant (ALG), University of Padjadjaran Contract no: 855/UN6.3.1/PL/2017.

References

[1] Li H and Bubeck C 2013 *Macromol. Research* **21** 290

[2] Fitrilawati, Syakir N, Aprilia A, Liu Z, Feng X and Bubeck C 2015 *Material Science Forum* **827** 317

[3] Fitrilawati, Perkasa M B, Syakir N, Aprilia A, Safriani L, Saragi T, Risdiana, Hidayat S, Bahtiar A, Siregar R, Sihombing R R, Nugroho A 2017 *IOP Conf. Series: Materials Science and Engineering* **196** 012027

[4] Risdiana, Fitrilawati, Siregar R E, Hidayat R, Nugroho A A, Tjia M O and Watanabe I 2010 *Physica B* **405** S381

[5] Risdiana, Fitrilawati, Hidayat R, Siregar R E, Tjia M O and Watanabe I 2010 *Journal of Physics: Conference Series* **200** 052024

[6] Risdiana, Fitrilawati, Siregar R E, Hidayat R, Nugroho A A, Tjia M O, Ishii Y and Watanabe I 2010 *Journal of Physics: Conference Series* **225** 012003

[7] Safriani L, Risdiana, Bahtiar A, Aprilia A, Siregar R E, Hidayat R, Saragi T P I, Kawasaki I and Watanabe I 2014 *Advanced Materials Research* **896** 477

[8] Riccò M, Pontiroli D, Mazzani M and Choucair M 2011 *Nano Lett.* **11** 4919

[9] Gaboardi M, Bliersbach A, Bertoni G, Aramini M, Vlahopoulou G, Pontiroli D, Mauron P, Magnani G, Salviati G, Züttel A and Ricco M 2014 *J. Mater. Chem. A* **2** 1039

[10] Kadono R, A. Matsushita A, Macrae R M, Nishiyama K and Nagamine K 1994 *Phys. Rev. Lett.* **73** 2724

Synthesis and characterization of $La_{2-y}Y_yCuO_4$ with T'-structure

A I Hanifah[1], S Saito[2], T Saragi[1], T Kawamata[2], Y Koike[2] and Risdiana[1*]

[1]Department of Physics, Faculty of Mathematics and Natural Sciences, Universitas Padjadjaran, Jl. Raya Bandung-Sumedang km 21 Jatinangor, Sumedang, West Java, Indonesia 45363
[2]Department of Applied Physics, Graduate School of Engineering, Tohoku University, °Sendai 980-8579 Japan

[*]risdiana@phys.unpad.ac.id

Abstract. Superconductivity is a phenomenon when the resistivity becomes zero below the critical temperature (T_c). The appearance of superconductivity in undoped bulk samples of T'-$La_{1.8-x}Eu_{0.2}M_xCuO_4$ (M = Sr, Ca; x = 0.05, 0.10, 0.15) has altered the existing phase diagram in superconducting cuprates and raised big questions about the mechanism of superconductivity. Here, new materials of T'-$La_{2-y}Y_yCuO_4$ with y = 0.05, 0.10, 0.15 have been synthesized by the low-temperature synthesis method in order to grow undoped superconductors with the single phase of the T'-structure. The powder x-ray diffraction analysis was performed to identify of the crystal structure, while magnetic susceptibility measurements were carried out using a SQUID magnetometer at low temperature down to 2 K. It has been found that the T'-structure is stabilized when it was annealed at 500 – 600°C. The c-axis length decreases with increasing Yttrium-concentration. For y = 0.10, the single phase of the T'-structure was obtained by the oxidation annealing at 300°C and the succeeding reduction annealing at 500°C for 24 hours. From magnetic susceptibility measurements, it has been found that no trace superconductivity is observed in all samples, which is probably due to the remaining excess oxygen suppressing the appearance of superconductivity in the sample.

1. Introduction

The mother compounds of high-T_c cuprates have long been considered to be antiferromagnetic insulators. The superconductivity appears through hole and electron doping into insulating mother compounds of Ln_2CuO_4 (Ln: lanthanide element) with the K_2NiF_4-type (so-called T-type) and Nd_2CuO_4-type (so-called T'-type) structures, respectively. There are a lot of studies reported on the crystal growth, magnetic and transport properties of T-type samples [1-3], which are contrarery to those of T'-type samples [4].

In adequately reduced thin films of T'-$Nd_{2-x}Ce_xCuO_4$, however, it has been reported that superconductivity appears in a wide range of x and even in the undoped mother compound of $x = 0$ [5]. Moreover, Takamatsu *et al.* have succeeded in synthesizing undoped and hole-doped bulk samples of T'-$La_{1.8-x}Eu_{0.2}M_xCuO_4$ (M = Sr, Ca; x = 0, 0.05, 0.10) by the low-temperature synthesis method [6, 7]. These surprising results suggest a new phase diagram of superconductivity and attract great interest on the mechanism of emerging superconductivity in T'-Ln_2CuO_4.

The hole and electron doping are usually achieved by the partial substitution of Sr^{2+} and Ce^{4+} for Ln^{3+}, respectively. The Ionic radius of Ln in Ln_2CuO_4 has a big deal to decide the crystal structure. It prefers to the T'-structure instead of the T-structure when the ionic radius is small [8]. Furthermore, in the T'-$La_{1.8}Eu_{0.2}CuO_4$ (T'-LECO) system, however, it has been known that Ce cannot be substituted for La because of the production of a stable second phase of T'-$Eu_{2-x}Ce_xCuO_4$. Where the ECCO itself, has been reported has the T' tetragonal structure [9]. Therefore, in order to investigate both hole and electron doping dependences of the superconductivity around $x = 0$ using the same T'-Ln_2CuO_4 system, we have chosen T'-$La_{2-x-y}Y_yM_xCuO_4$ (M = Sr, Ce), in which Y is used instead of Eu. In the present work, accordingly, we have tried to prepare superconducting bulk samples of the undoped mother compound of T'-$La_{2-y}Y_yCuO_4$ (T'-LYCO).

2. Experiments

First, samples of T-LYCO with y = 0.05, 0.10, 0.15 were prepared by the conventional solid-state reaction method. The raw materials were mixed using a ball mill and sintered in O_2 gas flow at 1050°C for 24 hours.

Next, the samples of T-LYCO were mixed with reductant CaH_2 in a molar ratio of T-LYCO:CaH_2 = 1:2 in an argon-filled glove box. Then, the mixtures were sealed in an evacuated Pyrex tube, heated at 275°C for 24 hours, and washed with saturated NH_4Cl in anhydrous ethanol, so that samples of $La_{2-y}Y_yCuO_{3.5}$ with the $Nd_4Cu_2O_7$-type (so-called N-type) structure were prepared. Then, the obtained samples of N-$La_{2-y}Y_yCuO_{3.5}$ were oxidized at 200 – 400°C to obtain samples of T'-LYCO. Since the obtained samples of T'-LYCO included excess oxygen, they were reduced in vacuum at various temperatures to remove the excess oxygen. All the samples were characterized by the powder X-ray diffraction (XRD) analysis at room temperature. Magnetic susceptibility measurements were carried out using a superconducting quantum interference device (SQUID) magnetometer.

3. Results and Discussion

From the powder XRD analysis, as shown in figure 1, it has been found that single-phase samples of T-LYCO with y = 0.05 and 0.10 are obtained by the conventional solid-state reaction method, while there remains a small amount of Y_2O_3 for y = 0.15. The peak width of samples sintered at 1050°C is found to be sharp, indicating T-LYCO samples sintered at 1050°C are of high crystalline quality. This condition is reasonable. The T-type structure is stable at high temperatures, because the size mismatch between the larger CuO_2 layer and the smaller LaO layer due to the substitution of Y^{3+} smaller than La^{3+} is relaxed at high temperatures owing to the thermal expansion of the ionic La-O bond larger than that of the covalent Cu-O bond [10].

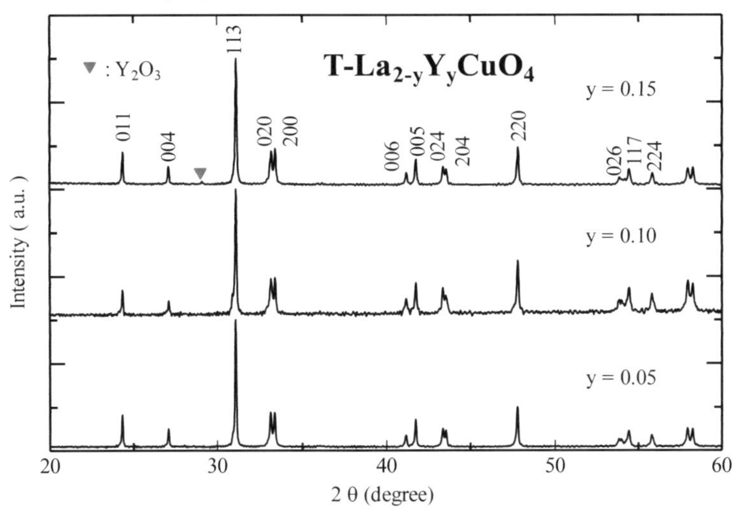

Figure 1. Powder XRD patterns of T-$La_{2-y}Y_yCuO_4$ samples with y = 0.05, 0.10, 0.15 heated in O_2 at 1050°C for 24 hours.

After the reduction using CaH_2, it has been found that single-phase samples of N-$La_{2-y}Y_yCuO_{3.5}$ with y = 0.05 are obtained, while there remains a small amount of Y_2O_3 for y = 0.10 and 0.15. Figure 2 shows the powder XRD patterns of T'-LYCO samples (y = 0.05, 0.10, 0.15). All of the pattern can be indexed as the T'-structure. It is found that T'-LYCO is obtained by the oxidation at 200 - 400°C. These results reveal that as-grown samples of T'-LYCO with y = 0.05 are of the single phase for all oxidation temperatures. Next, with increasing Y-concentration, it is found that a small Y_2O_3 peak appears for both samples with y = 0.10 and 0.15. Furthermore, an unknown peak is also observed for both sample with y = 0.10 and 0.15 around 25°.

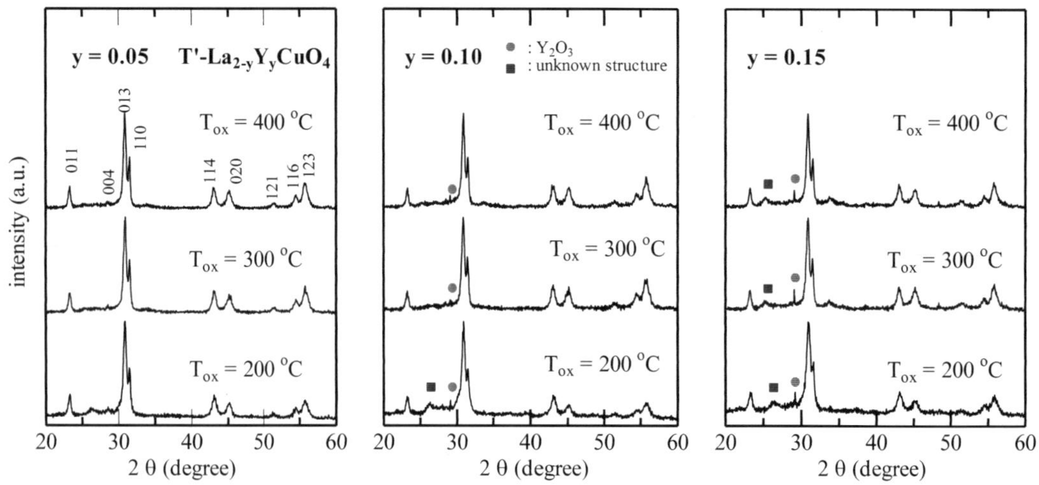

Figure 2. Powder XRD patterns of T'-$La_{2-y}Y_yCuO_4$ samples obtained by the reaction between T-$La_{2-y}Y_yCuO_4$ and CaH_2 and the oxidation at various temperatures. (a) y = 0.05, (b) y = 0.10, (c) y = 0.15. T_{ox} indicates the oxidation temperature.

Figure 3 shows the powder XRD pattern of the samples obtained by the reduction annealing. After the reduction annealing at various temperatures in vacuum for 24 hours for the removal of excess oxygen, it is found that single-phase samples of T'-LYCO are obtained only by the reduction annealing at 500°C and 550°C. While Y_2O_3 and La_2O_3 are found to appear together with T'-LYCO with increase of the Y-concentration and the reduction annealing temperature above 600°C, respectively. This may be due to the decomposition on the surface of crystalline grains.

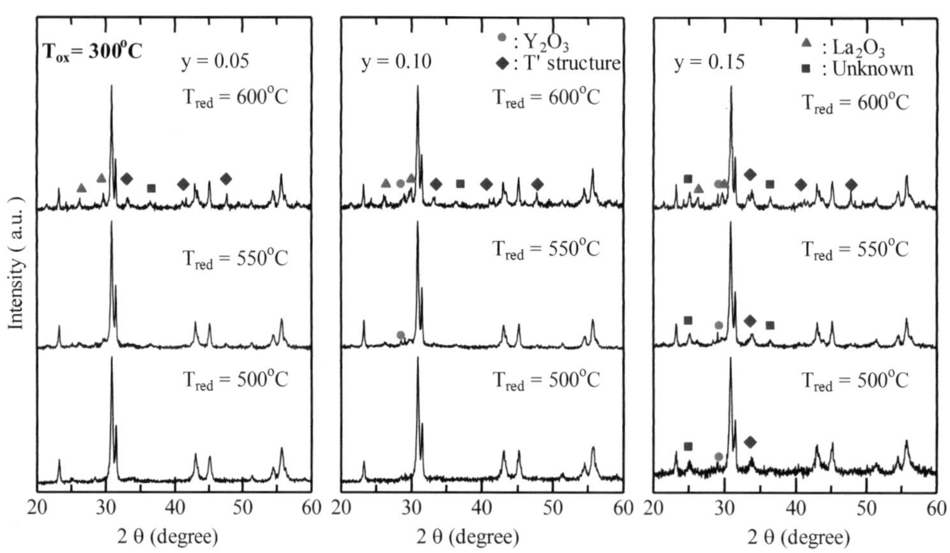

Figure 3. Powder XRD patterns of T'-$La_{2-y}Y_yCuO_4$ samples oxidized at 300°C and reduced at various temperatures. (a) y = 0.05, (b) y = 0.10, (c) y = 0.15. T_{red} indicates the reduction annealing temperature.

The T'-structure has been found to be broken with increasing reduction annealing temperature T_{red}, indicating that the T'-structure is unstable at high temperatures on account of the removal of indispensable oxygen by the strong reduction. The broken sample can be confirmed visually, because a part of the powder sample turns into brown. This is consistent with the former report [11] where it has been described that T'-La_2CuO_4 is stable only below 600°C.

Figure 4 shows the y dependence of the c-axis length of T'-LYCO samples reduced on various conditions. It is found that the c-axis length slightly decrease with increasing y. This is reasonable, because the ionic radius of Y^{3+} (0.92) Å is smaller than that of La^{3+} (1.16) Å. However, in the case of samples of y = 0.15, the c-axis lengths tend to be larger than those of y = 0.10. This may be caused by the presence of impurities of Y_2O_3 for y = 0.15.

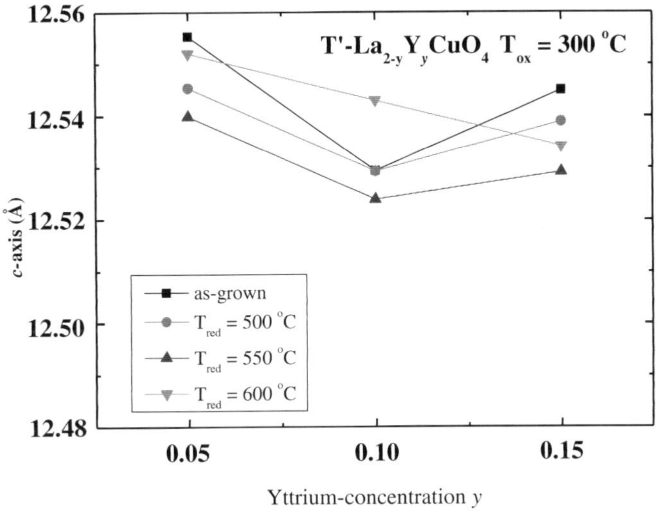

Figure 4. Yttrium-concentration dependence of the c-axis length of T'-$La_{2-y}Y_yCuO_4$ samples oxidized in O_2 at 300°C for 12 hours and reduced at various temperatures.

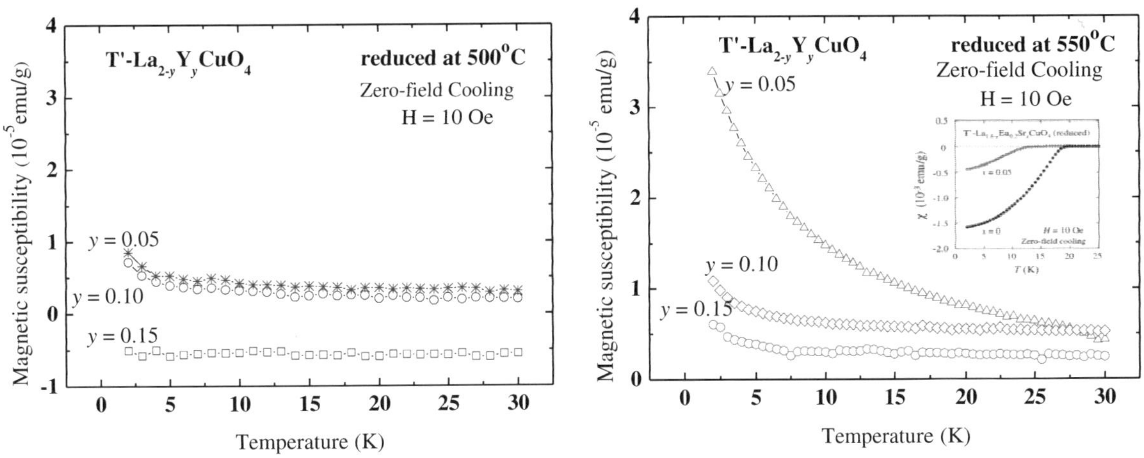

Figure 5. Temperature dependence of the magnetic susceptibility in a magnetic field of 10 Oe for T'-$La_{2-y}Y_yCuO_4$ samples (y = 0.05, 0.10, 0.15) reduced at 500°C and 550°C. The inset shows the temperature dependence of the magnetic susceptibility of superconducting T'-$La_{1.8}Eu_{0.2}CuO_4$ for reference [6].

Figure 5 shows the temperature dependence of the magnetic susceptibility of T'-LYCO. It is found that no superconductivity appears in any reduced samples. The magnetic susceptibility of superconducting T'-LECO is also shown in the inset of figure 5 for reference. It is known that undoped superconductivity is observed in samples with little excess oxygen whose *c*-axis length is not too large [6]. Meanwhile, the present result shows that the *c*-axis length of T'-LYCO is ~12.54 Å and longer than that of superconducting T'-LECO (12.458 Å), suggesting that there remains excess oxygen yet in T'-LYCO samples and that the excess oxygen suppresses the appearance of superconductivity.

4. Conclusions

LYCO (y = 0.05, 0.10, 0.15) samples with the T'-structures have successfully been synthesized by the low-temperature synthesis method at various temperatures below 600°C. The *c*-axis length decreases with increasing Yttrium-concentration. For y = 0.10, the single-phase sample of the T'-structure has been obtained by the oxidation annealing at 300°C and the succeeding reduction annealing at 500°C for 24 hours, with the *c*-axis length of 12.53 Å. However, no superconductivity has been observed in all the samples which is probably due to the existance of excess oxygen in the samples.

Acknowledgments

The author would like to thank M. Kato, T. Noji, and H. Sato for their technical support in the synthesis and characterisation measurement at Department of Applied Physics, Tohoku University, Japan. These works were supported by Junior Year Program in English (JYPE) Tohoku University, 2016-2017, Academic Leadership Grant, Universitas Padjadjaran 2017 No. 872/UN6.3.1/LT/2017 and also partially supported by International Research Collaboration and International Publication Grant No. 718/UN6.3.1/PL/2017.

References

[1] Suzuki K, Adachi T, Tanabe Y, Koike Y, Kawamata T, Risdiana, Suzuki T and Watanabe I 2010 *Phys. Rev.* B **82** 054519
[2] Tanabe Y, Adachi T, Risdiana, Kawamata T, Suzuki T, Watanabe I and Koike Y 2009 *Physica B* **404** 717
[3] Koike Y *et al.* 2005 *Physica C* **426-431** 189
[4] Risdiana, Adachi T, Oki N, Koike Y, Suzuki T, Watanabe I 2010 *Phys. Rev.* B **82** 014506
[5] Matsumoto O, Utsuki A, Tsukada A, Yamamoto H, Manabe T, Nito M 2009 *Physica C* **469** 924
[6] Takamatsu T, Kato M, Noji T, Koike Y 2012 *Appl. Phys. Express* **5** 073101
[7] Takamatsu T, Kato M, Noji T, Koike Y 2014 *Phys. Procedia* **58** 46
[8] Sawa H *et al.* 1989 *Nature* **337** 347
[9] Risdiana, Saputri M, Sobari M F, Hanifah A I, Somantri W A and Saragi T 2017 *IOP Conf. Series: Materials Science and Engineering* **196** 012012
[10] A. Manthiram and J. B. Goodenough, J 1991 *Solid State Chem* **92** 231
[11] Chou F C, Cho J, Uchida Y and Kato K 1990 *Phys. Rev. B* **42** 6172

Characteristics of crystalline silica (SiO_2) particles prepared by simple solution method using sodium silicate (Na_2SiO_3) precursor

I M Joni[1,2*], L Nulhakim[2], M Vanitha[2] and C Panatarani[1,2]

[1]Department of Physics, Faculty of Mathematics and Natural Sciences, Universitas Padjadjaran, Jl. Raya Bandung-Sumedang km 21 Jatinangor, Sumedang, West Java, Indonesia 45363
[2]Nanotechnology and Graphene Research Centre, Universitas Padjadjaran, Jl. Raya Bandung-Sumedang km 21 Jatinangor, Sumedang, West Java, Indonesia 45363

*imadejoni@phys.unpad.ac.id

Abstract. The crystalline silica (SiO_2) particle is successfully prepared using simple solution method from sodium silicate (Na_2SiO_3) precursor. The FTIR spectrum of the sample confirms the presence of SiO_2. The X-ray diffraction (XRD) shows that the sample is cristobalite type of SiO_2 which is comparable with ICSD ref. number of 01-076-0941. The crystallite size is about 28 nm as calculated by Scherrer method and the average particle diameter was about 697 nm as confirmed by Particle Size Analyzer (PSA) measurement. The Scanning Electron Microscope (SEM) image reveals that the sample reveal micro-flake with irregular rod-shaped morphology. The purity of sample is examined by X-ray Fluorescence (XRF) which shows about 93.1 mass % of pure SiO_2 whereas, the purity of the raw precursor Na_2SiO_3 before synthesis is about 60.5 Mass %. The synthesized SiO_2 particles can be used for several applications, where SiO_2 is used in its crystalline phase.

1. Introduction

In the last few years, there has been growing interest in the preparation of SiO_2 particle for a variety of applications where the purity of materials is one of the important key parameters. Pure SiO_2 particles with crystalline phase can be applied as the material source in photovoltaic [1], semiconductor electronic devices, catalysis, film substrates, ceramics, humidity sensors, etc. [2, 3]. Synthesis of the SiO_2 particle can be done by several methods such as chemical vapour deposition, plasma and combustion synthesis, hydrothermal method and also sol-gel processing. Among these methods, the simple solution method as a part of sol-gel has advantages of low cost in processing and also ease in controlling the properties of SiO_2, i.e., purity and homogeneity or modification of material composition. The report of simple solution method can be seen in Ref. [4-6]. In recent reports, the silica particle was prepared from an expensive raw material such as tetraethyl orthosilicate (TEOS) [7-8]. However, a simple method using inexpensive source materials is required for the mass production of the SiO_2 particle. In this work, the SiO_2 particle was prepared using simple solution method from an economic raw material, i.e., Na_2SiO_3 and its result were described.

2. Experiments

The starting material for synthesizing of SiO_2 is Na_2SiO_3 solution. Firstly, the precursor (Na_2SiO_3) was mixed with distilled water (with fraction 1:2) and then 2 M HCl was added and stirred at room temperature (RT) until a clear homogeneous mixture was obtained. The white powder was recovered by filtration followed by cleaning with distilled water in order to neutralize the acid and dried at a temperature of 60°C for about 12 h. In order to form crystalline phase of SiO_2, 2 g of the obtained white powder is mixed with PEG 6000 and distilled water in RT. The amount of SiO_2: H_2O: PEG 6000 fraction was about 1: 50: 5. The mixture of material was calcined at 900°C for about 2 h. Finally, the powder sample was characterized by various sophisticated techniques. The presence of SiO_2 in the sample was characterized by Fourier-transform infrared (FTIR) analysis (Shimadzu prestige 21). The structure and phase of silica were examined by X-ray Diffraction (XRD) Rigaku with Cu Kα with λ = 1.54060 Å. The average of particle diameter was measured by Particle Size Analyzer (PSA) Beckman Coulter LS 13 320. The purity of SiO_2 and its traces (Mass %) was measured by X-ray Fluorescence (XRF) Rigaku NEX CG. The morphology and elemental analysis of the sample were characterized by using Scanning Electron Microscopy (SEM) equipped with energy dispersive X-ray (EDX), FEI-Inspect 550, EDAX Ametek.

3. Results and Discussion

In order to confirm the presence of SiO_2, the sample was characterized using FTIR analysis. The spectrum of the sample after calcination at a temperature of 900 °C is shown in figure 1. It can be seen that two main characteristics peak were observed at around 794 cm^{-1} and 1079 cm^{-1} which is attributed to Si-O bending vibration band and asymmetric stretching vibration of the siloxane bonds (Si-O-Si) [9, 10], respectively. The band at about 473 cm^{-1} was associated with bending vibration modes in O-Si-O network [11]. On the other hand, the band at around 2357 cm^{-1} is concomitant to Si-C stretching (interaction of C from PEG with Si) and the band at around 3450 cm^{-1} is concomitant with Si-OH, characteristics of -OH group which commonly exists in water [12]. This peak still prevails even though the sample was calcined at high temperature. The reason for Si-OH existence is the physically absorbed water molecules from the environment by silica.

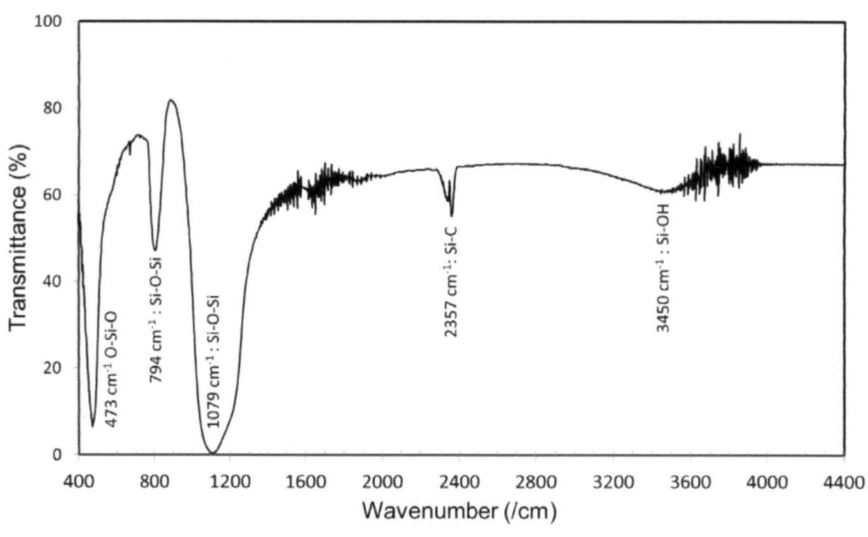

Figure 1. The FTIR Spectra of crystalline SiO_2 prepared by simple solution method.

Secondly, the structural property of sample was examined by XRD. The results showed that the SiO_2 exhibits a crystalline nature (figure 2(a)), representing cristobalite low type of SiO_2 (figure 2(b)). It matches with the reference of 01-076-0941 number (Calculated from ICSD using POWD-12++,

1997) [13]. The crystal phase of this type of silica is tetragonal with a lattice parameter of a-axis about 4.99 Å, b-axis 4.99 Å, and c-axis 7.02 Å, respectively. The alpha, beta and gamma were 90 degrees, respectively. However, the addition of PEG and calcination in high-temperature influences the crystallinity and structural phase of the material. The crystallite size (D) of the sample was calculated using Debye-Scherer's equation (1).

$$D = \frac{0.9\lambda}{\beta \cos \theta} \tag{1}$$

where λ is the wavelength of X-ray beam (λ=1.5406 Å), β is the full width at half maximum (FWHM) of the intense peak and θ is the Braggs angle. Based on the above calculation, crystalline SiO_2 has crystallite size of about 28 nm.

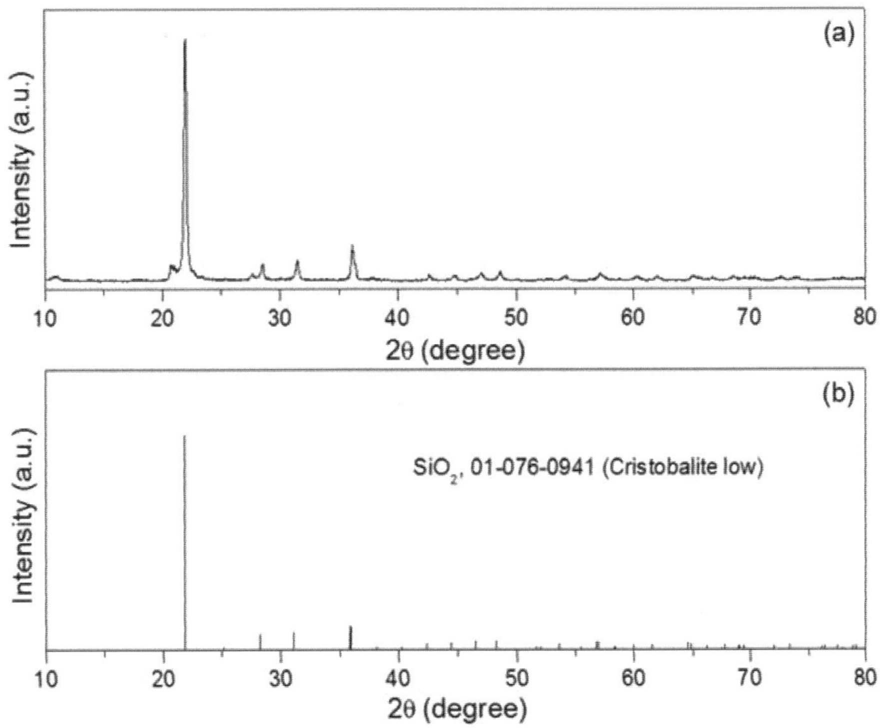

Figure 2. The XRD pattern of (a) SiO_2 powder, and (b) the standard SiO_2 pattern, cristobalite low type.

On the other hand, the particle diameter of the sample was measured by PSA analysis. The result (figure 3) showed that the average particle diameter was about 697 nm while volume median diameter of d_{50} (represent 50% of the total volume) was about 305 nm. This result was still larger compared to the crystallite size. Theoretically, it should be noted that the particle diameter will always be higher than that of crystallite size. We predict that this particle diameter was influenced by the existence of many agglomerations of particles on the sample which was detected by PSA. The existence of agglomeration and morphology of sample can be seen in figure 4.

The SEM characterization showed the surface morphology of silica particle at different magnification, (a) 100.000×, (b) 50.000×, (c) 20.000×, and (d) 150.000× (figure 4). The morphology of sample depicts mostly micro-flake and irregular rod-shape with agglomeration.

IOP Conf. Series: Journal of Physics: Conf. Series **1080** (2018) 012006 doi:10.1088/1742-6596/1080/1/012006

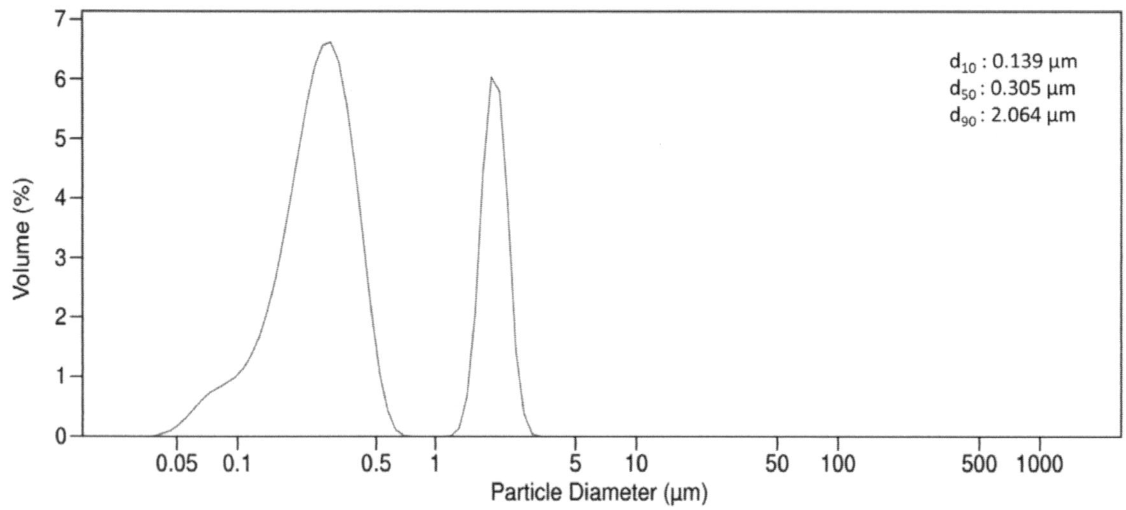

Figure 3. The particle diameter for crystalline SiO_2 prepared by simple solution method.

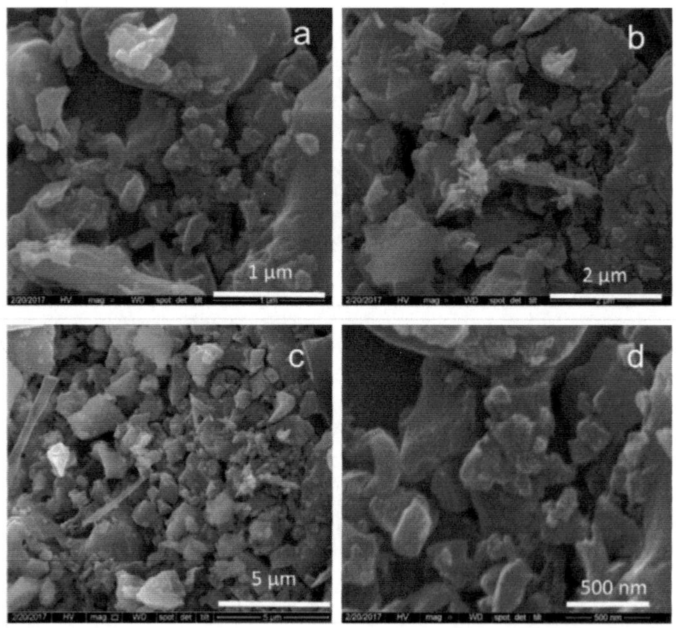

Figure 4. The SEM image of SiO_2 with various magnifications, (a) 100.000×, (b) 50.000×, (c) 20.000×, and (d) 150.000×.

The elemental analysis of SiO_2 particle is shown in figure 5. The presence of silica is confirmed by elemental analysis. The presence of oxygen in the EDX result indicates that the formation of silica as silicon dioxide. The element of oxygen is about 62 at % and Si is about 38 at%.

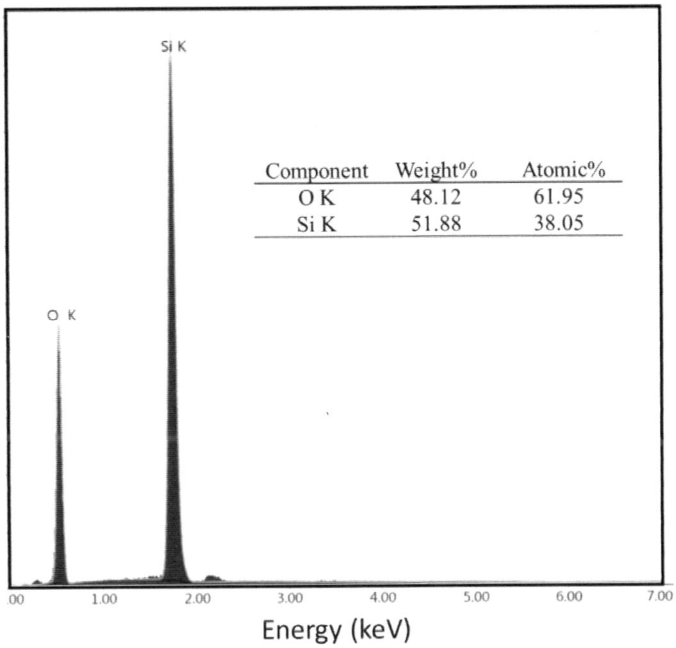

Figure 5. The result of EDX measurement of SiO_2 sample

Furthermore, not only the structural, morphology, and elemental analysis of sample but the purity of this SiO_2 particle was also observed. The purity of sample was examined by the XRF analysis. Table 1 shows the mass percentage of elements in sodium silicate as a precursor and the sample of crystalline silica. The purity of SiO_2 with sodium silicate as a raw precursor was about 60.5 % and the second highest is Na of about 38.6 % followed by other components as shown in table 1. On the other hand, the crystalline SiO_2 prepared by simple solution method has a high purity of SiO_2 of about 93.1 % with traces in a very less percentage (for each component).

Table 1. The mass percentage of elements in sodium silicate precursor and sample of crystalline silica prepared.

Component	Precursor (Mass%)	SiO_2 Particles (Mass%)
SiO_2	60.5	93.1
Na	38.6	1.13
ZrO_2	0.38	1.13
Al	0.34	0.41
TiO_2	0.04	1.18
Cl	0.04	-
S	0.03	-
Ca	0.025	0.63
Fe	0.012	0.43
K	0.0103	1.55
Cu	0.0011	-
ZnO	0.0007	1.07

4. Conclusions

Crystalline silica has been successfully prepared by using simple solution method and economic raw materials (Na_2SiO_3). The FTIR and EDX results confirmed the presence of silica in the form of silicon dioxide. The XRD showed that the silica prepared has characteristics crystalline phase with an average crystallite size of about 28 nm. This silica showed micro-flakes and irregular rod-shaped morphology with agglomeration as confirmed by SEM analysis. The XRF characterization displayed the purity of SiO_2 to be about 93.1%. In general, these characteristics can be used as an initial reference on the mass production of crystalline silica powder from an economic raw material for industrial and wide-scale applications.

Acknowledgements

The authors would like to acknowledge PUPT Research Scheme, Ministry of Research, Technology and Higher Education of the Republic of Indonesia, contract no. 30/E/KPT/2017 for funding this research.

References

[1] Rizky N P, Herny A B,. Agus M H, Sekartedjo and Doty D R 2017 *Proc. Eng.* **170** 93
[2] Bogus G H, Tracy M A and Zukosi C F 1988 *J. Non-Cryst Solids* **104** 95
[3] Sadeghi M, Dorodian M and Rezaei M 2013 *J. Adv. Chem.* **6** 917
[4] Panatarani C, Anggoro D and Faizal F 2010 *AIP. Conf. Proc. Eng.* **1284** 77
[5] Panatarani C, Anggoro D and Joni I M 2012 *AIP Conf. Proc.* **1454** 227
[6] Panatarani C and Joni I M 2013 *AIP Conf. Proc.* **1554** 109
[7] Guo Q, Huang D, Kou X, Cao W, Li L, Ge L and Li J 2017 *Ceram. Inter.* **43** 192
[8] Dubey R S, Rajesh Y B R D and More M A 2015 *Mat. Today: Proc.* **2** 3575
[9] Bansal V, Rautaray D, Bharde A, Ahire K, Sanyal A, Ahmad A and Sastry M 2005 *J. Mater. Chem.* **15** 2583
[10] Patel B H and Patel P N 2014 *Research J. Chem. Sci.* **4** 52
[11] Kalapathy U, Proctor A and Shultz J 2002 *Bioresource Technol.* **185** 285
[12] Moncada E, Quijada R and Retuert J 2007 *Nanotech.* **18** 335606
[13] Peacor D R and Kristallogr Z 1973 *Kristallgeom. Kristallphys. Kristallchem.* **138** 274

IOP Conf. Series: Journal of Physics: Conf. Series **1080** (2018) 012007 doi:10.1088/1742-6596/1080/1/012007

Analysis of structural properties of X-ray diffraction for composite copper-activated carbon by modified Williamson-Hall and size-strain plotting methods

Heryanto, Hendri, B Abdullah and D Tahir*

Department of Physics, Hasanuddin University, Makassar, 90245 Indonesia

*dtahir@fmipa.unhas.ac.id

Abstract. Structural properties of composite carbon with metal Cu (CCu) were analysis from the X-ray Diffraction (XRD) data by modified Williamson-Hall (W-H) and size-strain plotting methods. The effect of temperature shows influenced to increasing the resistance from 37.3% at room temperature to 66.4% for 600°C. By using the modified equation of Williamson-Hall, we found that the crystallite size about 40.78 nm. Modified Williamson-Hall (W-H) plotting and Gaussian distribution shows good correlation and accuration for determining the crystallite size from the XRD data in this study.

1. Introduction

Determination crystallite size from XRD data for composite metal and carbon quite rarely reported [1]. A convenient method is needed for determine the crystallite size from XRD data. Several reported for determine crystallite size usually by Scherrer equation but still quite different with result from transmission electron microscope (TEM) [2].

$$D_S = \frac{k\lambda}{\beta_S \cos(\theta)} \quad (1)$$

where D is crystallite size, k is the Scherrer constant (usually 0.98 for nanoparticle), λ is the wavelength (nm), β_s is full width half maximum (FWHM) of the diffraction peak in radian. β_s parameter should be corrected because instrumental effect by using standard sample. According to reference [1-8], the expression for instrumental effect correction β_s [3], where β_{exp} and β_{std} are FWHM from the width of sample in this study and width of the standard sample, respectively.

$$\beta_S = \sqrt{\beta_{exp}^2 - \beta_{std}^2} \quad (2)$$

Relation between the crystallite size and microstrain (ε) [4] as follows:

$$\varepsilon = \frac{\beta_S}{4\sin(\theta)} \quad (3)$$

where β_s is broadening due to microstrain and crystallite size, therefore, we can writte as follows:

$$\beta_S = \frac{k\lambda}{D_S \cos(\theta)} + 4\varepsilon \tan(\theta) \quad (4)$$

by multiplied cos(θ), equation (4) became:

$$\beta_s \cos(\theta) = \frac{k\lambda}{D_s} + 4\varepsilon_s \sin(\theta) \quad (5)$$

Equation (5) is called as Williamson-Hall (W-H) equation. The Crystallite size and microstrain are obtained from the intercept and slope value between $\beta_s \cos(\theta)$ as function of 4 $\sin(\theta)$, respectively. In this study, we modified the β_s value with interpolation-convolution and Gaussian distribution. We used linear interpolation techniques [5] as follows:

$$y = y_1 + \frac{(\varphi-\varphi_1)}{(\varphi_2-\varphi_1)}(y_2 - y_1) \quad (6)$$

The result of interpolation is y, (φ_1, y_1) and (φ_2, y_2) are the bottom and upper limit of a data, respectively . The φ axis as 2theta and intensity as y axis, interpolated data used for convolution as expressed in equation (7).

$$y = y_1 + \frac{(\varphi-\varphi_1)}{(\varphi_2-\varphi_1)}(y_2 - y_1) \quad (7)$$

Symbol (^{c}) is convolution operator and (η) is dummy variable. Result convolution is h(φ), f(φ) and g(φ) are input data and kernel convolutions, respectively. By convolution, the displacement line broadening (β_c) determined from the standard and instrumental line broadening effect. Gaussian distribution use for calculation the FWHM from the convolution method. According to reference [6] Gaussian formula (P(φ,μ,σ)) expressed in equation (8).

$$P(\varphi, \mu, \sigma) = \frac{1}{\sqrt{2\pi}\sigma} e^{-\frac{(\varphi-\mu)^2}{2\sigma^2}} \quad (8)$$

The mean is μ and the variance is σ. According to the derivative of Gaussian distribution the FWHM is about 2,35σ. Knowledge of crystallite size of composite metal is essential for search which suitable in the characteristic material for different applications. In this study, we focus our attention for calculation of crystallite size by using the modified W-H ploting from the x-ray diffraction data.We applied several method to determine the β_c, β_s, and FWHM by Gaussian methods (β_g) for accurate crystallite size.

2. Preparation and Instrumentation

In this study, composite (C-Cu) was prepared by mixing Cu powder (CAT 2703) with activated carbon (AC) 10% by Retsch MM for 30 minutes at frequency 10 Hz, which was controlled by XRF. The final mass of the sample is 15 grams and annealing in the free oxygen environment at 600°C for 30 minutes. The sample was measured by using X-ray diffraction (XRD) Shimadzu X-RD 7000. X-RD instrument with CuKα radiation using qualitative analysis and report of β values FWHM at 2θ values with the voltage input 40kV and current input 30 mA. Observation angle from are 20° to 80° with step size 0,02°/minutes. We compared our data with standard data from the Join Committee Diffraction Standar (JCPDS: 96-901-1605) [7] to identify the crystal orientations of the main peaks ([111], [200] and [202]).

3. Method and Analysis

Result of interpolation are density data then we convolute the diffraction data between standard Cu CAT 2703 and (CCu). FWHM determine from the Gaussian distribution but relatives larger compare with result from convolution (β_c) and Scherrer equations (β_s) . By using Williamson Hall (W-H) plots by using uniform deformation model (UDM) for explains distortion of the strain and imperfections of crystal induced broadening from the equation (3) [8]. The addition of the Scherrer equation after equation (3) we will get an equation (4). We modified β_s value in equation (4) with β_g and β_c. FWHM modified (β_m) is in between β_s, β_k and β_c as:

$$\beta_m \cos(\theta) = \frac{k\lambda}{D_m} + 4\varepsilon_m \sin(\theta_m) \tag{9}$$

D_m is crystalline size from W-H modified and ε_m is strain with modified which was determine from the intercept fitted data. We obtain D_m component and ε_m component by plotting β_m Cos(θ) versus 4sin(θ). In this study, We applied W-H plots UDM method [4][8].

4. Result

X-RD spectra are shown in Figure 1 for Cu, Cu_2O and CuO. We determine the diffraction peaks of Cu for determining the crystallite size. We compared between Cu and composite Cu with 10% C in figure 1. Carbon effective to increase the resistance of Cu by thermal treatment, from 37,3% to 66,4%. Composite of Cu at 600°C allows us to estimate crystallite size only for the peaks [111], [200] and [202].

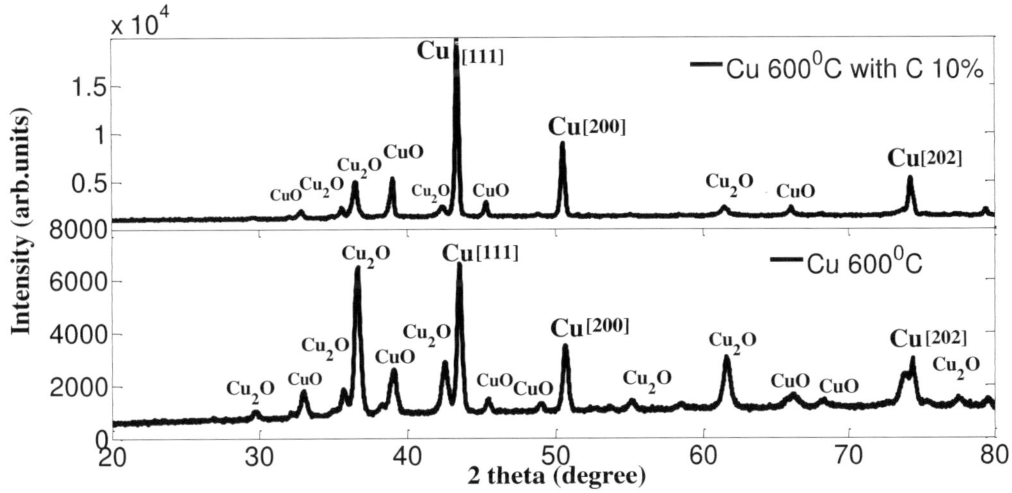

Figure 1. X-ray diffractogram pattern of Cu.

Figure 2 and figure 3 are convolution proses and calculation FWHM by Gaussian method. B_g value is in between β_s and β_c indicated that the weakness of the Gaussian distribution method due to assumption is broadening peak is ideal.

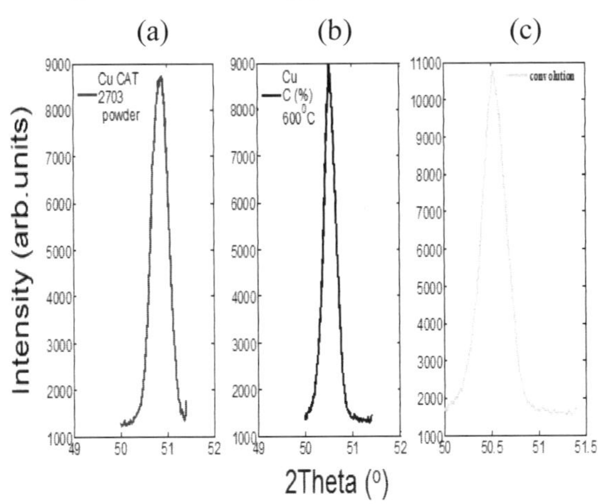

Figure 2. (a) is Cu CAT Standard data, (b) Carbon 10% and Cu 90%, and (c) result of convolution.

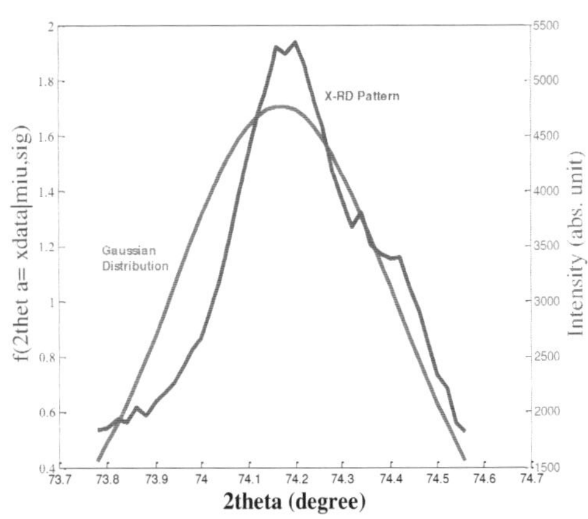

Figure 3. Comparison of the peak diffraction and Gaussian distribution in determine FWHM value.

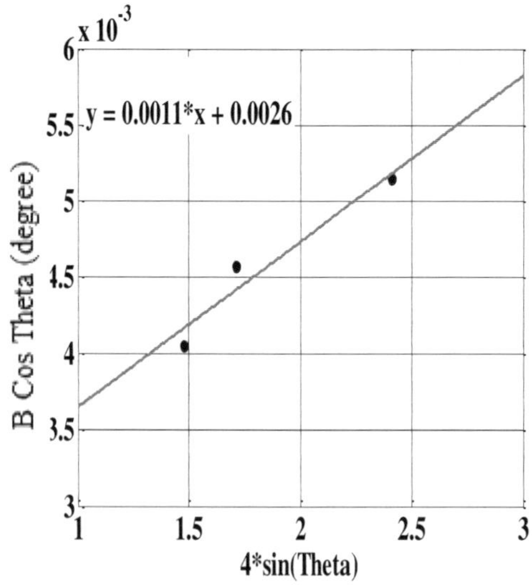

Figure 4. Uniform density model (UDM) without modified W-H plotting.

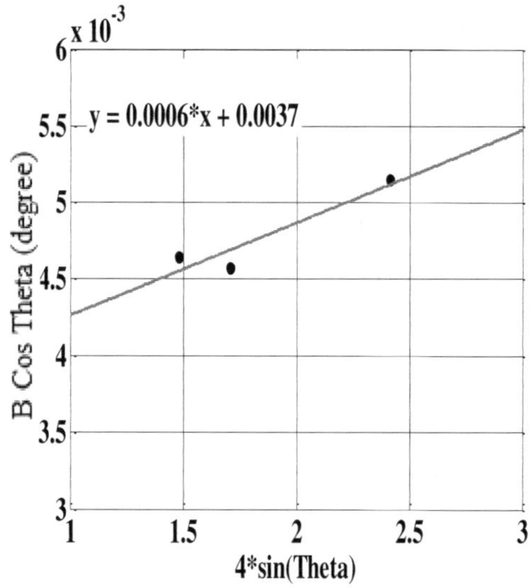

Figure 5. Uniform density model (UDM) with modified W-H plotting.

Figure 4 shows the slope (b=0.0006) and intercept (a=0.0037) by modified W-H plotting. Figure 5 shows the slope (b=0.0011) and intercept (a=0.0026) by W-H plotting (without modification). Residual error by fit line is shows in equation (10).

$$r_i^2 = \sum_{i=1}^{n}(y_i - (a + b\varphi_i)) \tag{10}$$

For n data points (φ,y) the slope and intercept shows in equation (11).

$$b = \frac{(\sum_{i=1}^{n}\varphi_i y_i) - n(\bar{\varphi}\bar{y})}{(\sum_{i=1}^{n}\varphi_i^2 - n\bar{\varphi}^2)} \quad and \quad a = \frac{\bar{y}(\sum_{i=1}^{n}\overline{\varphi_i^2}) - \bar{\varphi}\sum_{i=1}^{n}\varphi_i y_i}{\sum_{i=1}^{n}\varphi_i^2 - n\overline{\varphi^2}} \tag{11}$$

Algebra form of equation (11) can be written by equation (12).

$$na + b\sum_{i=1}^{n}\varphi_i = \sum_{i=1}^{n}y_i \ and \ a\sum_{i=1}^{n}\varphi_i + b\sum_{i=1}^{n}\varphi_i^2 = \sum_{i=1}^{n}\varphi_i y_i \tag{12}$$

Where y and $\bar{\boldsymbol{\varphi}}$ the average values of the y_i and $\boldsymbol{\varphi_i}$. If we compute for both equations can be written in matrix form (equation (13)).

$$\begin{bmatrix} n & \sum\varphi_i \\ \sum\varphi_i & \sum\varphi_i^2 \end{bmatrix}\begin{bmatrix} a \\ b \end{bmatrix} = \begin{bmatrix} \sum y_i \\ \sum\varphi_i y_i \end{bmatrix} \tag{13}$$

Equation (13) used for extracting crystalline size from intercept and microstrain extracted from slope value. Table 1 shows the crystalline size by Scherrer equation, interpolation-convolution, Gaussian distribution and modified Williamson-Hall plotts.

Table 1. Cystallite size for Cu.

Hkl	Scherrer equation (nm)	Scherrer with interpolation-convolution (nm)	Scherrer with Gaussian Distribution (nm)	Williamson Hall with modification (nm)	Williamson Hall without modification (nm)
[111]	32,5	37,2	25,89		
[200]	34,2	32,9	23,11	40,78	58,04
[202]	39,7	29,3	20,76		

5. Conclusion

We successful analysis the crystallite size for composites CCu with several methods from the X-RD spectra shows is effective way to increase the thermal resistance from 37,3% to 66,4%. In this study shows effective and efficient method for determine crystallite size from Interpolation-convolution, Gaussian distribution and modified W-H plotts.

Acknowledgments

We would like to thanks to the Hasanuddin University, Indonesia by PKLN-UH 2018 program for financial support of this research

References

[1] Wijaya S R, Tjandrawinata R and Parangtopo 1996 *Prosiding Pertemuan Ilmiah Sains Materi.* (Indonesia: PPSM)
[2] Monshi A, Foroughi M R and Monshi M R 2012 *World Journal of Nano Science and Engineering* **2** 154
[3] Goncalves N S, Carvalho J A, Lima Z M, Sasaki J M 2012 *Materials Letter* **72** 26
[4] Mote V D, Purushotham Y and Dole B N 2012 *Journal of Theoritical and Applied Physics* **6:6**
[5] Stoyan G 1978. *Applied Mathmetical Science 27 in Journal of Applied Mathematics and Mechanics Springer-Verlag* **XXIV392 S** DM 32,50
[6] Ribeiro M I 2004 *Gaussian Probability Density Function: Properties and Error Characterzatio* (Lisboa Portugal : Institute for System and Robotics Instituto Superior Tcnico 1049-001)
[7] Davey W P 1925 *Physical Review* **25** 753
[8] Prabhu Y T, Rao K V, Kumar V S S, Kumari B S 2014 *India. World Journal of Nano Science and Enginering* **4** 21

IOP Conf. Series: Journal of Physics: Conf. Series **1080** (2018) 012008 doi:10.1088/1742-6596/1080/1/012008

Crystallite size determination of barium hexaferrite nanoparticles using WH-plot and WPPM

M Manawan[1,2,4*], T Saragi[1], A Sukandi[2], Fachrudin[2], B Kurniawan[3], A Manaf[3], E P Boedijono[4] and Risdiana[1]

[1]Department of Physics, Universitas Padjadjaran, Sumedang 45363, Indonesia
[2]Energy Engineering, Politeknik Negeri Jakarta, Depok 16424, Indonesia
[3]Department of Physics, Universitas Indonesia, Depok 16424, Indonesia
[4]Integrated Laboratory, Esa Unggul University, Jakarta 11510, Indonesia

[*]maykeltem@gmail.com

Abstract. Scherrer formula and Williamson-Hall plot or WH-plot are the common methods to determine microstructure information from diffraction pattern, a technique with the expenses of the physical meaning of the result. Eventually a Whole Powder Pattern Modeling (WPPM) has been proposed, allowing physical information to be extracted from the diffraction data with the one-step refinement of the experimental pattern. In this paper, we reported the comparison between Williamson-Hall plot and Whole Powder Pattern Modeling to determine the crystallite size of nanoparticle barium hexaferrites which prepared by mechanical alloying and direct ultrasonic destruction.

1. Introduction

Recently nanostructured magnetic materials have become a hot topic due to their unusual properties raised from the surfaces of neighboring grains. The reason is that because of a great surface to volume ratio in the nanostructures system, then any physical phenomena taking place on the grain surfaces would contribute significant effects to the bulk properties. The unusual properties of permanent magnets were claimed for the first time by McCallum *et al.* [1] that they have developed the isotropic permanent magnet based on nanostructured materials with a remanent magnetization value significantly exceeding the conventional limits based on Stoner and Wohlfarth theory [2]. The enhancement in remanent magnetization is due to the interaction phenomenon since the orientation of the interacting spins becomes partly independent of the different crystallographic directions in each crystallite [3]. In fact, the interaction phenomenon that raised remanent enhancement occurred not only in the single phase nanostructured material [4] but also in the nanocomposite magnet system [5]. The basic principle to gain the effects of exchange grain interaction is the crystallite sizes of magnetic phase in the material must be in nanometer scale size.

Based on this fact, the determination of crystallite size becomes critical. Nanocrystalline barium hexaferrite (BHF) has been a subject of increasing interest due to the potential applications in the various areas like permanent magnets, microwave devices and fast magnetic recording devices [6-8]. Accurate determination of crystallite size is very important for a BHF as a hard magnetic material, leading to reduction in coercivity as it identity [8-11].

IOP Conf. Series: Journal of Physics: Conf. Series **1080** (2018) 012008 doi:10.1088/1742-6596/1080/1/012008

As the latest Scanning Electron Microscopes (SEMs) and Transmission Electron Microscopes (TEMs) easily achieve sub-nanometer resolution, determination of nanocrystalline materials becomes obvious. However, some limitations in analysis were still found such as limitation of analysis to only a few grains [12]. In addition to SEM and TEM, an indirect method like x-ray diffraction would also be possible for crystallite size analysis from the respective powder diffraction pattern. In years, Scherrer formula [13, 14] and Williamson-Hall plot [13, 15] methods have been used to determine the microstructural of the nanocrytalline material. Scherrer observed that the small crystallite size give rise to line broadening and derived a well-known equation for relating the crystallite size to broadening, which is called the Scherrer Formula [14]. Stokes and Wilson also observed that the crystal imperfection and distortion can cause line broadening [16]. Further, Williamson and Hall proposed a method for deconvoluting size and strain contribution by looking the peak width as a function of 2θ which is written in equation 1 [15].

$$\beta_{hkl} cos\theta = \frac{k\lambda}{D} + 4\varepsilon\, sin\theta \tag{1}$$

where β_{hkl} is the full width at half maximum of the line broadening, θ is the Bragg agle, k is the Scherrer constant depends on the how the line profile width is determined and usually not known, λ is the source wavelength, D is the crystallite size and ε is microstrain represents displacements of atoms from their ideal positions, produced by any lattice imperfection (dislocations, vacancies, interstitials, substitutionals, and similar defects).

It is clearly seen that the size and strain broadening are independent to each other, which is allowed to be separated when both occur together. The crystallite size was estimated from y-intercept and the strain from the slope. However, Scherrer formula (SF) and Williamson-Hall (WH) plot are strictly valid for Lorenzian peaks only, which are seldom meet a practice condition because some information also contain in Gaussian peaks. The average crystallite size is work well for isotropic domain, but not good enough for anisotropic domain. The microstrain contribution only provides a general term of defects, but does not identify their source [17]. The real obstacle comes from where SF and WH plot are extracted. Structure (Rietveld) refinement and whole powder pattern fitting (WPPF) are two distinct methods implementing the SF and WH-plot in the algorithms. The common tool for Rietveld and WPPF diffraction pattern analysis is shown in equation 2 [12].

$$y_{ic}(2\theta) = y_{ib} + S \sum_k G_{ik}\, M_k L_k |F_k|^2 P_k A_k E_k \tag{2}$$

where y_{ic} is the net intensity calculated at point i in the pattern, y_{ib} is the background intensity, G_{ik} is the normalized peak profile function contained size and strain effects, M_k is the multiplicity factor, L_k Lorentz polarization factor, F_k is the structure factor, P_k is the preferred orientation, A_k is the absorption correction and E_k is the extinction correction. The profile functions are introduced by modeling the peak profiles with an arbitrary bell-shaped function and further deduce the physical information. This is a two-step method where the result cannot be evaluated a priori.

The diffraction peak profiles are a complex combination of physical and instrumental effects. Constrains by the given analytical function can introduce a systematic (model) errors in correlated with structural and non-structural parameters. As a consequence, the results can be biased and difficult to assess. It also comes to the practical issues that the refinement is very sensitive to minimizing the least-squares of observed and calculated data.

Alternatively, there is a technique proposed to cover all Gauss/Lorentz-type combinations and one of those, called Whole Powder Pattern Modeling (WPPM) [17]. This technique directly refines the entire pattern with the individual parameter corresponding to individual source of broadening. It is able to model such information with a consistency and remarkable agree with the microscopy result [17-19]. It starts from Fourier approach and works in convolutive way where diffraction profiles can be seen in equation 3 [17].

$$I_{\{hkl\}}\left(d^*, d^*_{\{hkl\}}\right) = k(d^*). \sum_{hkl} w_{hkl} \int_{-\infty}^{\infty} C_{hkl}\,(L) \exp(2\pi i L.\, s_{hkl})\, dL \tag{3}$$

where d^* is the reciprocal space coordinate (d^*_{hkl} in Bragg condition), $k(d^*)$ groups of known geometrical and structural terms, w_{hkl} is the weight function (dependent on defects present in the material), L is the length in real space (inversely proportional to $d^*_{\{hkl\}}$), s_{hkl} is the distance from peak centroid in the reciprocal space and C is the Fourier transform (FT) multiplication as explained in equation 4 [17].

$$C_{\{hkl\}} = T_{pV}^{IP} \cdot A_{\{hkl\}}^{S} \cdot A_{\{hkl\}}^{D} \cdot \left(A_{hkl}^{F} + iB_{hkl}^{F}\right) \cdot A_{\{hkl\}}^{APB} \cdot \left(A_{hkl}^{GSR} + iB_{hkl}^{GSR}\right) \cdot \left(A_{hkl}^{SF} + iB_{hkl}^{CF}\right) \qquad (4)$$

where T_{pV}^{IP} is the FT of instrumental profile, $A_{\{hkl\}}^{S}$ is the FTs for lognormal, Γ or York distribution depends on grain shape and size, $A_{\{hkl\}}^{D}$ is the lattice distortion due to dislocation, $(A_{\{hkl\}}^{F} + iB_{\{hkl\}}^{F})$ is the stacking fault broadening, $A_{\{hkl\}}^{APB}$ is the anti-phase broadening, $(A_{\{hkl\}}^{GSR} + A_{\{hkl\}}^{GSR}$ is the grain surface relaxation broadening, $(A_{\{hkl\}}^{SF} + iB_{\{hkl\}}^{SF})$ is the stoichiometric fluctuations broadening and additional line broadening source that can be included due to Fourier Transform of the various effects are multiplied. It is one-step procedure that removed the arbitrary hypotheses by direct modeling the whole diffraction pattern in terms of physical models of the microstructure.

2. Experimental

The nanoparticles BHF was prepared through two successive steps. In the first step, the mechanical alloying of the solid-state reaction method was employed. In this method, stoichiometric quantities of the analytical-graded precursors $BaCO_3$ and Fe_2O_3 from Sigma-Aldrich with 99% purity level were mille in a ball mill apparatus with a ball-powder mass ratio of 10:1 using a jog speed of 160 rpm for 20 hours. Milled powders resulted from mechanical milling were compacted into a cylindrical die of 25 mm diameter under 10 tons load force. The green compact was then sintered at 1100°C for 2 hours leading to fully crystalline bulk sample. The crystalline bulk sample was re-mille for 10 hours and then washed with a light HCl solution to eliminate impurities. In the second step, the fully crystalline re-milled powders were then further treated with direct ultrasonic probe with 5% Vol/Vol of water, at a frequency 20 kilohertz and 60 μm transducer amplitude. X-ray diffraction was performed under a Philips diffractometer using Co-Kα radiation due to the fluorescence effect of iron. The crystallite size of powder diffraction data was determined using High Score Plus (HSP) software implementing WH-plot and PM2K software implementing WPPM. Powder XRD data of 5 hours ultrasonically irradiated sample were used in refinements after determining the instrument broadening with Si-standard (SRM-640). The average particle size distributions of powder specimens were determined using Dynamic Light Scattering (DLS) Particle Size Analyzer and Scanning Electron Microscope (SEM).

3. Result and Discussion

Fitted powder XRD patterns of ultrasonic irradiated samples in figure 1(a) shows a typical single phase M-type hexagonal structure of BHF with space group P63/mmc. Strong preferred orientation on *hhl* -planes in bulk pattern is a typical of mechanical alloyed BHF sintered between 1000-1250°C [20, 21]. As the ultrasonic irradiation time increased the intensity of diffracted peaks also decreased due to the more random distribution of destructed crystals. The decrease in intensity was followed by line broadening of diffracted peaks. In figure 1(b), broadened peaks of (114) for ultrasonic irradiated powders are compared. It shows the Full Width Half Maximum (FWHM) increased as the ultrasonic irradiation time increased. The longer irradiation time, the higher value of FWHM. It is concluded that the ultrasonic irradiation can reduce the crystallite size. In the previous work, Karina *et al.* [22] have reported their comprehensive studies on the use of high power ultrasonicator to refine further the mechanically alloyed powders. Prior to irradiation ultrasonically the crystalline mechanically milled SrO.6 (Fe_2O_3) powders which initially have mean particle and crystallite sizes respectively 723 nm and 179 nm, both were progressively reduced the size, though with a different reduction rate after irradiated under an ultrasonic transducer of 55 μm amplitude. After 5 hours irradiation time the polycrystalline particles reduced the size towards monocrystalline particles of about 87 nm. Hence, crystallites in the particle can be further refined.

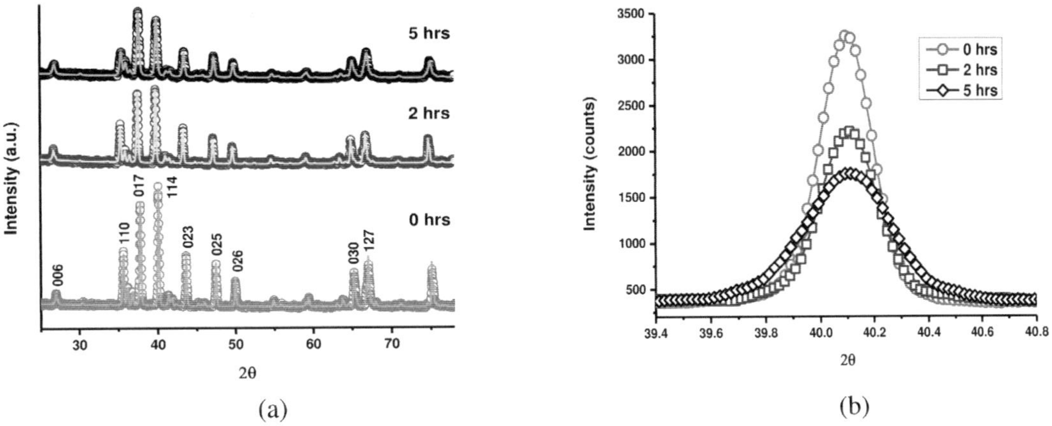

Figure 1. PXRD patterns of: (a) stacked ultrasonic assisted samples and (b) comparison of 023 peaks.

Figure 2(a) is a plot of particle size distribution of powder sample evaluated by particle size analyzer after ultrasonically irradiated for 5 hours. It shows a bi-modal distribution with a mean size of 51.2 nm and 276.3, respectively. The high mean distribution occurred due to effect of fine particles that tend to agglomerate. It can be seen that the particles of a low mean particle size occupied the largest volumetric distribution by 64 % and that of a high mean particle size occupied by 36%.
SEM micrograph of such agglomerated particles is shown in figure 2(b). It can be clearly seen that the agglomerated particles present as a homogenous spherical shape with size range from a few ten nanometers to a few hundred nanometers. The mean crystallite size and crystallite size distribution of such particles can be evaluated from the respective XRD data.

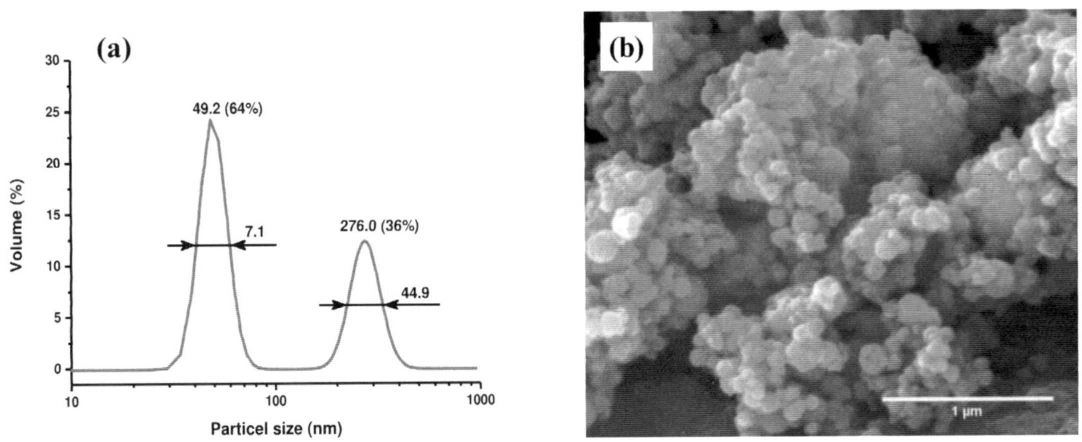

Figure 2. (a) Particle size distribution and (b) SEM micrograph of ultrasonic irradiated sample.

WH-plot and WPPM analysis of XRD data for the 5 hours ultrasonically irradiated powder sample is shown in figure 3. Such plot was made of refined XRD data of figure 1(a), from which the refinement result of WH-plot has Rexp 9.0, Rwp 9.2 and GoF 1.1. It indicates a good agreement between data and model. Linear fitting Williamson-Hall plot in figure 3(a) has resulted in a linear relation of $\beta * cos\theta = 0.209 + 0.070 * sin\theta$. Referring to equation (1), it confirms the presence of microstrain 0.018 (3) % of the particles. The mean crystallite size of particles is 42.434(2) nm. The dash-dot linear plot shows 12% (± 7 nm) deviation with similar weighted residue to continuous line. The modelled pattern of WPPM has Rexp 5.2, Rwp 5.5 and GoF 1.1 and the refinement parameters are lesser than Rietveld method. It was found the perfect match between data, model and the size distribution was modeled by a lognormal distribution as shown in figure 3(b). The result superimposed second peak from DLS graph. Mean size of 49.1 nm is very close to DLS result (49.2 nm), concludes that each particle contained single crystallite. The curve width is about 3 nm narrower than DLS result (7.1 nm), shows a difference between the techniques.

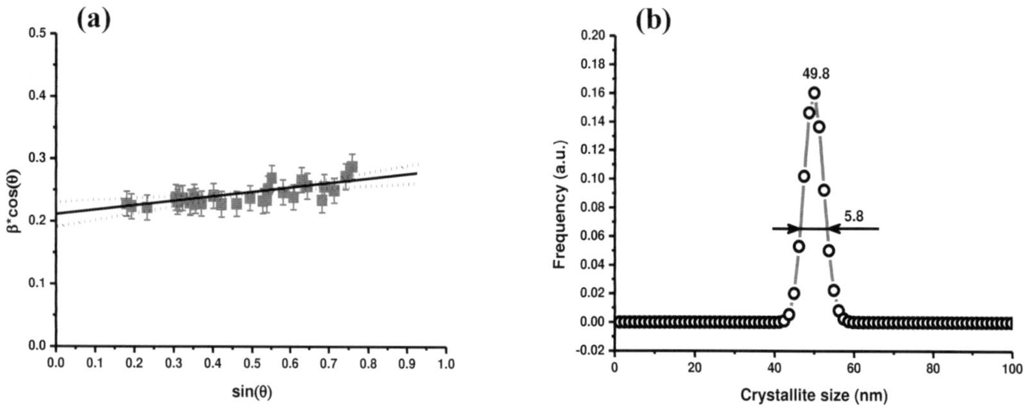

Figure 3. (a) Rietveld/WPPF result as WH-plot and (b) WPPM result as a lognormal distribution.

4. Conclusions

A comparison between WH-plot and WPPM method of crystallite size determination at nano particle BHF has successfully performed. It is found that the average crystallite size of WH-plot was very close to the mean crystallite size distribution from WPPM. The size of WH-plot was about 42.4 nm and for WPPM about 49.1 nm. Both results have good agreement with DLS and SEM analysis. WPPM give the advantage of size distribution which is very important information in nano material applications.

Acknowledgment

Maykel Manawan, Togar Saragi and Risdiana were supported by Academic Leadership Grant of Padjadjaran University 2017, No: 872/UN6.3.1/LT/2017. The authors gratefully acknowledge the support of the Physics Department of the Universitas Indonesia for the research facilities. We thank Professor Matteo Leoni (Università degli Studi di Trento - University of Trento Trento, Italy) for valuable discussion and permission to use PM2K software.

References

[1] McCallum R W, Kadin A M, Clemente G B, Keem J E 1987 *J. Appl. Phys.* **61** 3577
[2] Stoner E C and Wohlfarth E P 1948 *Phil. Trans. Roy. Soc.* **A-240** 599
[3] Giselher Herzer 2005 *NATO Science Series II: Mathematics, Physics and Chemistry* **184**
[4] Eds. B. Idzikowski, P. Švec, M. Miglierini (Kluwer Academic, Dordrecht) p 15
[5] Manaf A, Leonowicz M, Davies H A, Buckley R A 1992 *Materials Letters* **13** 194
[6] Manaf A, Buckley R A and Davies H A 1993 *J. Magn. Magn. Mater.* **128** 307
[7] Pullar R C 2012 *Prog. Mat. Sci.* **57** 1191
[8] Sharrock M P and Carlson L W 2002 *IEEE Trans. Magn.* **31** 2871
[9] Manawan M, Manaf A, Soegijono B, Yudi A 2014 *Adv. Mat. Res.* **896** 401
[10] Manaf A, Hafizah M A E, Belyamin, Nainggolan B, Manawan M 2017 *Int. J. Tech.* **2** 37
[11] Ataie A and Zojaji S 2007 *J. Alloys Comp.* **431** 331
[12] Chaudhury S, Rakshit S, Parida S, Singh Z, Mudher K S, Venugopal V 2008 *J. Alloys Comp.* **455** 25
[13] Leoni M 2012 *Information on Imperfections, in Uniting Electron Crystallography and Powder Diffraction* (Trento: Springe p 173)

IOP Conf. Series: Journal of Physics: Conf. Series **1080** (2018) 012008 doi:10.1088/1742-6596/1080/1/012008

[14] Alexander L E and Klug H P 1974 *X-ray Diffraction Procedures for Polycrystalline and Amorphous Materials 2nd ed.* (New York: Wiley)
[15] Scherrer P 1918 *Nachrichten von der Gesellschaft der Wissenschaften, Gottingen. Mathematisch-Physikalische Klasse* **2** 98
[16] Williamson G K and Hall W H 1953 *Acta Metal* **1** 22
[17] Stokes A R and Wilson A J C 1944 *Proc. Phys. Soc.* **40** 197
[18] Scardi P and Leoni M 2001 *Acta Crys.* **A58** 190
[19] Leoni M, Maggio R D, Polizzi S, Scradi P 2004 *J. American Cer. Soc.* **86** 1133
[20] Abdellatief M, Abele M, Leoni M, Scardi P 2013 *J. Thin Solid Films* **530** 44
[21] Wang S, Ding J, Shi Y, Chen Y 2000 *J. Magn. Magn. Mater.* **219** 206
[22] Dho J, Lee E K, Park J Y, Nur N H 2005 *J. Magn. Magn. Mater.* **285** 164
[23] Fitriana K N, Hafizah M A E and Manaf A 2017 *Int. J. Tech.* **4** 644

Influences of Al dopant atoms to the structure and morphology of Al doped ZnO nanorod thin film

A Aprilia[1*], H Fernando[1], A Bahtiar[1], L Safriani[1] and R Hidayat[2]

[1]Department of Physics, Faculty of Mathematics and Natural Sciences, Universitas Padjadjaran, Jl. Raya Bandung-Sumedang km 21 Jatinangor, Sumedang, West Java, Indonesia 45363

[2]Department of Physics, Faculty of Mathematics and Natural Sciences, Institut Teknologi Bandung, Jl. Ganesha 10 Bandung 40132, West Java, Indonesia

[*]a.aprilia@phys.unpad.ac.id

Abstract. We report our work about dependency of Al dopant atoms to the growth process of Al doped ZnO (AZO) nanorod structure. ZnO is one of type metal oxide, which is classified as semiconducting material with direct wide band gap (~ 3.4 eV). This metal oxide is widely used in optoelectronic devices, such as solar cell, gas sensing, and photocatalyst. To produce effective photoelectrode in solar cell devices and high sensitive sensor in gas sensing application, many researchers have been modifying the nanostructure, such as inserting dopant ions, variation in deposition technique and surface modifications. By inserting some dopant ions/atoms, the structure and morphology of ZnO nanorod can be controlled. Nucleation process of nanorod structure which is growth by chemical solution method, are depend on seed layer morphology. We have prepared ZnO nanorod thin film (growth by self-assembly method) with inserting aluminum as dopant atom both in seed layer and nanorod growth solution. The morphology of ZnO nanostructured significantly changes as the existence of aluminum atoms. Based on X-ray diffraction pattern, the wurtzite structures of AZO nanorod still possessed and only affect to the small shift of lattice crystal. A high surface roughness of seed layer can produce wide radius of nanorod, since the nucleation process initiated by seed layer particle size as a template of nanorod arrangement.

1. Introduction

Nanostructured of semiconducting material possess convenient and useful in physical, electrical, and optoelectronic properties in accordance with device applications. For emerging photovoltaic devices, ZnO photoanodes with variety of nanostructures have been confirmed can increase solar cell performance [1, 2]. ZnO 'trees like' structure called as nanoforest, with high density and long branched hierarchical crystalline phase could significantly improve the efficiency of DSSC (dye sensitized solar cell) [2]. This specific structure can increase the photogenerated current due to greatly augmented surface area supporting higher dye loading and light harvesting. An appropriate nanostructured can reduce the charge recombination through direct conduction along nano-crystalline ZnO. Feng Xu et.al have been synthesised ZnO nanoarchitectures with two-step process and produced nanowire-nanosheet assembly which is increases power conversion nearly twice as high as that of the DSSC using the primary ZnO nanosheet arrays [3].

In other electronic devices such as gas sensing, ZnO was widely used as an active component to detect the existence some toxic gases [4]. Compared to TiO_2, ZnO shows a good catalytic reaction in the gas sensor due to existing various defects such as oxygen vacancies, zinc interstitials, and oxygen interstitials. The performances Gas sensor-based metal oxides can be enhanced by modifying the morphology that increasing the surface-to-volume ratio, creating more centres for gas interaction, and changing the energy band structures [5]. One of effort in modifying ZnO structure is by inserting dopant ions/atoms [4]. Some literatures have reported that Al doping can increase the sensitivity and conductivity of ZnO respects to accelerating chemical reactions in the gas sensor [4]. Moreover, Al doped ZnO based materials such as $ZnAl_2O_4$ is also suitable for ultraviolet electronic applications for the example is photocatalyst [6].

A low surface roughness of ZnO seed layer shows tendency to produce smaller diameter of ZnO nanorod [7]. In this work, we study the existences of aluminium content to the ZnO structure. In order to observe the influence of aluminum dopant to the formation of AZO nanorod structure, aluminum chloride as doping agent was inserted in precursor solution of ZnO seed layer and nano-rod growth. Intercalation of aluminum atom to ZnO nanorod structure was estimated occurred during deposition processes via self-assembly.

2. Experimental Methods

All chemicals used for preparation both of Al doped ZnO seed layer and nanorod thin film were analytical grade without further purifications. As the purposes for optoelectronic device applications, Al doped ZnO nanorod was deposited on FTO (fluorine doped tin oxide) substrate. Before used FTO substrates were cleaned and washed by ethanol and acetone in ultrasonic bath. 0.5 mM Zinc acetate dehydrate ($Zn(CH_3COO)_2.2H_2O$-Merck) in 2-methoxyethanol and diethanolamine (Sigma-Aldrich) as stabilizer was used as precursor of ZnO seed layer (ZnO-SL). Aluminum chloride (Sigma-Aldrich) (0.5 wt%) was added for inserting dopant atoms in ZnO-SL. ZnO-SL was deposited on cleaned FTO substrate by spin coating method. The preparation of Al doped ZnO seed layer was followed our previous work [7, 8].

Undoped ZnO and Al doped ZnO nanorods (ZnO-Nrs) were grown by self-assembly method in isolated glass bottle. The Al doped ZnO seed layer on FTO substrate (FTO/Zn-SL) was positioned facing down at 45° inside glass bottle. The solution for growing ZnO nanorods was prepared by dissolving equimolar zinc nitrate hexahydrate ($Zn(NO_3)_2.6H_2O$-Merck) and hexametylenetetramine in deionized water [9]. 0.5 wt% and 1 wt% of Aluminum chloride respect to the zinc nitrate hexahydrate was added for inserting Al dopant atoms on the growth process. Growing temperature was fixed at 100°C for 150 min inside the regularly laboratory oven. After that, the resulted undoped ZnO (0-0 wt%) and Al doped ZnO nanorods thin film was washed in ethanol and deionized water several times and subsequently annealed up to 500°C for 30 minutes on isolated hotplate.

The influences of Al dopant ion to the both of ZnO seed layer and nanorod optical properties were also investigated using UV-Vis spectroscopy. Structure and morphology of Al doped ZnO thin film were observed by X-ray diffraction (XRD) and scanning electron microscopy (SEM), respectively. The grain crystal size of Al doped ZnO thin films were calculated using Debye Scherer equation. To ensure the nanorod moprhology, tunneling electron microscope (TEM) was also carried out.

3. Results and Discussion

The ultraviolet visible (UV-Vis) absorbance spectra of Al doped ZnO nanorod thin film is shown in figure 1 ZnO seed layer (ZnO-SL) has highly transparent in visible wavelength and shows UV absorbance peak at 367 nm. Among the other samples, undoped ZnO nanorod (ZnO-Nr) thin film has highest UV absorbance peak which identified at 378 nm. The sample with Al dopant atoms (0.5 wt%) both in seed layer and nanorod growth solution is denoted by AZO-Nr 0.5-0.5. Whereas, AZO-Nr 0-0.5 is use undoped ZnO thin film as seed layer whilst employing 0.5 wt% Al in nanorod growth solutions. The Al atoms in AZO-Nr thin films can reduce the absorbance spectrum in visible wavelength although there is a small increment at 400-470 nm (for AZO-Nr 0.5-0.5). AZO-Nr thin film relatively shows high transparency compared to ZnO-Nr in visible wavelength relates to the smaller nanorod size. The secondary absorbance peaks below 370 nm which are identified for both of AZO-Nr thin films, were estimated relates by the existences of smaller size of ZnO. The smaller particles can decrease the onset absorbance curve due to quantum confinement size effects [10]. The influence of inserting Al dopant in ZnO particles is already report in many references [11]. Aluminium atom has been proved can reduce ZnO particle size which affect the optical properties, such as absorbance/transmittance and also photoluminescence [12]. The size reduction was played by Al atoms which has smaller atomic radius (~118 pm) compared than Zn (142 pm); that occupy the Zn sites (Zn vacancy) [13]. The identification of reduction particle size and the changes of ZnO structure by Al dopant are also clarified by SEM image and XRD results respectively, which are shown in figure 2 and 3.

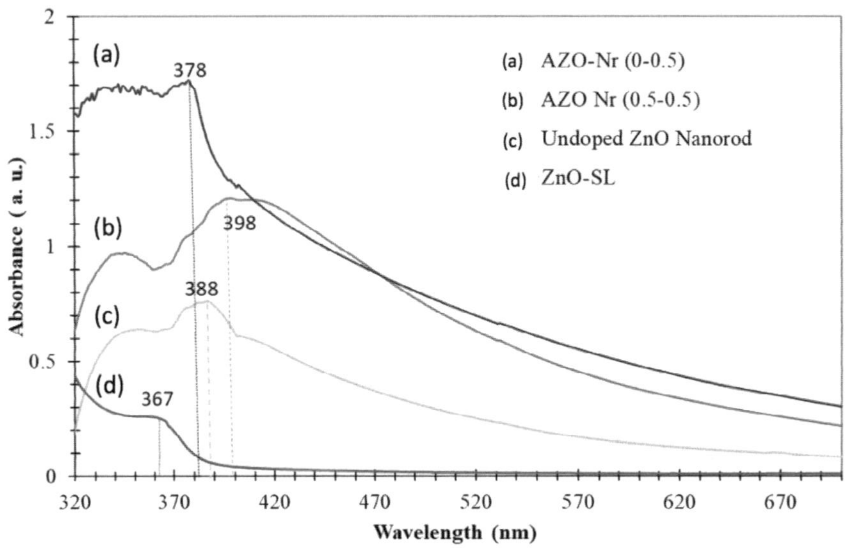

Figure 1. Absorbance spectra of undoped ZnO nanorod, Aluminum doped ZnO nanorod (AZO-Nr), and ZnO seed layer (ZnO-SL) thin films.

The influence of Al dopant to the morphology of ZnO nanorod was observed by scanning electron microscopy (SEM) that shown in figure 2. The surface and cross section morphology of each layer are provided in the figure. All samples present vertically aligned hexagonal rods structure. The diameter are varied and relate to the existences of Al dopant. The rod diameter is significant depend on the existences of Aluminum dopant. Undoped ZnO-Nr has a largest diameter (~ 450 nm) whilst all of samples AZO-Nrs have varied size which is not show significant different from each other (40 ~ 130 nm). The different result between AZO-Nrs thin film are identified in rod well alignment and empty space between rods on each sample. Al dopant is not only affects the rod diameter ($d = 2r$) but is also effect to the length (L). Inserting 0.5wt% of Al dopant both in seed layer and growth solution will reduce significantly the length of rod structure. Subsistence of Al atoms can increase the aspect ratio (L/r) of ZnO nanorod structure [14]. The table 1 shows aspect ratio comparison of all samples. Higher aspect ratio is obtained by 0.5-0.5 wt% which is imply that aluminium atom shows tendency to reduce in diameter rather than length of rod structure.

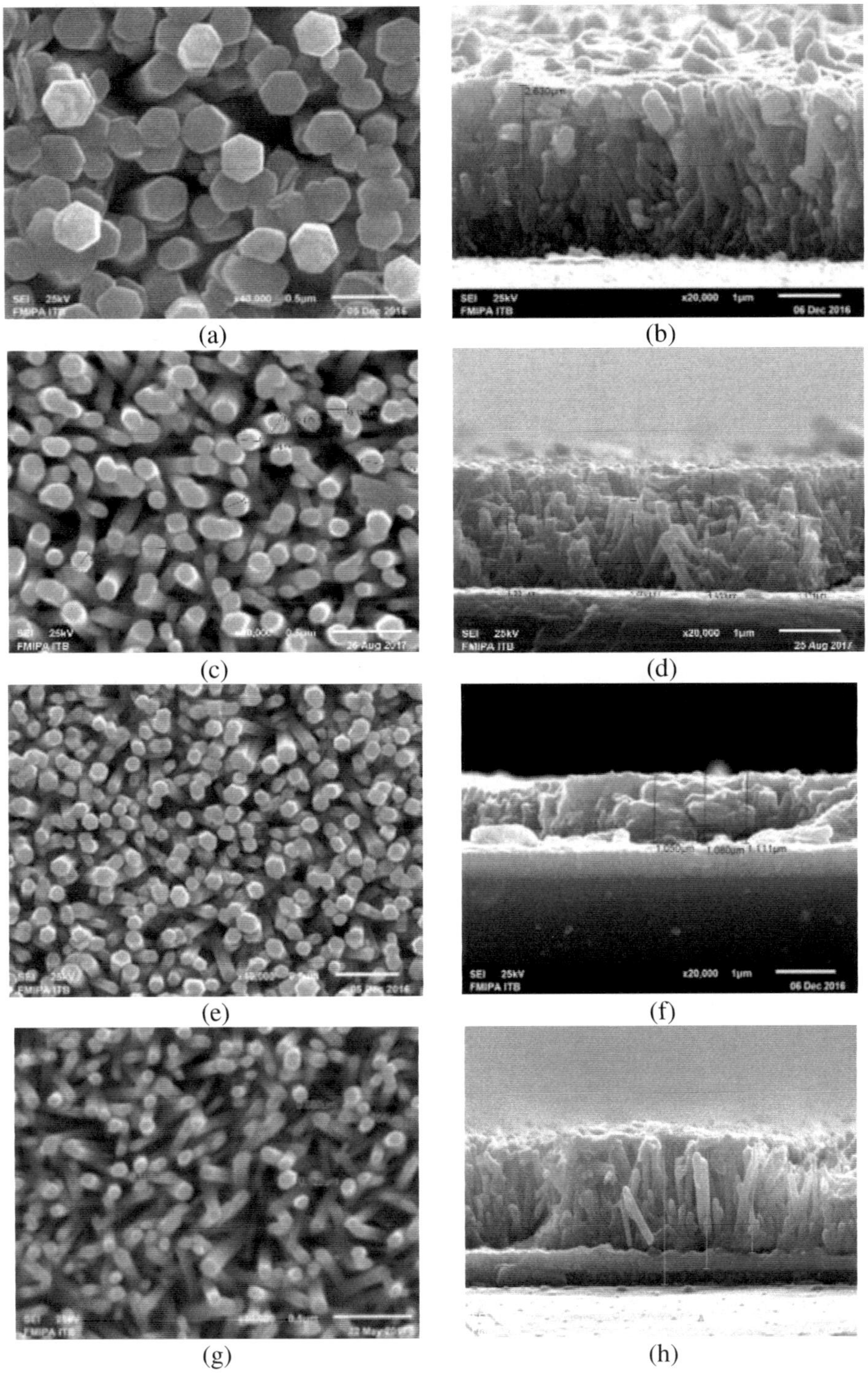

Figure 2. SEM images of ZnO-Nr and AZO-Nr thin film. (a) - (b) ZnO-Nr [7], (c) - (d) AZO-Nr (0-0.5), (d) - (e) AZO-Nr 0.5-0, (g) - (h) AZO-Nr 0.5-0.5.

Figure 3 shows X-ray diffraction pattern results of ZnO-Nr and aluminium doped ZnO (AZO-Nr). All samples show hexagonal wurtzite structure with c-axis (002) plane preferred orientation. Surprisingly, the (101) plane of metallic zinc are identified in each sample ($2\theta \cong 43^{\circ}$). It is probably relate to the existence and agglomeration of Zn-interstitial in all samples during growing process and annealing treatment. Debye-Scherer formula was used to determine crystal size using the full width at half maximum (FWHM) of the strongest (002) diffraction peaks. The resulting calculation of average crystallite size is tabulated in table 2. Seed layer morphology was play an important rule to control crystal size of ZnO [15]. The used of undoped ZnO seed layer (0-0 wt%) produced a largest crystal size among the others. The small grain size of seed layer tends to be produced relatively small diameter of rod and vice versa. To ensure the morphology of rod structure, the exfoliated thin film of AZO-Nrs 0.5-0.5 was observed by TEM (tunneling electron microscopy) as presented in figure 4. Although AZO-Nr seems overlap but the size and morphology of rod structure are clearly identified.

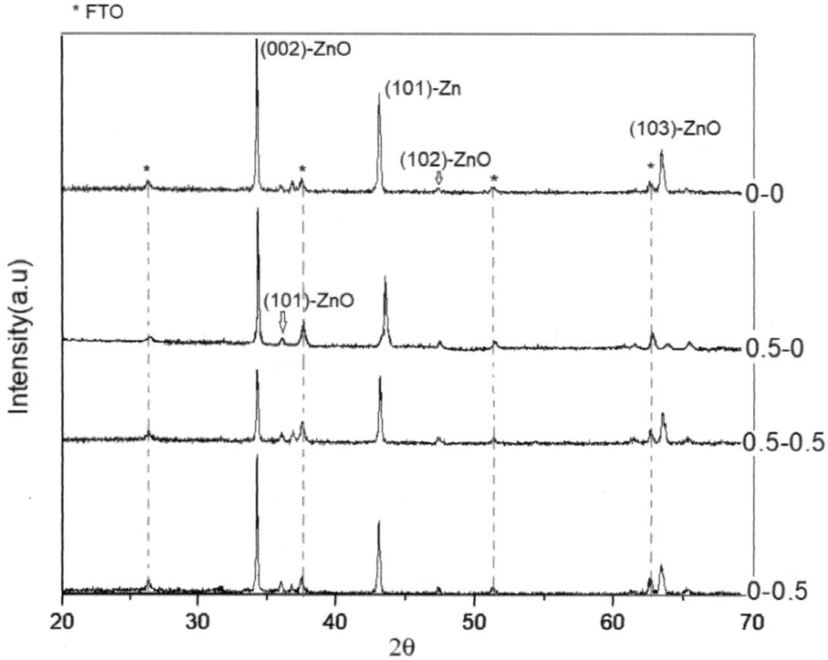

Figure 3. X-ray diffraction pattern of ZnO-Nr and Al doped ZnO (AZO-Nr).

Table 1. Aspect ratio (L/r) of ZnO and Azo-Nrs. The length (L) and diameter (2*r*) of nanorod were obtained from SEM images.

Aluminum Content Seed layer – nanorod	Length (μm)	Radius (μm)	Aspect Ratio (L/r)
Undoped (0-0 wt%)	2.63	0.225	11.69
0.5-0 wt%	1.95	0.07	27.86
0-0.5 wt %	1.05	0.065	16.15
0.5-0.5 wt%	1.7	0.06	28.33

Table 2. Comparison of crystal lattice and crystallite size of ZnO-Nr and AZO-Nr.

Aluminum Content inside ZnO SL-ZnO Nanorods	2θ (002)	*c* (Å)	*D* (nm)
0-0 wt %	34.226	5.24	105.13
0.5-0 wt%	34.258	5.23	71.39
0-0.5 wt %	34.215	5.24	86.27
0.5-0.5 wt%	34.235	5.23	68.39
ZnO (JCPDS)	**34.42**	**5.2**	**-**

Figure 4. Tunneling electron microscopy (TEM) image for exfoliated of AZO-Nr (0.5-0.5) thin film. The rod structure clearly identified with a diameter approximately of ~80 nm and length of around 300 nm and 650 nm.

4. Conclusions

In summary, aligned growth of ZnO and Al doped ZnO nanorods have been successfully prepared on FTO substrates using self-assembly method. The inserting of aluminum dopant (0.5 wt%) in seed layer and nanorod growth solution have been proved can reduce significantly the nanorod size (both of diameter and length). The existences of Al in seed layer, can reduce the surface morphology due to reducing ZnO particle size. As initiator layer, grain size of seed layer is strongly affect the nanorod growth. The smallest crystal size and largest aspect ratio (28.33) is obtained by AZO-Nrs 0.5-0.5. This situation can be explained by an occupying aluminum atom at Zinc sites in rod structure which is reducing the nanorod size. Aluminum has a smaller atomic radius (118 pm) compare with Zn atom (142 pm). Nevertheless, in this case the Al substance likely tendency to produce random rod aligned due to smaller formed rods.

Acknowledgment

This work was supported by Kemenristek-Dikti Republic of Indonesia under Penelitian Fundamental/RDUPT (Riset Dasar Unggulan Perguruan Tinggi) research grant, contract number 718/UN6.3.1/PL/2017.

References

[1] Lori E G, Matt L, Dawud H T, Max M, Josh G, Gabor S and Peidong Y 2005 *Nano Lett.* **5** 1231
[2] Seung H K, Daeho L, Hyun W K, Koo H N, Joon Y Y, Suk J H, Costas P G and Hyung Jin S 2011 *Nano Lett.* **11** 666
[3] Feng Xu, Min D, Yinong L and Litao S 2010 *J. Phys. Chem. C.* **114** 2776
[4] R Sankar G, M Navaneethan, G K Mani, S Ponnusamy, K Tsuchiya, C Muthamizhchelvan, S Kawasaki and Y Hayakawa 2017 *Journal of Alloys and Compound* **698** 555
[5] Baochang C, Z Ouyang, B Tian, Y Xiao and S Lei 2013 *Ceramics International* **39** 7379
[6] Suellen B, Caroline R, Eric da C S, Marcio A M, Raquel C K, Andre G and Edson L F 2014 *Materials Research* **17** 734

IOP Conf. Series: Journal of Physics: Conf. Series **1080** (2018) 012009 doi:10.1088/1742-6596/1080/1/012009

[7] Safriani L, Nurrida A, Mulyana C, Susilawati T, Bahtiar A and Aprilia A 2017 *IOP Conf. Serie: earth and Environmental Science* **75** 012004
[8] Aprilia A *et al.* 2017 *Journal of Physics: Conf. Series* **877** 01023
[9] Aprilia A *et al.* 2017 *IOP Conf. Series: Materials Science and Engineering* **196** 012046
[10] Nurul Azri K A and Mohd Rafir J 2012 *Journal of Nanomaterials* **2012** 439010
[11] Sio A D, Chakanga K, Sergeev O, Maydell K V, Parisi K, hauff E V 2012 Solar Energy Materials & Solar Cells **98** 52
[12] Zhang D, Ma X, Zhang Q, Xie Y, Shi Q, Wang Q and Wang W 2016 *Journal of Alloys and Compounds* **688** 581
[13] Aprilia A, Wulandari P, Suendo V, Herman, Hidayat R, Fujii A and Ozaki M 2013 *Solar Energy Materials & Solar Cells* **111** 181
[14] Polsongkram D, P Chamninok, S Purkird, L Chow, O lupan, G Chai, H Khallaf, S Park and A Schulte 2008 *Physica B* **403** 3713
[15] Hyuk Jae L, Jun-Ho S, Jisu C, Jung Bin K, tae-Heui K and Kyung B P 2013 *Electronic Materials letters* **9** 357

IOP Conf. Series: Journal of Physics: Conf. Series **1080** (2018) 012010 doi:10.1088/1742-6596/1080/1/012010

Effect of localized surface plasmon resonance from incorporated gold nanoparticles in PEDOT:PSS hole transport layer for hybrid solar cell applications

E Susanti, P Wulandari* and Herman

Physics of Magnetism and Photonics Research Division, Faculty of Mathematics and Natural Sciences, Institut Teknologi Bandung, Jl. Ganesha 10 Bandung 40132, West Java, Indonesia

*wulan@fi.itb.ac.id

Abstract. We report optical and structural properties of poly (3,4-ethylene dioxythiophene):poly(styrene sulfonate) (PEDOT:PSS) incorporated gold nanoparticles capped by citrates (Au-citrates) as hole transport layer for hybrid organic-inorganic solar cell applications. PEDOT:PSS and Au-citrates both are water solution and the mixed are done in 2.378 wt%, 2.978 wt% and 3.144 wt% of Au-citrates. Plasmonic peak in the wavelength area of ~525 nm for Au-citrates is found quite stable after addition of iso-propanol for wettability purpose. In our result, enhanced absorbance spectra of PEDOT:PSS incorporated Au-citrates indicates the effect of localized surface plasmon resonance from Au-citrates and it is estimated to enhance performance of hybrid solar cell efficiency. Optical properties of the thin films in hybrid solar cell device are done in the structure of ITO/ZnO/P3HT:PCBM(active layer)/(PEDOT:PSS:Au-citrates). The complete layer structure shows the enhancement of optical property in all UV-vis region in comparison to that structure without Au-citrates.

1. Introduction

Hybrid organic-inorganic solar cell have recently studied due to simple fabrication processes, large area and flexibility, where the structure integrates the advantage of organic material with high hole mobility and inorganic material with high electron mobility [1]. Ji *et al.* demonstrated that power conversion efficiency (PCE) is enhanced in hybrid ZnO-polymer solar cell by varied the size and morphology of ZnO nanostructures [2]. In our previous study, we investigated influences of dopant concentrations in sol-gel method derived Al doped ZnO (AZO) layer on the performance of inverted hybrid solar cell. Doping at high concentration produces higher charge carrier density, but not for charge carrier mobility and its electrical conductivity, which may be influenced by the surface morphology [3]. In hybrid solar cell structure, poly (3,4-ethylenedioxythiophene):poly (styrene sulfonate) (PEDOT:PSS) is well known using as good hole transport layer material but water based PEDOT:PSS has a wettability problem when it is introduced on top of active hydrophobic polymer layer by used of spin-coating technique. Modification of PEDOT:PSS through addition of surfactant such as FS-31 successfully reduces the contact angle of PEDOT:PSS on the hydrophobic photoactive layer and the wetting property is sufficiently developed [4].

On the other case, metal nanoparticles such as gold (AuNPs) and silver (AgNPs) have been used in organic solar cells due to strong localized surface plasmon resonance (LSPR) wave, which contributes to the enhancement of local electromagnetic field and it is believed to be able to improve light absorption and to promote exciton dissociation or charge transfer generation in organic solar cells. The effect of AuNPs incorporated into the active layer of polymer solar cell experimentally and theoretically enhance the light absorption in the active layer because LSPR near field mainly distributes laterally along the active layer [5]. Incorporation of AuNPs or AgNPs in the hole transporting buffer layer of PEDOT:PSS, which is formed on an indium tin oxide (ITO) surface by the spin-coating of PEDOT:PSS-Au or AgNPs composite solution exhibit an 8% improvement in their efficiency due to the enlarged surface roughness of the PEDOT:PSS, which lead to an improvement in the charge collection and in the short-circuit density and fill factor [6].

In this research study, we investigated optical and structural properties of AuNPs, which has capping material of citrates (Au-citrates) incorporated into PEDOT:PSS as hole transport layer (HTL) material for hybrid organic-inorganic solar cell applications. Localized surface plasmon resonance (LSPR) effect from Au-citrates embedded into PEDOT:PSS will be expected to improve charge carrier generation and transport in hybrid solar cell. We studied in detail in how to modify both PEDOT:PSS and Au-citrates solution by used of iso-propanol solvent since it has been known that they are water solution and have wettability problem on hydrophobic layer of active polymer layer in the structure of hybrid solar cell.

2. Experiments

2.1. Preparation of PEDOT:PSS incorporated Au-citrates solution

Au-citrates in water solution with spherical shape and 15 nm of diameter size (Sigma-Aldrich) and *poly (3,4-ethylenedioxythiophene):poly(styrene sulfonate)* (PEDOT:PSS) in water solution (Heraeus Clevios GmbH) both were used as received without additional processes. First, 0.0027 mg/ml of Au-citrate were mixed with 50 μl, 100 μl, 200 μl and 500 μl of iso-propanol (IPA) in order to check the stability of Au-citrates in solution. Incorporation of Au-citrates with PEDOT:PSS solution was done by mixed several concentration of Au-citrates (2.378 wt%, 2.978 wt% and 3.144 wt%) into PEDOT:PSS solution and stirred for 60 min at room temperature. Thin film preparation was done by dropping PEDOT:PSS:Au-citrates solution on clean glass substrate and spin-coated at 1000 rpm for 10 s and annealed at 150°C for 15 min.

2.2. Preparation of hybrid solar cell layers

Thin films for hybrid organic-inorganic solar cell application was prepared on indium tin oxide (ITO) glass which were consisted of ZnO/P3HT:PCBM/PEDOT:PSS:Au-citrates layers. ITO glass was cleaned in ultrasonic bath by de-ionized (DI) water and continued by used of mixed iso-propanol and aceton and finalized by ethanol. ITO glass was dried by N_2 gas before used. Next, ZnO layer was prepared by following our previous study (sol-gel method) [3] in which 0.0878 g of *zinc acetate-dihydrate* ($Zn(CH_3COO)_2.2H_2O$, Merck) was dissolved in iso-propanol as solvent and stirred for 15 min at temperature of 60°C. Next, *diethanolamine* ($HN(CH_2CH_2OH)_2$, Sigma-Aldrich) as stabilizer agent was gently added into zinc acetate (Zn-Ac) solution and kept stirring for additional 12 h in room temperature. Deposition of Zn-Ac on ITO glass was done by used spin-coating technique at 4000 rpm for 10 s and the thin film was pre-heated at 300°C for 10 min. This process was repeated for 3 times and then post-heated at 500°C for 60 min to form ZnO film. Active layer was prepared by mixed *poly (3-hexylthiophene* (P3HT, Sigma-Aldrich) and *phenyl C61 butyric acid methyl ester* (PCBM, Sigma-Aldrich) with weight ratio of 1:1 in chlorobenzene and stirred at 35°C for 12 h. Deposition of active layer on ZnO thin film was done by spin-coated at 2000 rpm for 10 s. PEDOT:PSS with and without incorporated Au-citrates was prepared as explained in section 2.1.

2.3. Measurements

Optical properties of Au-citrates, PEDOT:PSS incorporated Au-citrates including thin films in hybrid solar cell structure were characterized by used of Ultra Violet-visible spectrometer (UV-vis Agilent Technologies 8453). Chemical structures of PEDOT:PSS incorporated Au-citrates was investigated by used of Fourier Transform Infrared spectrometer (FT-IR Prestige 21 Shimadzu, Japan). Information of morphology and thickness of PEDOT:PSS thin film was recorded by used of Scanning Electron Microscope (SEM Hitachi Model SU3500).

3. Results and Discussions

3.1. Optical properties of PEDOT:PSS incorporated Au-citrates

In figure 1, typical plasmonic peak of Au-citrates solution is appeared clearly at wavelength of 525 nm together with sharpen peak shape, which is indicated distribution of particles size is relatively homogenous around 15 nm in diameter as shown also in our previous work for synthesis of Au-citrates [7]. In this study, 0.027 mg/ml of Au-citrates in water solution is quite stable with addition of 50 μl, 100 μl and 200 μl of IPA as shown in absorbance spectra in comparison to the spectrum of Au-citrates in water solution as reference. A small broadening peak is found in the 500 μl addition of iso-propanol into Au-citrates solution indicated the particle starts to aggregate.

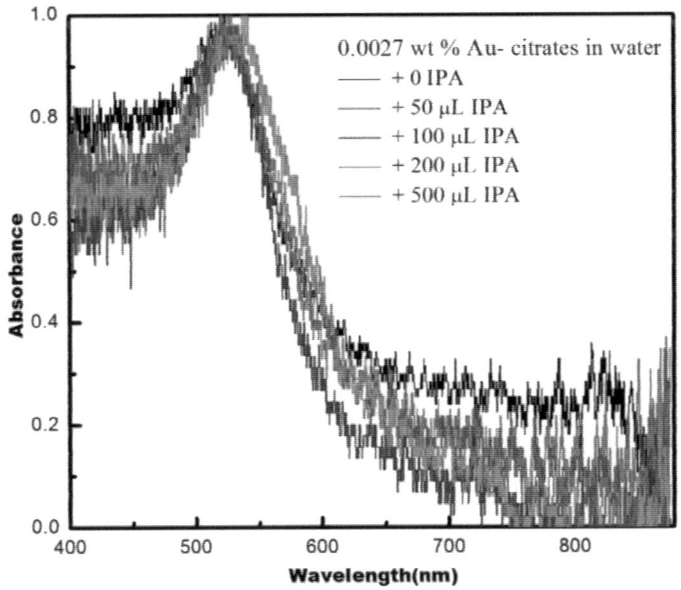

Figure 1. UV-Vis spectra of Au-citrate in water and iso-propanol.

Figure 2(a.2)–(a.4) shows solution of PEDOT:PSS in water mixed with 2.378 wt%, 2.978 wt% and 3.114 wt% of Au-citrates compare to the PEDOT:PSS solution without addition of Au-citrates (figure 2(a.1)). This result indicates that Au-citrates is quite stable in PEDOT:PSS solution. Moreover, UV-vis spectroscopy result as shown in figure 2(b) suggests that the enhancement of the absorbance peaks in the region of PEDOT:PSS spectra caused by LSPR effect of Au-citrates. In our experiment, we investigated the enhancement of absorbance peaks at LSPR peak position for Au-citrates as indicated by black dash line in the figure 2(b). The absorbance peak intensities are found much enhanced for PEDOT:PSS solution contain 2.378 wt% and 2.978 wt% of Au-citrates. In case of incorporation 3.114 wt% Au-citrate in PEDOT:PSS solution, the increased intensity is not significant, which indicated Au-citrates starts to aggregate even though the precipitate does not appear yet in the solution (figure 2(a.4)).

IOP Conf. Series: Journal of Physics: Conf. Series **1080** (2018) 012010 doi:10.1088/1742-6596/1080/1/012010

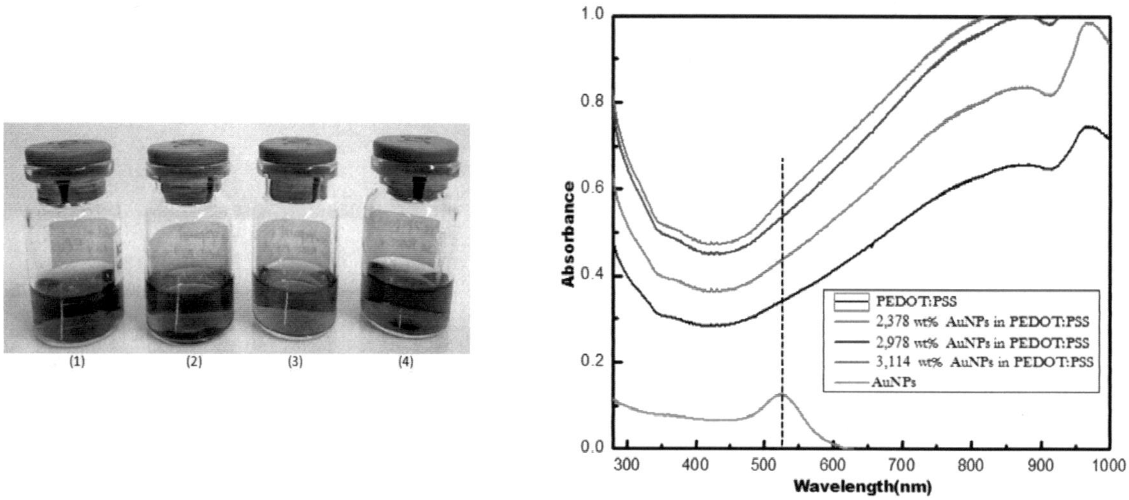

(a) (b)

Figure 2. (a) PEDOT:PSS solution incorporated Au-citrates (1) 0 wt%, (2) 2.378 wt%, (3) 2.978 wt% and (4) 3.144 wt% and (b) Absorbance spectra of PEDOT:PSS incorporated Au-citrates in solution.

Optical property of PEDOT:PSS:Au-citrates thin film is presented in figure 3, which indicates similar phenomena to the sample solutions. The increased of absorbance peak in the wavelength area of ~300 nm for PEDOT:PSS : Au-citrates (2.978 wt%) in comparison to the absorbance peaks of PEDOT:PSS without Au-citrates indicates the strong effect of LSPR from Au-citrates.

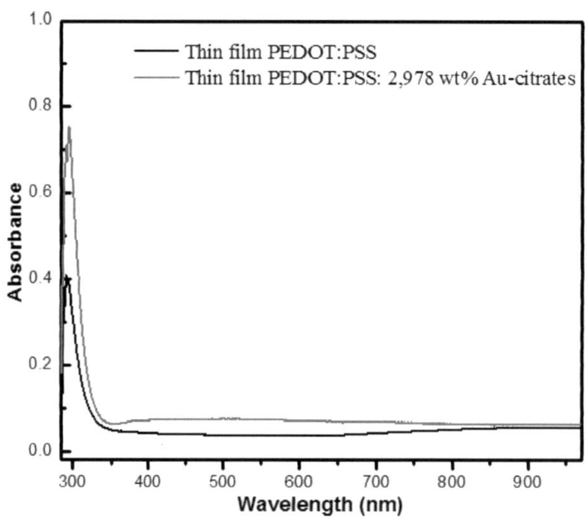

Figure 3. Absorbance spectra of PEDOT:PSS with and without incorporated Au-citrates in thin film.

3.2. Structural properties of PEDOT:PSS incorporated Au-citrates

Chemical structures of PEDOT:PSS incorporated Au-citrates was investigated by used of FT-IR spectroscopy. The samples were prepared by dried-up the solution of PEDOT:PSS with and without Au-citrates in room temperature and mixed with *potassium bromide* (KBr) to form a pellet. The differences in FT-IR peaks of PEDOT:PSS with and without Au-citrates are shown in figure 4. The peaks related to C-C, C-O, S-H and C-S bonds. Peak of C-S at wavenumber 831 cm^{-1} in PEDOT:PSS are reduced drastically in the PEDOT:PSS:Au-citrates, whereas C-O-C bond at wavenumber of 1388 cm^{-1} is disappeared. These results indicate that the changes intensities are caused by new bonding formation in the sample especially between PEDOT:PSS and citrates capping of Au nanoparticles. The shifted of peak intensity in FT-IR spectra to higher and lower frequencies are still in our investigation which could be caused by the effect of LSPR from Au-citrates.

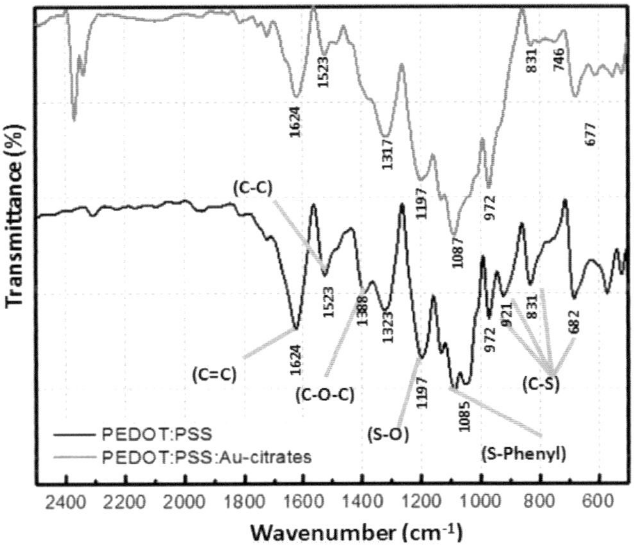

Figure 4. FTIR spectra of PEDOT:PSS and PEDOT:PSS:Au-citrates.

Morphology of PEDOT:PSS:Au-citrates thin film on glass substrate resulted from spincoating technique at 1000 rpm for 10 s is clearly shown in figure 5(a) and its cross section view shows the thickness around ~174 nm. Annealing process at 100°C after spin-coated process is important to eliminate the solvent and to form homogenous thin film.

Figure 5. SEM images of thin film PEDOT:PSS:Au-citrate on glass substrate (a) on top view and (b) cross section view.

3.3. Layer by layer optical properties of fabricated hybrid solar cell

In order to investigate more detail of the LSPR effect from Au-citrates in PEDOT:PSS as hole transport layer in hybrid solar cell applications, we measured optical properties layer by layer structure in hybrid solar cell by used of UV-vis spectroscopy. Figure 6 shows typical absorbance spectra from each layer structure of ITO as cathode, ZnO as electron transport layer (ETL), P3HT:PCBM as photoactive layer and PEDOT:PSS as hole transport layer (HTL) with and without Au-citrates. Complete layers structure of hybrid solar cell incorporated Au-citrates shows enhancement of absorption in all UV-vis region compared to structure without Au-citrates. All enhancement of optical properties which are suggested by the effect of LSPR from Au-citrates is expected to be improved the efficiency of inverted type hybrid solar cell.

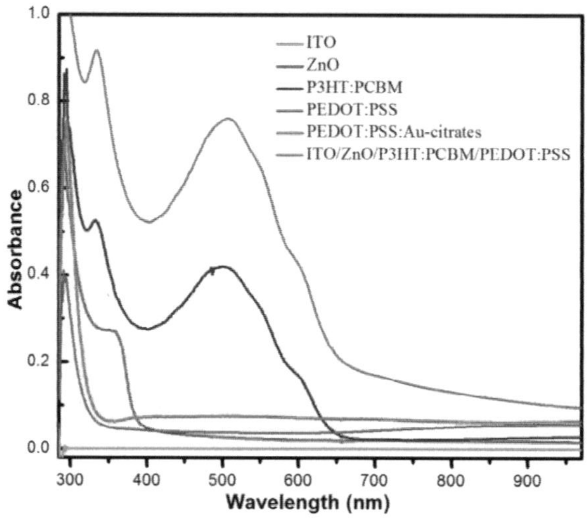

Figure 6. Absorbance spectra of layer by layer hybrid solar cell structure incorporated Au-citrates.

4. Conclusions

Optical properties of PEDOT:PSS incorporated Au-citrates with concentration of 2.378 wt%, 2.978 wt% and 3.144 wt% were investigated by use of UV-vis spectroscopy. Enhanced absorbance spectra of PEDOT:PSS incorporated Au-citrates in solution and thin film are caused by the effect of localized surface plasmon resonance (LSPR) from Au-citrates. In this case, we conclude that Au-citrates is quite stable, when it is embedded into PEDOT:PSS and iso-propanol solution. FTIR characterization confirm that there are peaks intensity changes and peaks shifted between PEDOT:PSS and PEDOT:PSS incorporated Au-citrates due to chemical bonding between PEDOT:PSS and citrates from Au nanoparticles. In our result, enhancement of optical properties in complete layer structure of hybrid solar cell incorporated Au-citrates in all UV-vis region can explain that the generation of LSPR is expected to be improve the efficiency of inverted type hybrid solar cell.

References

[1] Kudo N, Honda S, Shimazaki Y, Ohkita H, Ito S and Benten H 2007 *Appl. Phys. Lett.* **90** 183513
[2] Ji L W *et al.* 2013 *Solar Energy Material & Solar Cells* **111** 181
[3] Yi M, Hong S, Kim J R, Kang H, Lee J, Yu K, Kee S, Lee W and Lee K 2016 *Solar Energy Materials & Solar Cells* **153** 117
[4] Wang C C D, Choy W C H, Duan C, Fung D D S, Sha W E I, Xie F X, Huang F and Cao Y 2012 *J Mater. Chem.* **22** 1206
[5] Woo S, Jeong J H, Lyu H K, Han Y S and Kim Y 2010 *Nanoscale Research Letters* **7** 641
[6] Wulandari P, Li X, Tamada K and Hara M 2008 *J Nonlinear Opt. Phys. Mater.* **17** 185

Charge carrier transport in blend of P3HT and ZnO nanoparticles at low temperature studied by μSR

L Safriani[1*], Risdiana[1], Fitrilawati[1], M Manawan[1], A Bahtiar[1], A Aprilia[1], D P Sari[2,3], J Angel[2,4] and I Watanabe[2,3,4]

[1]Department of Physics, Faculty of Mathematics and Natural Sciences, Universitas Padjadjaran, Jl. Raya Bandung Sumedang km 21 Jatinangor, Sumedang, West Java, Indonesia 45363
[2]RIKEN Nishina Center, 2-1 Hirosawa, Wako, Saitama 351-0198, Japan
[3]Department of Physics, Osaka University, Toyonaka, Osaka 560-0043, Japan
[4]Department of Physics, Hokkaido University, Sapporo, Hokkaido, 060-0810, Japan

[*]lusi.safriani@phys.unpad.ac.id

Abstract. Recently, hybrid solar cell, which is a combination of organic and inorganic material, has been developed to produce better performance especially in power conversion efficiency of solar cell. Organic material based on conjugated polymer such as Poly(3-hexylthiophene) (P3HT) attracted much attention because P3HT shows the highest hole mobility among the series of Poly(3-alkylthiophene). On the other hand, it is well known that ZnO nanoparticles are an inorganic material with high electron mobility that can be acted as electron acceptor to dissociate excitons formed in organic material. The existence of ZnO nanoparticles in active layer of solar cell will lead to dissociation of excitons in P3HT via rapid electron transfer to ZnO. To understand the mechanism of charge carrier transport in blend of P3HT and ZnO nanoparticles, we have been conducted an experiment using muon spin relaxation (μSR) at low temperature of 10 K with variation of longitudinal field from 2 to 395 mT. We also investigated the effect of light irradiation to the charge carrier transport in blend of P3HT and ZnO nanoparticles. Without light irradiation, it is clearly observed that depolarization rate (λ) was proportional to linier field of $H^{-0.5}$, indicating one-dimensional intra-chain transport of charge carrier. With light irradiation, it is found that asymmetry data changed slightly indicating the trace of any changing of charge carrier transport to increase in the amount of exciton in samples causing by light irradiation.

1. Introduction

Organic solar cells have attracted much attention because of their potential in the low-cost production of solar panels, featuring a flexible, lightweight, and ultrathin device. The power conversion efficiency (PCE) of organic solar cells is now reaching up to 10% [1] and it could be improved by modifying the active material of solar cells. Poly-(3-hexylthiophene) (P3HT) is a polymer, which is widely used as active material in photovoltaic devices [2]. P3HT has considerable attracted research interest because P3HT shows the highest hole mobility compared to other conjugated polymers [3].

Recently, many researchers give attention to hybrid solar cells to improve performance of solar cells whilst still maintain low cost processability [4]. Hybrid solar cells consist of organic materials, normally conjugated polymer that absorb light as the donor and inorganic materials as the acceptor and electron transporter in the structure. ZnO as inorganic material has some advantages such as having high electron mobility, becoming one of candidates to be combined with organic material of P3HT in active layer of hybrid solar cell. The advantage of P3HT mixed with ZnO nanoparticles give a promising material for better performance of hybrid solar cells [5]. ZnO prepared as nanoparticles resolve the problem of small diffusion range of P3HT [6]. The existence of ZnO in active layer will support charge transfer from P3HT to electrode of solar cell. The phenomena could be realized due to the conduction band of ZnO is lower than that of low unoccupied molecular orbital (LUMO) of P3HT as shown in figure 1.

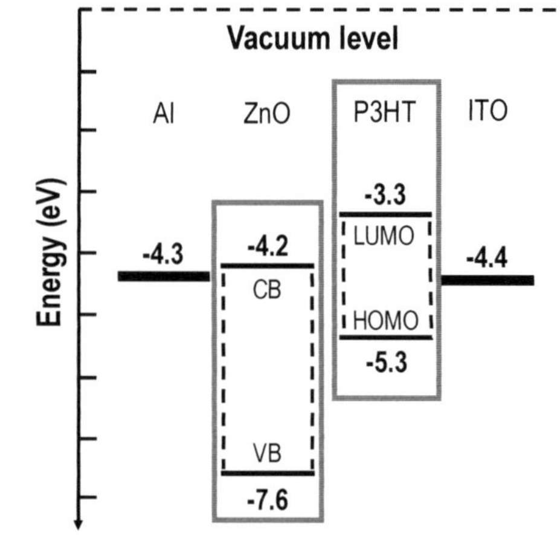

Figure 1. Energy level of P3HT and ZnO.

We have studied the microscopic intrinsic charge carrier dynamics in active material of P3HT:ZnO blend of hybrid solar cells along and perpendicular to the chain by using the longitudinal field (LF) μSR method. Figure 2 shows the longitudinal field dependence of λ_1 in P3HT:ZnO [7]. We found that the charge carrier mobility changes from intra-chain diffusion that charge diffused along the polymer chain to inter-chain diffusion that charge diffused in between the polymer chain. One-dimensional intra-chain diffusion is observed in the samples at temperatures below 10 K, while three-dimensional inter-chain is observed at temperatures above 25 K. Compared with μSR data of P3HT that shows dimensional crossover at 25 K [8], it is apparent that the dimensional crossover from one-dimensional to three-dimensional in P3HT:ZnO is observed at lower temperature. ZnO nanoparticles facilitate the electron transfer from P3HT in blend of P3HT:ZnO more easily than in P3HT only.

In solar cells application, light was irradiated to initiate the charge carrier from P3HT as donor to ZnO nanoparticles as electron acceptor. In this paper, we performed light irradiation into sample of P3HT:ZnO to investigate the intra and inter-chain charge carrier dynamics by using longitudinal-field (LF) μSR. The results of charge carrier dynamics in active layer of P3HT:ZnO solar cell could be directly implied in fabrication of high performance of power conversion energy device.

2. Experimental methods

Conjugated polymer P3HT was obtained from Sigma-Aldrich without further purification. Sol-gel method was used to synthesize ZnO nanoparticles [9]. Hybrid organic-inorganic sample consisted of P3HT and ZnO nanoparticles were mixed by dissolving in chloroform and 3% methanol to produce a homogeneous solution. The solution was evaporated at 150°C to produce bulk of hybrid polymer P3HT:ZnO nanoparticles. For μSR measurement, P3HT:ZnO nanoparticles bulk sample was mounted on a silver plate in cryostat. The LF-μSR measurements were performed in magnetic field ranging from 0 to 395 mT at low temperature of 10 K. μSR measurements was carried out at the RIKEN-RAL Muon Facility, Rutherford-Appleton Laboratory, UK using a pulsed positive surface muon beam [10, 11]. White light source was irradiated to P3HT:ZnO nanoparticles bulk sample, all the time during μSR measurements. The longitudinal asymmetry parameter of μSR signal $A(t)$ at a time t is defined in equation 1.

$$A(t) = [F(t) - \alpha B(t)]/[F(t) + \alpha B(t)] \tag{1}$$

where $F(t)$ and $B(t)$ are total muon events counted by the forward and backward counters, respectively, and α is the calibration factor reflecting the relative counting efficiencies between the forward and backward counters [12].

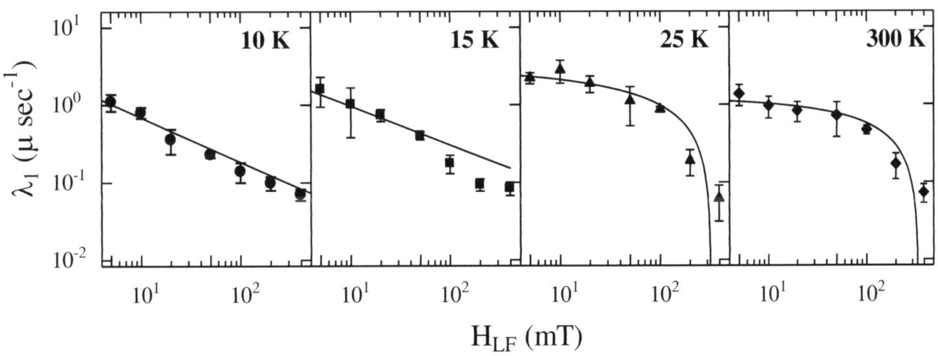

Figure 2. Longitudinal field dependence of λ_1 of P3HT:ZnO without light irradiation [7].

3. Results and Discussions

Figure 3 shows the longitudinal-field (LF) dependent variation of raw asymmetry at low temperatures of 10 K. Red circle data is asymmetry data with light irradiation and black circle data is asymmetry data without light irradiation. It is clearly shown that all initial asymmetry shifted to the higher values with increasing field due to repolarization of the muonium state [13].

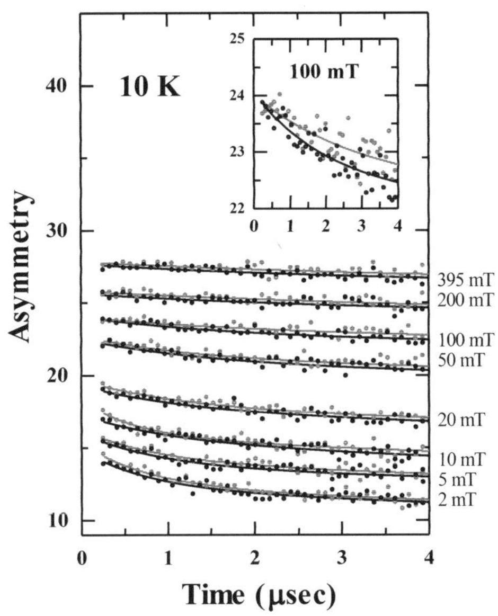

Figure 3. The asymmetry data of P3HT:ZnO nanoparticles at low temperatures of 10 K for various longitudinal magnetic field values (• with light irradiation and • without light irradiation).

The information of electron mobility on the polymer chain of P3HT was obtained from the spectra that were analyzed using two-component function as presented in equation 2.

$$A(t) = A_1 e^{\lambda_1 t} + A_2 e^{\lambda_2 t} \quad (2)$$

where A_1 and A_2 are the initial asymmetries, λ_1 and λ_2 are corresponding to depolarization rates associated with the fast and slow components, respectively. In figure 3, it is shown that the raw asymmetries are well fitted indicated by the solid line. It is also clearly seen that using light irradiation, the asymmetry shifted slightly as shown in the insert of figure 3. The shifting of asymmetry by light irradiation indicated the change of charge carrier transport in the samples.

The longitudinal field (H_{LF}) dependence of λ_1 in P3HT:ZnO is presented in figure 4. The LF dependence of λ reflects the direction of charge diffusion in the polymer chain. That is, one-dimensional intra-chain diffusion is indicated by λ proportional to linier field of $H^{-0.5}$ and three-dimensional inter-chain diffusion is characterized by λ ~ C-$H^{0.5}$ curve [14]. With light irradiation, for the low temperature of 10 K, λ_1 was proportional with C-$H^{0.5}$ curve, indicating three-dimensional inter-chain diffusion. The similar tendency also observed for H_{LF} dependence of λ_2. Comparing with the previous μSR data of P3HT:ZnO which the experiment was done without light irradiation, one-dimensional intra-chain diffusion is still observed in the samples at low temperature of 10 K, while three-dimensional inter-chain is observed at high temperature above 25 K [7]. With light irradiation, the crossover temperature from one-dimensional to three-dimensional of P3HT:ZnO bulk sample was obtained at lower temperature of 10 K compared to previous result at 25 K without light irradiation. This is probably related to process irradiation which initiate the production of exciton that increase number of charge carrier in sample and transport not only along the chain of polymer (one-dimensional) but also perpendicular to other chain of polymer (three-dimensional). Even though qualitatively the amount of exciton produced by light irradiation could not be determined yet, the result of μSR measurement shows that light irradiation to P3HT:ZnO effect its charge carrier transport property.

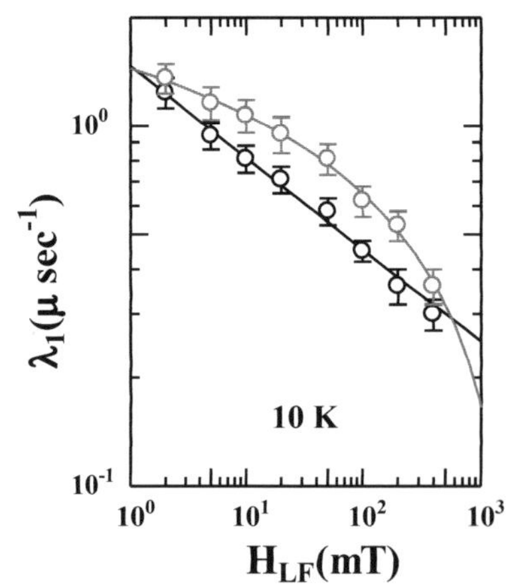

Figure 4. The longitudinal-field dependence H_{LF} of relaxation rate λ_1 of P3HT:ZnO at low temperature of 10 K (o with light irradiation and o without light irradiation).

4. Conclusions

We have studied charge carrier transport of blend P3HT and ZnO nanoparticles using muon spin relaxation method at low temperature with light irradiation. We found small changes of asymmetry for sample with and without light irradiation at low temperature of 10 K. Without light irradiation, λ_1 was proportional with $H^{-0.5}$, which indicating one-dimensional intra-chain diffusion. While with light irradiation, λ_1 was proportional with C-$H^{-0.5}$ curve which indicating three-dimensional inter-chain diffusion. With light irradiation, the charge carrier transport increased initiating by exciton production.

Acknowledgement
This work was financially supported by Padjadjaran University under research HIU (Hibah Internal Unpad) with contract No. 431/UN6.3.1/PL/2017.

References

[1] Nguyen T L *et al.* 2014 *Energy Environ. Sci.* **7** 3040
[2] Günes S, Neugebauer H and Sariciftci N S 2007 *Chemical Reviews* **107** 1324
[3] Babel A and Jenekhea S A 2005 *Synth. Met.* **148** 169
[4] Wright M and Uddin A 2012 *Solar Energy Materials & Solar Cells* **107** 87
[5] Safriani L, Risdiana, Bahtiar A, Aprilia A, Kawasaki I and Watanabe I 2015 *Materials Science Forum* **827** 131
[6] Ji L W, Shih W S, Fang T H, Wu C Z, Peng S and Meen M T H 2010 *J. Matter Sci.* **45** 3266
[7] Safriani L, Risdiana, Bahtiar A, Aprilia A, Siregar R E, Hidayat R, Saragi T P I, Kawasaki I and Watanabe I 2014 *Advanced Materials Research* **896** 477
[8] Risdiana, Fitrilawati, Hidayat R, Nugroho A A, Siregar R E, Tjia M O and Watanabe I 2010 *Journal of Physics Conf. Series* **200** 052024
[9] Eric M A 1998 *J. Phys. Chem.* B **102** 5566
[10] Matsuzaki T, Nagamine K, Tanase M, Kato M, Kurosawa K, Sugai H, Ishida K, Nakamura S N, Watanabe I and Eaton G H 2002 *Nucl. Instr. Meth.* A **480** 814
[11] Nagamine K, Matsuzaki T, Ishida K, Watanabe I, Kadono R, Eaton G H, Jones H J, Thomas G and Williams W G 1994 *Hyperfine Interactions* **87** 1091
[12] Risdiana, Fitrilawati, Hidayat R, Siregar R E, Tjia M O and Watanabe I 2010 *Physica* B **405** S381
[13] Pratt F L 1997 *Philos. Mag. Lett.* **75** 371
[14] Butler M A, Walker L R and Soos Z G 1976 *J. Chem. Phys.* **64** 3592

IOP Conf. Series: Journal of Physics: Conf. Series **1080** (2018) 012012 doi:10.1088/1742-6596/1080/1/012012

J-V characteristic of perovskite solar cells using lead(II) thiocyanate doped-methylammonium lead iodide ($MAPbI_3$) as active material

A Bahtiar[1,*], M Putri[1], E S Nurazizah[1], Setianto[1], T Saragi[1], Risdiana[1], M Yamashita[2], S Ikawa[2] and Y Furukawa[2]

[1]Department of Physics, Faculty of Mathematics and Natural Sciences, Universitas Padjadjaran, Jl. Raya Bandung-Sumedang km 21 Jatinangor, Sumedang, West Java, Indonesia 45363

[2]Department of Chemistry and Biochemistry, School of Advanced Science and Engineering, Waseda University, Shinjuku-ku, Tokyo 169-8555, Japan

[*]ayi.bahtiar@phys.unpad.ac.id

Abstract. Perovskite solar cells based on methylammonium iodide ($MAPbI_3$) as active material have been intensively investigated for low cost and high power conversion efficiency (PCE). Todays, the PCE beyond 20% have been obtained using different structures of solar cells. However, there are still many problems needed to be solved prior to mass production and commercialization, in particular the stability of solar cells. The PCE decreases significantly in high humidity due to the decomposition of organic methylammonium iodide with water vapor from air-humid into lead iodide. Many researchers are now paying a lot of attentions to improve the stability of perovskite solar cells, including device structures, interfaces engineering and materials engineering. Recently, perovskite solar cell based on methylammonium lead(II) thiocyanate [$MA(PbSCN)_2$] improved the humidity-stability of solar cells, due to higher binding energy of thiocyanate (SCN) ions to Pb ions as than that of iodide (I) ions to Pb ions. However, the PCE is lower than that of commonly used $MAPbI_3$ solar cells, because the large bandgap leads to low light absorption. In this work, we present our recent study on J-V characteristic of solar cells using lead(II) thiocyanate $Pb(SCN)_2$ doped lead(II) iodide as active materials of perovskite solar cells. We prepared perovskite films using two-step spin-coating method. We used two different structures of solar cells, mesoporous device and inverted planar structure. The best PCE of solar cells is 7% obtained with inverted planar structure. The mesoporous device structure shows lower performance than that of inverted planar structure, which is due to poor quality of electron transport layer, both titanium dioxide (TiO_2) both blocking and mesoporous layers.

1. Introduction

Perovskite solar cells have much great research interest in the last decade, due to their highly potential as high performance and low-cost solar cells [1]. Perovskite materials have crystal structure ABX_3, where the A is the organic cation, the B is the metal cation and the X is the halogen anion. This perovskite crystals have superior optoelectronic properties, including optimal band-gap (~1.5 eV) [2], high absorption coefficient [3], long carrier diffusion length (~1 μm) [4] and high carrier mobility (~25 $cm^2 V^{-1} s^{-1}$) [5]. Perovskite properties also can be widely tuned by various elements positioned at A, B and X, such as methyl ammonium iodide (MAI), formamidinium iodide (FAI), lead(II) iodide, lead(II) chloride, etc. [6–8]. Perovskite materials can be used not only as light absorber in solar cells but also as charge transport materials both for hole and electron. Power conversion efficiency (PCE) of perovskite solar cells is increased dramatically only less than a decade from the first report by Miyasaka with 3.9% in 2009 [9] to beyond 20% reported by several groups [10-16].

However, stability in particular to humidity or moisture is a main drawback for mass production being used as electricity generating devices. Perovskites based lead is easily to react with water and oxygen, which leads to degradation of perovskite solar cells. Recently, many researches are focussing on fabricating high stability of perovskite solar cells, in particular to humidity, including new stable perovskite materials, device structure optimization and interface engineering [17]. Moreover, morphology of perovskite layer is very crucial for obtaining high performance and high stability solar cells. The perovskite film has to possess homogeneous grains with size larger than 500 nm and pin-hole free [12, 18-19]. Pin-holes in the perovskite layer cause direct contact between hole- and electron-transport layers, and thus lead to high resistance value of device and low performance of solar cells.

Recently, Tai *et al.* [20] fabricated perovskite solar cell using methylammonium lead(II) thiocyanate [$MA(PbSCN)_2$] instead of methylammonium lead(II) iodide ($MAPbI_3$) as active material. The solar cell showed stable high stability, irrespective to the relative humidity-stability. The high stability of $MA(PbSCN)_2$ of solar cells is due to higher binding energy of thiocyanate (SCN) ions to Pb ions as than that of iodide (I) ions to Pb ions [21]. However, the maximum PCE (15%) is lower than that of $MAPbI_3$ solar cells, due to higher bandgap (> 1.5 eV) [21]. Moreover, $MA(PbSCN)_2$ has different crystal structure, therefore its optoelectronic properties are also different as $MAPbI_3$ [21].

In this paper, we report our recent study of $Pb(SCN)_2$ doped $MAPbI_3$ to form new perovskite $MAPbI_{3-X}(SCN)_X$ with the aim to improve moisture stability of perovskite while maintain the superior optoelectronic properties of $MAPbI_3$. We fabricated two different perovskite solar cells structure, namely mesoporous and inverted planar solar cells.

2. Materials and Experimental Methods

2.1 Materials

Titanium dioxide (TiO_2) blocking layer (BL-1) and mesoporous (18NR-T) were purchased from Dyesol. Methylammonium iodide (MAI) and spiro-OMeTAD were purchased from Luminescence Technology Inc. and used as received without further purification. Lead (II) iodide (PbI_2), lead(II) tyocyanate [$Pb(SCN)_2$], poly(3,4-ethylenedioxythiophene) polystyrene sulfonate (PEDOT:PSS), methanol, isopropanol (IPA), N-N'-dimetylformamide (DMF) were purchased from Sigma Aldrich.

2.2 Characterization

UV-Vis spectroscopy (T70+ PG Instrument) was applied to measure optical absorption of thin films of PbI_2, $Pb(SCN)_2$ precursor and perovskite $MAPbI_{3-X}(SCN)_X$. Crystal structures of films were measured by use of X-Ray diffractometer. Cross section images of perovskite solar cells were measured by scanning electron microscope (SEM) SU3500 (Hitachi).

2.3 Fabrication of Perovskite Solar Cells

We fabricated two different solar cells structures, namely mesoporous and inverted planar structures. The mesoporous structure consists of glass/FTO/TiO_2/perovskite/spiro-OMeTAD/Au, while inverted structure comprises of Glass/ITO/PEDOT:PSS/perovskite/C60/BCP/Ag.

2.3.1. Mesoporous Solar Cells Structure

Titanium oxide (TiO_2) layers consist of solid film as blocking layer (bl) and mesoporous layer (mp). TiO_2 blocking layer was prepared by screen printing of TiO_2 solution in methanol using mesh T71 at room temperature. The film was then annealed at 450°C with two-step process. First, the film was heated to 250°C for 15 minutes and then at 450°C for another 15 minutes. The increasing temperature from room temperature to 250°C and from 250°C to 450°C was adjusted to have slow increased temperature gradient in order to have good quality of TiO_2 layer. The mesoporous TiO_2 layer was prepared by spin-coated of TiO_2 mesoporous paste at 1500 rpm for 30 s. The film was then annealed at 450°C using the same procedure as TiO_2 blocking layer.

Perovskite film was prepared from mixture of PbI_2 and 5% wt $Pb(SCN)_2$ precursor in DMF solvent. The solution was previously prepared by mixing 461 mg PbI_2 and 23 mg $Pb(SCN)_2$ in 1 ml DMF to form $PbI_2(SCN)_2$ solution. The mixed solution was then stirred for 3 hours at 60°C to obtain yellow transparent solution. The perovskite film was spin-coated on TiO_2 layer at spinning speed 3000 rpm for 40 s, using two-step process, i.e. 500 rpm for 5 s and 3000 rpm for 40 s. The MAI solution (10 mg MAI in 1 ml IPA) was then spin-coated with 3000 rpm for 40 s on top perovskite precursor film to convert the precursor film into perovskite film and followed by annealing it at 80°C for 5 minutes.

Hole transport layer was prepared by spin-coated Spiro-OMeTAD solution at 3000 rpm for 40 s using two-step process, first 500 rpm for 5 s and 3000 rpm for 40 s. The spiro-OMeTAD solutions were prepared dissolving 90 mg spiro-OMeTAD in 1 ml chlorobenzene and stirred at 70°C for 60 minutes. Finally, 100 nm of metal electrode consisting nickel and gold (5 nm Ni/95 nm Au) was deposited by thermal evaporation using a shadow mask to pattern the electrodes.

2.3.2. Inverted Solar Cells Structure

Devices structure of inverted planar perovskite solar cells comprises p-i-n structure, where PEDOT:PSS was used as p-type or hole transport material. Hole transport layer was prepared by spin-coated of PEDOT:PSS solution at 2500 rpm for 60 s. The film was then annealed at 140°C for 10 minutes, forming a PEDOT:PSS layer of 45 nm. Perovskite solution was prepared similar way as in mesoporous structure. The perovskite film was spin-coated on PEDOT:PSS layer at spinning speed 3000 rpm for 40 s, using two-step process, i.e. 500 rpm for 5 s and 3000 rpm for 40 s. The MAI solution (40 mg MAI in 1 ml IPA) was then spin-coated on top perovskite precursor film to convert into perovskite film and followed by annealing it at 100° C for 5 minutes.

Electron transport layer was prepared by highly vacuum thermal evaporation of fullerene C60 layer. The thickness of C60 was set 40 nm. The buffer layer bathocuproine (BCP) with thickness 8 nm was evaporated onto C60 layer by vacuum evaporation. Finally, 100 nm silver (Ag) was thermally evaporated on top of the device to form the cell's back contact using a shadow mask to pattern the electrodes.

2.4. J-V Characteristic Measurement

The current density-voltage (J–V) curves of the photovoltaic devices were obtained by a Keithley 4200 source-measure unit with a scanning forward direction from –0.2 to 1.0 V and reverse direction from 1.0 to –0.2 V with a sweep rate of 0.2 V/s. The photocurrent was measured under simulated 100 mW/cm^2 AM 1.5 G irradiation using a Xe-lamp-based solar simulator [Oriel 300 W solar simulator (Thermo Oriel 91160–1000)]. The simulated irradiance was calibrated using a certified silicon diode, as determined by a standard silicon solar cell. The effective area of the cell was defined as approximately 0.10 cm^2 using a non-reflective photo mask.

3. Results and Discussion

Figure 1 shows UV-Vis spectra of thin precursor films of PbI_2, $Pb(SCN)_2$ and perovskite $MAPbI_{3-X}(SCN)_X$. Thin film $Pb(SCN)_2$ absorbs light at wavelength shorter than 400 nm, while PbI_2 absorbs light at wavelength shorter than 500 nm. Absorption of perovskite $MAPbI_{3-X}(SCN)_X$ cover wide range of wavelength ranging from 780 nm to UV light. This wide range absorption of light is expected to generate large number of excitons in active material of solar cells, thus to obtain the high PCE of solar cells.

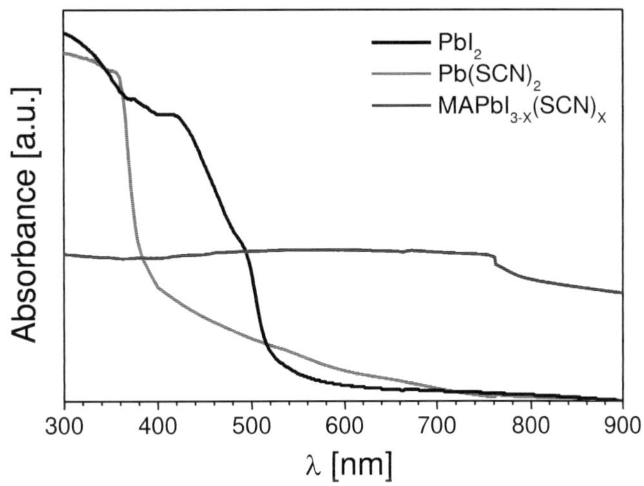

Figure 1. UV-Vis spectra of thin films of PbI_2, $Pb(SCN)_2$ and perovskite $MAPbI_{3-X}(SCN)_X$.

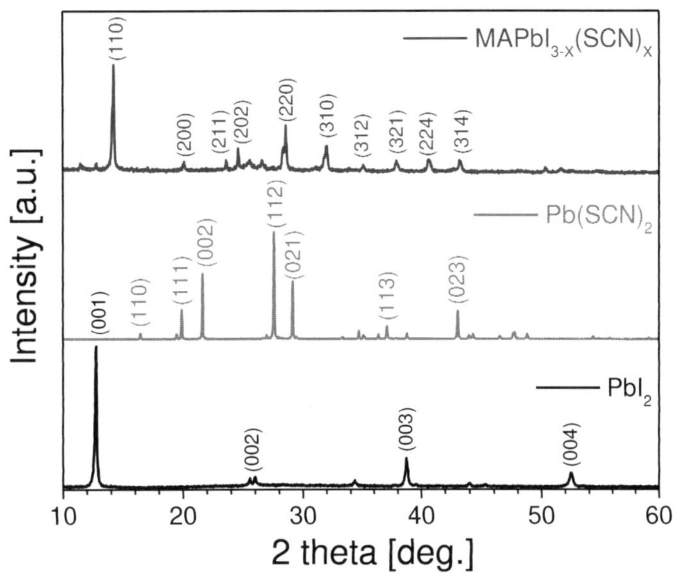

Figure 2. XRD pattern of PbI_2, $Pb(SCN)_2$ and perovskite $MAPbI_{3-X}(SCN)_X$ films.

XRD pattern of PbI_2, $Pb(SCN)_2$ and perovskite $MAPbI_{3-X}(SCN)_X$ films are shown in figure 2. Lead(II) iodide has four distinguished peak at 12.76°, 26.00°, 38.74° and 52.46° correspond to the crystal plane of (001), (002), (003) and (004), respectively. All crystal planes are oriented in z-axis similar as reported by other group [22]. PbI_2 film deposited by spin-coating from DMF solution crystallizes in the form of the hexagonal 2H polytype [22]. The XRD pattern of lead(II) thiocyanate shows also similar peaks reported in literature [23]. Lead(II) thiocyanate has monoclinic crystal structure with a = 9.661Å, b = 6.544 Å, c = 8.253 Å , $\alpha = \gamma = 90°$ and $\beta = 92.37°$, belonging to the space group C2/c [24]. We calculated the XRD pattern of $Pb(SCN)_2$ powder using its crystal structure to identify the crystal plane. The peaks at 16.47°, 19.91°, 21.64°, 27.58°, 29.14°, 37.08°, and 43.03° correspond to the crystal plane of (110), (111), (002), (112), (021), (113) and (023), respectively. No XRD peak is observed in perovskite film $MAPbI_{3-X}(SCN)_X$ originated both from PbI_2 and $Pb(SCN)_2$ peaks. It shows that both precursors are fully converted into perovskite crystal. The XRD pattern of perovskite shows tetragonal crystal structure [25-26]. The peaks at 14.25°, 20.12°, 23.60°, 24.63°, 28.57°, 31.98°, 35.07°, 37.88°, 40.67° and 43.26° correspond to the crystal plane of (110), (200), (211), (202), (220), (310), (312), (321), (224) and (314), respectively [25-26]. No significant PbI_2 peak at around 12° is observed, which shows that the perovskite $MAPbI_{3-X}(SCN)_X$ film is stable to high humidity as compared to the common perovskite material $MAPbI_3$ [19-20]. In the case of $MAPbI_3$, the perovskite is easily degraded into lead iodide PbI_2, even less than one hour during annealing of the film in ambient air with high relative humidity [27].

Figure 3 shows J-V curves of mesoporous and inverted planar perovskite solar cells in dark and illuminated conditions both forward and reverse measurements. The solar cell's parameters are displayed in table 1. In dark-condition, the mesoporous solar cell shows very good diode characteristic with voltage threshold's value of 0.48 V. It shows that p-i-n junction is well formed. However, in illuminated condition, the current in the device is very low that causes low value of short-circuit current (J_{SC}). Moreover, the open-circuit voltage (V_{OC}) is lower than that of in dark condition. The fill-factor (FF) value is also very low that resulted PCE less than 1%. On the other hand, the PCE of inverted planar structure is 6.56%. This PCE and also the value of V_{OC}, J_{SC} and FF are higher than that of mesoporous solar cells.

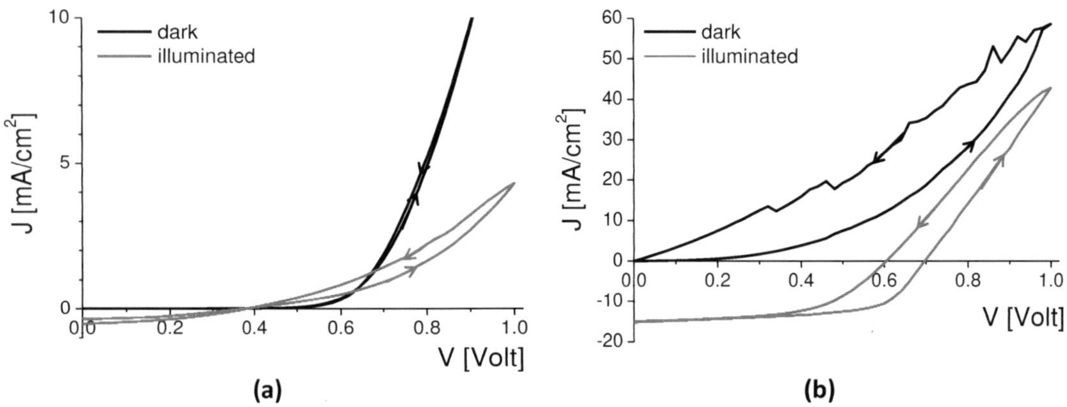

Figure 3. J-V Curve of perovskite solar cells (a) Mesoporous structure and (b) Inverted planar structure.

Table 1. Measured parameters of perovskite solar cells.

		V_{OC} (V)	J_{SC} (mA/cm^2)	FF	PCE (%)
Mesoporous	Forward	0.38	0.35	0.38	0.05
	Backward	0.38	0.52	0.37	0.07
Inverted planar	Forward	0.70	15.21	0.62	6.56
	Backward	0.60	14.97	0.56	5.00

The low values of solar cell's parameters of mesoporous structure are strongly related to the quality of TiO_2 layer as electron transport layer. Figure 4 shows the SEM images of solar cells cross section both mesoporous and inverted planar structure. It is clearly seen that the thickness of mesoporous TiO_2 layer, perovskite film and hole-transport spiro-OMeTAD layer is more than 2 μm which is too thick for optimum perovskite solar cells. The thickness of TiO_2 blocking layer must be less than 100 nm [28-29]. The layers in inverted planar structure, on the other hand, show homogeneous layer. The thicknesses of hole transport layer PEDOT:PSS (300 nm), electron transport layer C60/buffer layer BCP (60 nm) and electrode Ag (100 nm) are the optimum thickness of each layer in high reported PCE perovskite solar cells. The thickness of C60 needs to be carefully controlled. It needs to be thick enough to form a compact layer over the perovskite film, but a too thick C60 layer will increase the series resistance of the device. The thickness of perovskite film is around 800 nm which is two-times thicker than that the optimum layer for high PCE perovskite $MAPbI_3$ (300 – 400 nm) [30]. Liu et.al. [30] reported that for the devices with perovskite film thicknesses larger than 400 nm, the PCE decreases significantly. As the film thickness continues to increase, the series resistance of device is increased. The series resistance of the cells with the thickness of perovskite layers more than 500 nm is over an order of magnitude higher than for optimized devices.

Hysteresis in J-V curves of mesoporous and inverted planar solar cells might be caused by poor quality of perovskite film, in particular related to the "pin-holes" which lead to high series resistance [17]. However, mesoporous structure shows less hysteresis than inverted planar which might be due to a better p-i-n junction formation. Morphological control is highly needed to obtain high quality perovskite film, i.e. pin-hole free and homogeneous perovskite grains with average size larger than 500 nm [16-18]. The large hysteresis of J-V curve in inverted planar structure might be caused by poor contact between perovskite and PEDOT:PSS layer due to nature hydrophilic property of PEDOT:PSS.

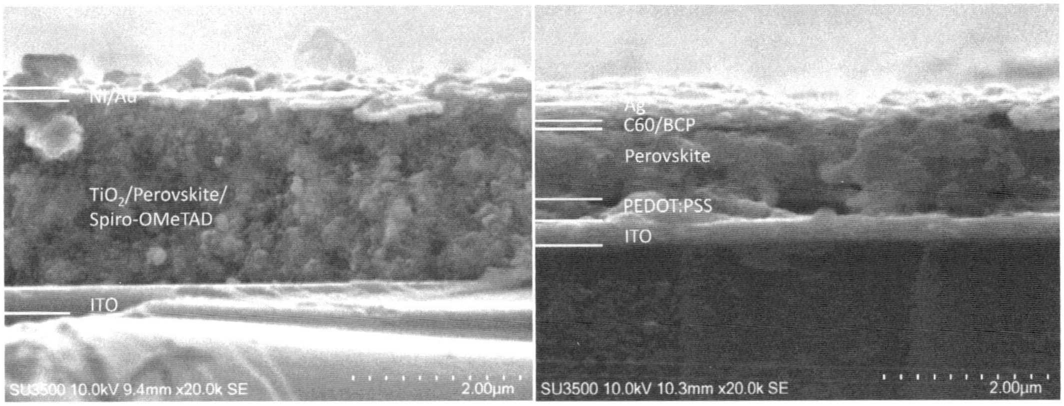

Figure 4. Cross section SEM images of (a) Mesoporous structure and (b) Inverted planar structure.

4. Conclusion

We have fabricated perovskite solar cells using lead(II) thiocyanate-doped lead(II) iodide as precursor materials to form perovskite $MAPbI_{3-X}(SCN)_X$ as active material both in mesoporous and inverted planar structures. The best PCE 6.56% is obtained with inverted planar solar cells. All parameters of mesoporous solar cells show lower values than that of inverted planar structure, which is caused by poor quality and too thick of electron transport layer, titanium dioxide (TiO_2) both blocking and mesoporous layers. However, mesoporous structure shows less hysteresis of J-V curve than that of inverted planar structure.

Acknowledgement

This work is financially supported by Ministry of Research, Technology and Higher Education Indonesia under Research Grant for International Collaboration and Scientific Publication 2017, Contract No. 718/UN6.3.1/PL/2017.

References

[1] Correa-Baena J P, Saliba M, Buonassisi T, Grätzel M, Abate A, Tress W and Hagfeld A 2017 *Science* **358** 739
[2] Kim H S *et al.* 2012 *Sci. Rep.* **2** 591
[3] Green M A, Ho-Baillie A and Snaith H J 2014 *Nat. Photon.* **8** 506
[4] Stranks S D, Eperon G E, Grancini G, Menelaou C, Alcocer M J P, Leijtens T, Herz L M,. Petrozza A and Snaith H J 2013 *Science* **342** 341
[5] Ponseca C S *et al.* 2014 *J. Am. Chem. Soc.* **136** 5189
[6] Noh J H, Im S H, Heo J H, Mandal T N and Seok S I 2013 *Nano Lett.* **13** 1764
[7] Eperon G E, Stranks S D, Menelaou C, Johnston M B, Herz L M and Snaith H J 2014 *Energy Environ. Sci.* **7** 982
[8] Park N G 2015 *Mater. Today* **18** 65

[9] Kojima A, Teshima K, Shirai Y and Miyasaka T 2009 *J. Am. Chem. Soc.* **131** 6050
[10] Yang W S, Noh J H, Jeon N J, Kim Y C, Ryu S, Seo J, Seok S I 2015 *Science* **348** 1234
[11] Saliba M *et al.* 2016 *Energy Environ. Sci.* **9** 1989
[12] Li X, Bi D, Yi C, Décoppet J D, Luo J, Zakeeruddin S M, Hagfeldt A and Grätzel M 2016 *Science* **353** 58
[13] Tan H *et al.* 2017 *Science* **355** 722
[14] Shin S S, Yeom E J, Yang W S, Hur S, Kim M G, Im J, Seo J, Noh J H, Seok S I 2017 *Science* **356** 167
[15] Yang W S *et al.* 2017 *Science* **356** 1376
[16] Arora N, Dar M I, Hinderhofer A, Pellet N, Schreiber F, Zakeeruddin S M and Grätzel M 2017 *Science* **358** 768
[17] Xu T, Chen L, Guo Z and Ma T 2016 *Phys.Chem.Chem.Phys.* **18** 27026
[18] Kim H D, Ohkita H, Benten H, Ito S 2016 *Adv Mater.* **28** 917
[19] Im J H, Kim H S and Park N G 2014 *APL Materials* **2** 081510
[20] Tai Q, You P, Sang H, Liu Z, Hu C, Chan H L W and Yan F 2016, *Nat. Commun.* **7** 11105
[21] Jiang Q, Rebollar D, Gong J, Piacentino E L, Zheng C and Xu T 2015 *Angew. Chem. Int. Ed.* **54** 7617
[22] Burschka J, Pellet N, Moon S J, Humphry-Baker R, Gao P, Nazeeruddin M K and Grätzel M 2013 *Nature* **499** 316
[23] Ke W *et al.* 2015 *Angew. Chem. Int. Ed.* **54** 7617
[24] Mokuolo J A A and Speakman J C 1975 *Acta Cryst.* **B31** 172
[25] Halder A, Chulliyil R, Subbiah A S, Khan T, Chattoraj S, Chowdhury A, Sarkar S K 2015 *J. Phys. Chem. Lett.* **6** 3483
[26] Oku T 2015 *Crystal Structures of $CH_3NH_3PbI_3$ and Related Perovskite Compounds Used for Solar Cells* (Solar Cells - New Approaches and Reviews ed L A Kosyachenko (InTech) chapter 3 77)
[27] Yang J, Siempelkamp B D, Mosconi E, De Angelis F and Kelly T L 2015 *Chem. Mater.* **27** 4229
[28] Qarony W, Jui Y A, Das G M, Mohsin T, Hossain M I, Islam S N 2015 *Am. J. Energy. Research* **3** 19
[29] Lu H, Ma Y, Gu B, Tian W and Li L 2015 *J. Mater. Chem. A* **3** 16445
[30] Liu D, Gangishetty M K and Kelly T L 2014 *J. Mater. Chem. A* **2** 19873

IOP Conf. Series: Journal of Physics: Conf. Series **1080** (2018) 012013 doi:10.1088/1742-6596/1080/1/012013

Photocatalytic activity of gadolinium doped TiO_2 particles for decreasing heavy metal chromium (VI) concentration

D R Eddy[1*], I Rahayu[1], S Wyantuti[1], Y W Hartati[1], M L Firdaus[2] and H H Bahti[1]

[1]Department of Chemistry, Faculty of Mathematics and Natural Sciences, Universitas Padjadjaran, Sumedang, West Java, Indonesia 45363
[2]Graduate School of Science Education, University of Bengkulu, Bengkulu, Indonesia 38371

[*]diana.rahmawati@unpad.ac.id

Abstract. Titanium dioxide (TiO_2) is among the most potential photocatalyst. However, TiO_2 only works under UV light due to the relatively wide band gap energy. To overcome this, TiO_2 is infiltrated (doped) by rare earth elements such as gadolinium (Gd). TiO_2 was synthesized using sol gel method, while Gd/TiO2 photocatalyst was synthesized by the mixing methods of sol-gel and impregnation. The crystals are then characterized using XRD and SEM-EDX. The XRD diffractogram demonstrated that the crystallinity synthesized of TiO_2 had an anatase structure similar to the structure of anatase P-25 as references. The SEM-EDX showed that Gd ion could not enter the crystal lattice, but only dispersed on the surface of TiO_2 particles. The optimum result of decreased Cr(VI) concentration was reported using Gd/TiO_2 1% w/w by a decrease of up to 94%, after irradiated with ultraviolet for one hour.

1. Introduction

Recently, titanium oxide has gained great attention due to its ability as a photocatalyst semiconductor. In photocatalysts, irradiation that exceeds the semiconductor band gap is absorbed to produce electron pair holes, which facilitate an important reaction such as water purification. The use of TiO_2 photocatalyst in the form of a dispersed powder has the potential to act as a catalyst that has a larger surface area than in film or coating. The photocatalyst reaction occurs near the surface of the catalyst, and the TiO_2 powder is used in an aqueous suspension for water purification [2].

TiO_2 has many advantages because it is cheap, non-toxic and has long term stability. There are three TiO_2 crystallographic phases, i.e. anatase, rutile, and brookite. Anatase TiO_2 is the most commonly applied photocatalyst for oxidative decomposition of organic compounds and the conversion of solar light energy, due to its high photoactivity. TiO_2 commonly used as a photocatalyst is TiO_2 with anatase and rutile structures [3]. Thermodynamically, anatase possesses chemical and physical properties that are more stable than the rutile phase. Furthermore, titania has a sufficiently high surface area which is play an important role in aqueous photocatalyst reactions [4]. Titanium dioxide can be applied as a photocatalyst because it has a suitable energy gap to support several chemical reactions. The energy gap of TiO_2 absorbs electromagnetic radiation in the near ultraviolet region. Decreasing band gap of TiO_2 can increase the photocatalyst performance. Compared to other rare earth elements such as Sm(III), Ce(III) and Nd(III), Gd(III) is selected as a dopant ion since it could reduce band gap of TiO_2 smaller than 3.2 eV [5]. We have already reported the used of TiO_2 [6, 7] and Gd-doped TiO_2 to decrease Chemical Oxygen Demand [8].

In this work, we prepared Gd(III) doped TiO_2 with various Gd(III) concentration which could affect photocatalyst activity. TiO_2 was prepared by the sol-gel method to produce TiO_2 nanoparticles with low calcination temperature and an affordable cost. In the environmental, if Cr(VI) has a high level concentration, it could cause irritation and corrosion [9]. We studied Gd/TiO_2 photoactivity with Cr(VI) as an object.

2. Materials and Methods

2.1 Materials

The materials used in this study are deionized water, distilled water, nitric acid p.a (HNO_3 65% b/b, Merck), ethanol p.a (C_2H_5OH, 99%, Merck), gadolinium(III)-oxide p.a (Gd_2O_3, 99%, Sigma Aldrich), titanium(IV)-oxide (TiO_2, P25 Degussa), titanium tetrachloride ($TiCl_4$, 99%, Merck), and potassium chromate (K_2CrO_4, 99%, Merck).

2.2 Methods

2.2.1 Synthesis of Gd/TiO$_2$

Synthesis of Gd/TiO_2 was performed by mixing 8 ml of ethanol dropwise to 10 ml of titanium tetrachloride. The color of mixed solution after two hours became yellow. A yellowish-color gel was formed, which was allowed to settle undisturbed for 24 hours. The gel was then heated in an oven at a temperature of 120°C for 3 hours. A yellowish powder was formed, which was crushed by a ball mill and calcinated at a temperature of 500°C for 2 hours. The TiO_2 powder has impregnation by $Gd(NO_3).6H_2O$ (prepared from Gd_2O_3 and HNO_3 concentrated) with various concentrations (1%, 3%, and 5% w/w). The mixture was dried in the oven at a temperature of 120°C for 24 hours, and then calcinated to 500°C for 3 hours. The Gd/TiO_2 powder was then characterized by XRD and SEM-EDX.

Gd/TiO_2 photocatalyst was synthesized by the mixing methods of sol-gel and impregnation. Titanium chloride was selected as a precursor due to a large amount of TiO_2 could be produced by hydrolysis. The hydrolysis process of TiO_2 is as followed.

$$TiCl_4 + 2C_2H_5OH \longrightarrow TiO_2 + 4HCl + 4CH_2 \quad (1)$$

The viscosity of the solution changes by adding ethanol one dropwise into gel. The resulting gel contains colloid particle suspension of titania which cause by the Brownian motion. The formed suspension was allowed to settle for 24 hours to gradually remove HCl gas [10].

Photocatalyst that was synthesized was dried in an oven at a temperature of 120°C for 3 hours. Then, it was ground with ball mill to expand the surface area and refine the powder. As the Gd(III) ion source, Gd_2O_3 could not be directly mixed because it is difficult to dissolve with TiO_2. Modification of TiO_2 by $Gd(NO_3).6H_2O$ was performed by the impregnation method. The residual impurities were removed by heated at a temperature 120°C for 24 hours, and then calcinated to 500°C to reach the anatase phase.

2.2.2 Activity Test of Gd/TiO$_2$ Photocatalyst

Metal Cr(VI) absorbance was measured using AAS (Systronics, model-2203). The Gd/TiO_2 powder was added into a photoreactor and irradiated with an Hg lamp. The samples (2 ml) were extracted every hour using a syringe membrane and subsequently measured by AAS. The same treatment was performed for TiO_2 without doping as a comparison.

3. Results and Discussion

3.1 Characterization

3.1.1 XRD Characterization

Crystalline phase and size of the synthesized materials were identified. Undoped TiO_2 and Gd/TiO_2 were characterized by XRD to analyze the effect of Gd(III) concentration solvent and calcination temperature to TiO_2 crystal structure.

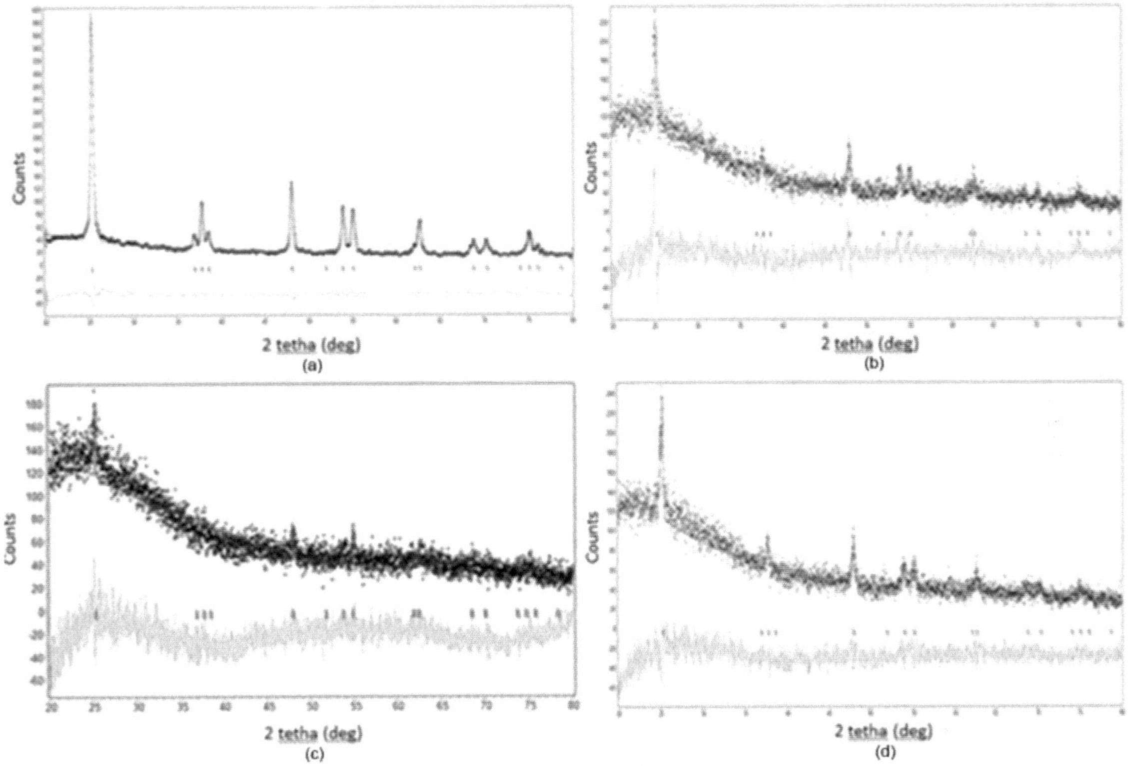

Figure 1. X-Ray diffraction of (a) TiO_2 without doping, (b) Gd/TiO_2 1%, (c) Gd/TiO_2 3%, (d) Gd/TiO_2 5%.

The use of ethanol as a solvent produces anatase and rutile phase of TiO_2, but anatase phase is more dominant. Figure 1(a) shows the crystalline phase of anatase TiO_2 which is assigned by high intensity of peak at $2\theta = 25°$. The difference in crystal structure which shown in figure 1(b-d) is due to the difference of concentration Gd(III) and solution acidity. The acid state in this system act as catalyst in the process of nucleation and crystallization. The crystalline phase from amorphous to anatase began at 350°C, therefore the calcination temperature at 500°C produced well-crystallized TiO_2.

Figure 1 also shows the Le Bail refinement of TiO_2 and Gd/TiO_2 at concentration of 1%, 3%, and 5%. These concentration were the optimum concentrations as reported by Zhang *et al.*, 2015 [11]. The XRD data were collected on the range of 2θ between 20°-80° in tetragonal group system, space group 141/amd with Z = 4. XRD data from measurement (black line) and calculation (red line) should match each other with small difference which presented as green line curve in figure 1 [8]. The three highest peaks (2θ) at 25°, 38° and 48° are indicated anatase phase while at 27°, 36° and 54° are rutile phase. These result have been confirmed with JCPDS Card no. 21-1272. Figure 2 shows the XRD pattern of pure Gd_2O_3 with dominant peak (2θ) at 29.8°.

XRD patterns of Gd/TiO_2 1% and 5% in figure 1 are similar to the anatase phase of TiO_2 patterns. The addition of Gd 1% dispersed well on the surface of TiO_2, which efficiently inhibit direct contact between TiO_2 grains, the agglomeration of TiO_2 particles and the growth of rutile phase. However, in

Gd(III) doped TiO_2, the peaks of Gd_2O_3 are less identified on the XRD pattern, because gadolinium only attach to TiO_2 surface.

Table 1. Crystallite size of TiO_2 and Gd/TiO_2 synthesized concentration variations 1%, 3%, and 5%.

Samples	Crystallite size (nm)
TiO_2	195.20
Gd/TiO_2 1%	244.64
Gd/TiO_2 3%	213.77
Gd/TiO_2 5%	220.77

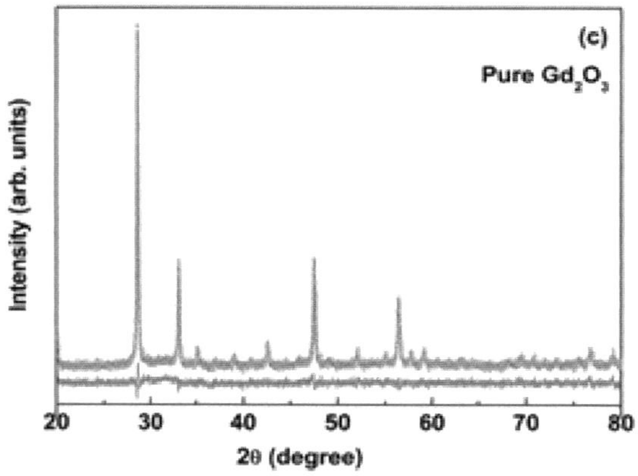

Figure 2. XRD pattern of Gd_2O_3 [12].

Table 1 shows the crystallite size of Gd(III)/TiO_2 at different concentration of Gd. The average crystallite size (D) was estimated from Scherrer's equation (equation (2)) from the highest peak of each sample. In doping process, not all gadolinium ion enters the lattice of TiO_2, therefore, we cannot observe the effect of concentration to the crystallite size.

$$D = \frac{k\lambda}{\beta \cos \theta} \quad (2)$$

3.1.2 SEM-EDX Analysis

The morphology, distribution and composition of Gd/TiO_2 were characterization by using SEM-EDX. Figure 3(b-d) shows that Gd/TiO_2 has agglomeration and part of Gd^{3+} ions are distributed on the surface of crystallites. Gd/TiO_2 1% has a better gadolinium distribution pattern on TiO_2.

Table 2 shows the EDX results of Gd/TiO_2 1%, and gadolinium composition is 0.44%.The small amount of gadolinium was caused by the loss of gadolinium during synthesis process. We assumed that gadolinium just attach to TiO2 surface rather than inserted in crystal lattice. This result has a good agreement with our previous research [8].

Based on XRD analysis, there is no distinctive peak of gadolinium that signifies TiO_2 doped slightly. The SEM-EDX analysis shows that in TiO_2, the catalyst results remain to have empty cavities, and gadolinium does not substitute titanium, but is only scattered on the surface.

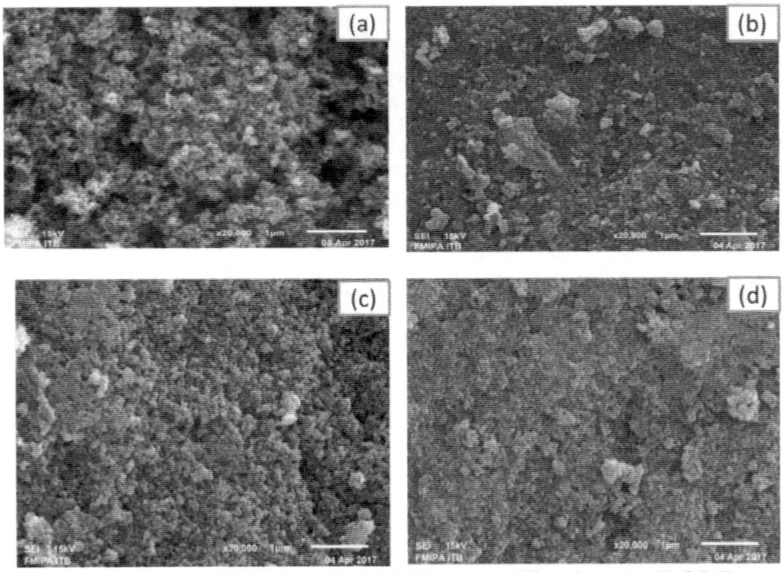

Figure 3. Surface morphology of SEM with 20,000x magnification (a) P-25 Degussa, (b) Gd/TiO_2 1%, (c) Gd/TiO_2 3% and (d) Gd/TiO_2 5%.

Table 2. EDX analysis of Gd/TiO_2 1%.

Element	(keV)	mass%	Error%	Atom%	K
O K	0.525	48.85	0.28	74.20	6.7916
Ti K	4.508	50.71	0.14	25.73	82.6930
Gd L	6.053	0.44	0.75	0.07	0.5255
Total		100.00		100.00	

Fitting Coefficient: 0.2263

3.2 Photocatalyst test of TiO_2 and Gd/TiO_2

A photocatalytic test was done to determine performance of the synthesized photocatalyst compared to P-25 Degussa. Cr(VI) is one of the heavy metals that found in river stream of industrial waste disposal. Cr(VI) degradation simulation was obtained from K_2CrO_4 solution.

Table 3. The decreasing of Cr(VI) concentration 2.2722 ppm by P-25, TiO_2 and Gd/TiO_2 using UV light.

Time/hour	Photocatalysts				
	P25	TiO_2	Gd/TiO_2 1%	Gd/TiO_2 3%	Gd/TiO_2 5%
1	0.1105	0.2008	0.1425	0.9865	0.2252
2	0.1105	0.1425	0.1519	0.2308	0.2308
3	0.1162	0.1218	0.1613	0.1820	0.2026

The analyst results using Atomic Absorption Spectroscopy (AAS) showed that there was a decrease in Cr(VI) concentration after the addition of photocatalyst by UV light (table 3). The reaction that occurs in the decrease of Cr(VI) concentration is a radical reaction in which the capture and release of electrons occurs when the oxidant is formed in the photocatalyst process. The highest decreasing of Cr(VI) is 1% that was irradiated for 1 hour. However, the decreasing in Cr(VI) concentration by TiO_2 is much higher than that of TiO_2 doped. Gd atom should substitute the Ti atom, which makes the charge unbalanced, and the hydroxides ion will be absorbed on the surface. Hydroxide ions act as trap holes that inhibit the recombination of electron pairs or holes.

The decrease of Cr(VI) concentration from different photocatalysts is shown in figure 4. It shows the Gd/TiO_2 1% photocatalyst effectively decreased Cr(VI) concentration. Gd/TiO_2 1% for 1 hour is the optimum result by decreasing Cr(VI) concentration with a value of 94 %, compared to the other TiO_2 doped composition. However, the addition of dopant in larger quantities does not necessarily increase the activity of photocatalysts. It occurs because too much Gd form the recombination center, where the photoinduction carrier is retained, causing the photocatalyst activity to be lower.

Table 4. The decreasing ability of Cr(VI) concentration 2.2722 ppm by P-25 and TiO_2.

Photocatalysts	Cr(VI) concentration (ppm)	% decrease
P25	0.1162	95 %
TiO_2	0.1218	91 %

Based on table 4, the decrease of Cr(VI) concentration by synthesized TiO_2 is similar to commercial P-25 Degussa. In accordance with the regulation of the Minister of Environtment No. 01 of 2010 on the Management of Water Pollution Control, the maximum limit of Cr(VI) level is 0.3 mg/L, therefore, the synthesized TiO_2 and Gd/TiO_2 1% w/w effectively decrease Cr(VI) concentration.

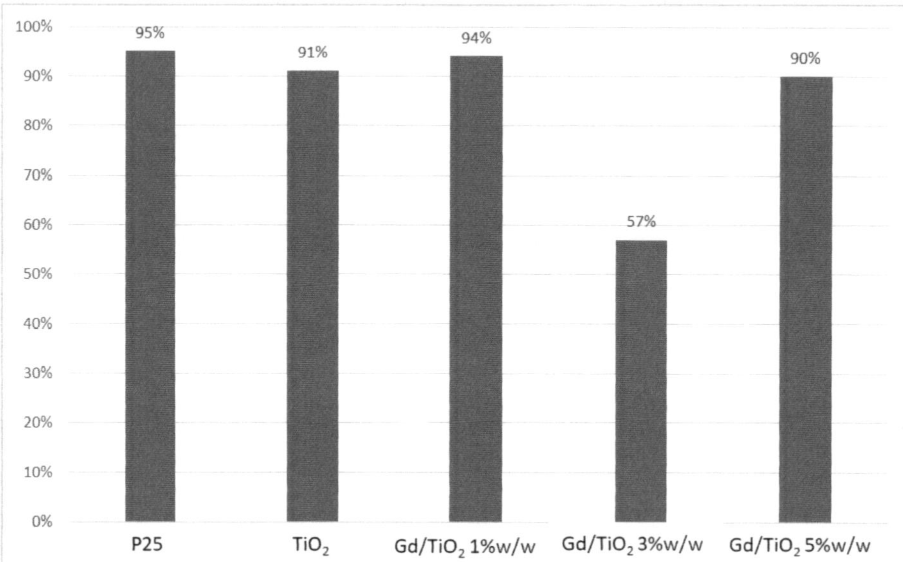

Figure 4. Decreasing concentration of Cr(VI) by different photocatalysts after irradiation for 1 hour.

4. Conclusion

Both XRD and SEM analysis indicate that Gd ion disperse on the surface of TiO_2 particles and form Gd cluster. Different Gd^{3+} doping level affect the crystallite size. Both TiO_2 and Gd/TiO_2 1% w/w effectively decrease Cr(IV) concentration. This work provides a strategy for designing and fabricating of high efficient TiO_2 based photocatalysts by medium concentration.

Acknowledgment

This research was supported by Academic Leadership Grant (ALG) project, Universitas Padjadjaran, Republic of Indonesia. We thank Kemal Hamdani for his helpful experiment. We also thank reviewers whose valuable suggestion improved the quality of the manuscript.

IOP Conf. Series: Journal of Physics: Conf. Series **1080** (2018) 012013 doi:10.1088/1742-6596/1080/1/012013

References

[1] Agorku E S, Mamba B B, Pandey A C and Mishra A K 2014 *J. Nanomaterial* **1** 1
[2] Hanaor D H, Chironi I, Karatchevtseva I, Triani G and Sorrell C C 2012 *Advances in Applied Ceramics* **111** 149
[3] Li W and Zeng T 2011 *PLoS ONE* **6** e21082
[4] Dastan D and Chaure N B 2014 *Inter. J Materials Mechanics and Manufacturing* **2** 21
[5] Štengl V, Bakardjieva S and Murafa N 2009 *Materials Chemistry and Physics* **114** 217
[6] Destiarti L, Tjokronegoro R, Eddy D R and Rudiyansyah 2015 *Makara J. Sci.* **19** 1
[7] Eddy D R, Puri F N and Noviyanti A R 2015 *Procedia Chemistry* **17** 55
[8] Eddy D R, Dwiyanti D, Rahayu I, Hastiawan I and Bahti H H 2017 *AIP Conf. Proc. on Mathematics, Science, and Computer Science Education (MSCEIS 2016)* **10** 03003-1
[9] Khalil L B, Mourad W E and Rophael M W 1998 *J. Appl. Catalysis B: Environmental* **17** 267
[10] Shahab M, Tabish T, Zaman B and Kamran M 2005 *J. Inter. Engineering* **3** 313
[11] Zhang J, Wu W, Yan S, Chu G and Zhao S 2015 *J. Appl. Surface Science* **344** 249
[12] Sarkara B J, Debb A K and Chakrabarti P K 2016 *RSC Adv.* **6** 6395

IOP Conf. Series: Journal of Physics: Conf. Series **1080** (2018) 012014 doi:10.1088/1742-6596/1080/1/012014

The compositional observation of synthesized ZrO_2-Al_2O_3-SiO_2 as dental composite fillers

V Takarini[1], Z Hasratiningsih[1], X Lee[2], Sunarso[1] and A Cahyanto[1,3*]

[1]Department of Dental Materials Science and Technology, Faculty of Dentistry, Universitas Padjadjaran, Jl. Raya Bandung Sumedang km 21 Jatinangor 45363, Indonesia
[2]Vanadia Utama Ltd., Komplek Sentra Industri Terpadu, Pantai Indah Kapuk, Blok G1 No. 5 Jakarta 14470, Indonesia
[3]Oral Biomaterials Study Center, Faculty of Dentistry, Universitas Padjadjaran, Jl. Sekeloa Selatan 1, Bandung 40134, Indonesia

[*]arief.cahyanto@fkg.unpad.ac.id

Abstract. One aspect of improvements in the field of restorative dental material properties is focusing on reinforcement of ceramic fillers. Indonesian researchers are encouraged to synthesize and fabricate dental materials product to promote national dental materials independency. This research aims to synthesize ZrO_2-Al_2O_3-SiO_2 ceramic fillers and observed the composition of the particles using Scanning Electron Microscope (SEM) and Spectrum-Line Energy Dispersive Spectroscopy (EDS). The results show that the average particle size ranges between 0.5 – 150 μm. Zirconia formed as homogenous particles and the highest percentage among other elements. This composition shall provide the transformation toughening mechanism of ceramic dental composite as restorative materials.

1. Introduction

Physical and mechanical properties of restorative dental materials can be improved by filler reinforcement. These factors affect the clinical performance that will change the compressive and tensile strength, modulus of elasticity, and toughness. Dental composite is one of restorative dental materials used filler to reinforce the matrix using ceramic particles, such as glass, silica, and other metal oxides [1].

Ceramic fillers such as zirconia can be used on a variety of restoration needs such as crowns, veneers, bridge frameworks as well as reinforced materials. Zirconia (ZrO_2) is a widely used ceramic material in various fields from electronic equipment to the field of dentistry. Zirconia is a crystalline dioxide from zirconium [2, 3]. Its mechanical properties are similar to metals and their colour is similar to tooth. Zirconia crystals consist of three types depending on temperature i.e. monoclinic (M), cubic (C), and tetragonal (T). Pure zirconia crystals at room temperature are in monoclinic form, however, from 1170°C to 2370°C the crystalline structure becomes tetragonal before transforming to cubic at above 2370°C until its melting point [2-4].

At room temperature, tetragonal and cubic crystalline structures are unstable and tend to transform into monoclinic. ZrO_2 mixtures with metal oxides such as Al_2O_3, SiO_2, MgO, CaO, and Y_2O_3 can produce stable molecules. Metal oxide will serve as a stabilizer that supports tetragonal structures to be more stable at room temperature, therefore more effective to inhibit crack propagation. When there is a stress application, a transition between the different crystal (reticulation) arrays due to force on the zirconia surface occurs, resulting in a volumetric change in the crystal. This condition causes partially stabilized zirconia to have high strength and fracture toughness at the tip of the crack. The ability to inhibit the crack propagation is known as the transformation toughening mechanism [4-7].

To promote national independency, Indonesia researchers also encouraged to synthesize and fabricates filler components to develop dental materials products. Therefore, our research focusing on ZrO_2-Al_2O_3-SiO_2 synthesis as ceramic fillers to reinforces restorative dental materials, which compositional observations of morphological interest at a micron and sub-micron level features as well as to study the chemical composition of synthesized ZrO_2-Al_2O_3-SiO_2 ceramic fillers in terms of the amount of each element present characterized using Scanning Electron Microscopy (SEM) coupled with Spectrum-Line Energy Dispersive Spectroscopy (EDS). Those are two analysis techniques that are widely used to study morphology and elements of solid samples, from inorganic to biological [9].

2. Materials and Methods

2.1 Synthesis of ZrO_2-Al_2O_3-SiO_2

Zirconia-alumina-silica was synthesized using sol-gel technique. Aluminum nitrate nonahydrate $[Al(NO_3)_3 9H_2O]$, tetraethyl orthosilicate (TEOS; $Si(OC_2H_5)_4$), and zirconium chloride ($ZrCl_4$) were used as precursors. Distilled water and 1% w/v chitosan were used as solvent and dispersant, respectively. Briefly, 11.652 g of $ZrCl_4$, 9.378 g of Al_2NO_3 and 38.785 ml of TEOS were dissolved in 500 ml distilled water, followed by the addition of 100 ml of 1 w/v% chitosan solution. Subsequently, the solution was ultra-turaxed at 10 rpm for 5 minutes and repeated for 6 times. The as-prepared filler of composite was homogenized using ultrasonic homogenizer. After drying at 100°C, the filler particles of composite was calcined at 550°C for half an hour followed by calcined in 700°C for 2 hours in an electric furnace (Cress Electric Furnace C1228/935, Carson City, Nevada).

2.2 Characterization of ZrO_2-Al_2O_3-SiO_2

The as-synthesized powder was characterized using Scanning Electron Microscope (SEM) and Energy Dispersive Spectroscopy (EDS). For EDS analysis, both spectrum point and line scanning were performed.

SEM as samples characterization technique can be used widely for hard materials, such as metals or ceramics, and soft materials, such as polymers or biological tissues. Electron beam is finely focused and scanned over the analysis area and signals are produced due to the electron-matter interactions on the surface of the specimen is the basic work principle of SEM. Such signals include secondary and backscattered electrons for the image formation and characteristic x-ray from the elements present in the sample for composition studies [9]. In order to evaluate the composition of the obtained ZrO_2-Al_2O_3-SiO_2 particles, EDS analysis was also performed. EDS analysis is a spectroscopic technique which allows to determine the presence and relative abundance of the elements that compose the surface of the specimen studied [9].

3. Results and Discussion

Figure 1 shows the SEM image of the ZrO_2-Al_2O_3-SiO_2 particles obtained by sol gel technique. The size of ZrO_2-Al_2O_3-SiO_2 particles measured by SEM ranges from 0.5 μm to 150 μm. The particles seem to be agglomerated. Based on the particle size the ZrO_2-Al_2O_3-SiO_2 filler can be classified as hybrid fillers.

Smaller filler particle size can increase the surface area, therefore surface smoothness and esthetics characteristic can be improved. A preliminary observation If the volume fraction of filler approximately 70%, abrasion and fracture resistance are raised to levels approaching those of tooth tissue, thereby increasing both clinical performance and durability. Ceramic fillers not only can be used for reinforcement materials, but also reduce polymerization shrinkage, reduction in thermal expansion, viscosity control, decreased water sorption, and imparting radio opacity [1, 10].

Sol gel technique used in this study is expected to produce ZrO_2-Al_2O_3-SiO_2 nanoparticles. However, nano-size particles could not be observed from SEM analysis. It is known that nanoparticles tend to agglomerate due to their high surface energy. Therefore, further experiments need to be carried out to grind down these agglomerated particles to nano-sized particles.

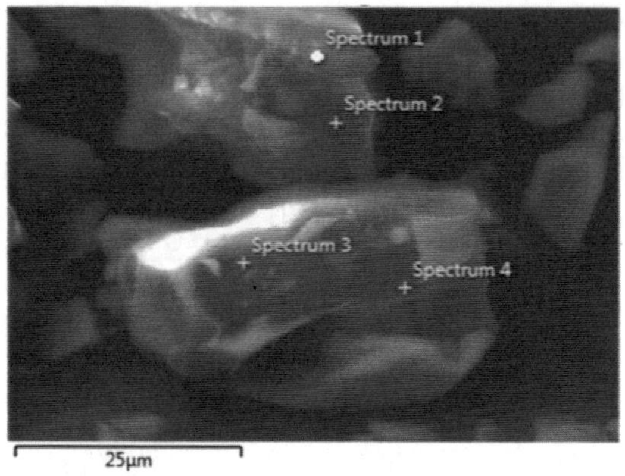

Figure 1. SEM image of the ZrO_2-Al_2O_3-SiO_2.

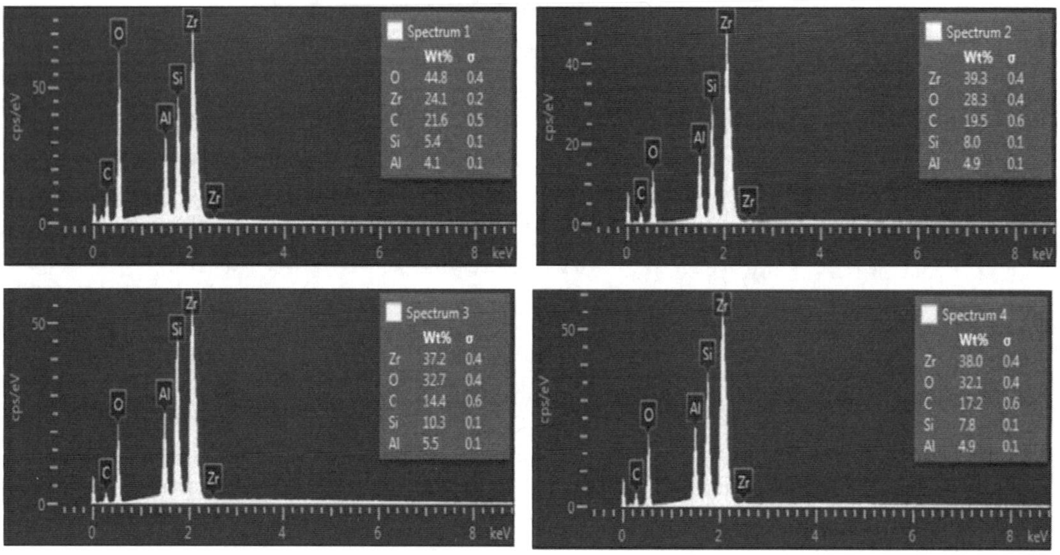

Figure 2. EDS analysis of ZrO_2-Al_2O_3-SiO_2 particles at 4 different points.

Figure 2 shows the relative amount of Zr, Al and Si elements in ZrO_2-Al_2O_3-SiO_2 particle at the given points in figure 1. The average amount of Zr, Si and Al elements is shown in table 1.

Table 1. Average amount of filler elements.

Element	
Zr	34.65±7.09 wt%
Al	4.85±0.57 wt%
C	18.17±0.57 wt%
O	34.47±0.4 wt%
Si	7.88±2.00 wt%

The EDS line scanning was also employed to evaluate whether the composition of Zr, Si, and Al is relatively similar throughout the particle. Figure 3 shows the EDS line scanning obtained from the ZrO_2-Al_2O_3-SiO_2 particles. EDS line scanning demonstrated that the composition of Zr, Si, and Al is relatively similar with that obtained in figure 2. The result indicated that the obtained particles have homogenous composition of Zr, Si, and Al.

The results of composition and morphological of ZrO_2-Al_2O_3-SiO_2 particles also can be influenced

IOP Conf. Series: Journal of Physics: Conf. Series **1080** (2018) 012014 doi:10.1088/1742-6596/1080/1/012014

by ceramic firing like the sintering temperature and time. The sintered particles would normally flow and fill up the pore spaces during sintering. Another study is being carried out to determine the relationship between sintering temperature/time and pore space generation. So far, a result from a porcelain version suggests that the sintering temperature may have distinguishable effects on pore generation [11].

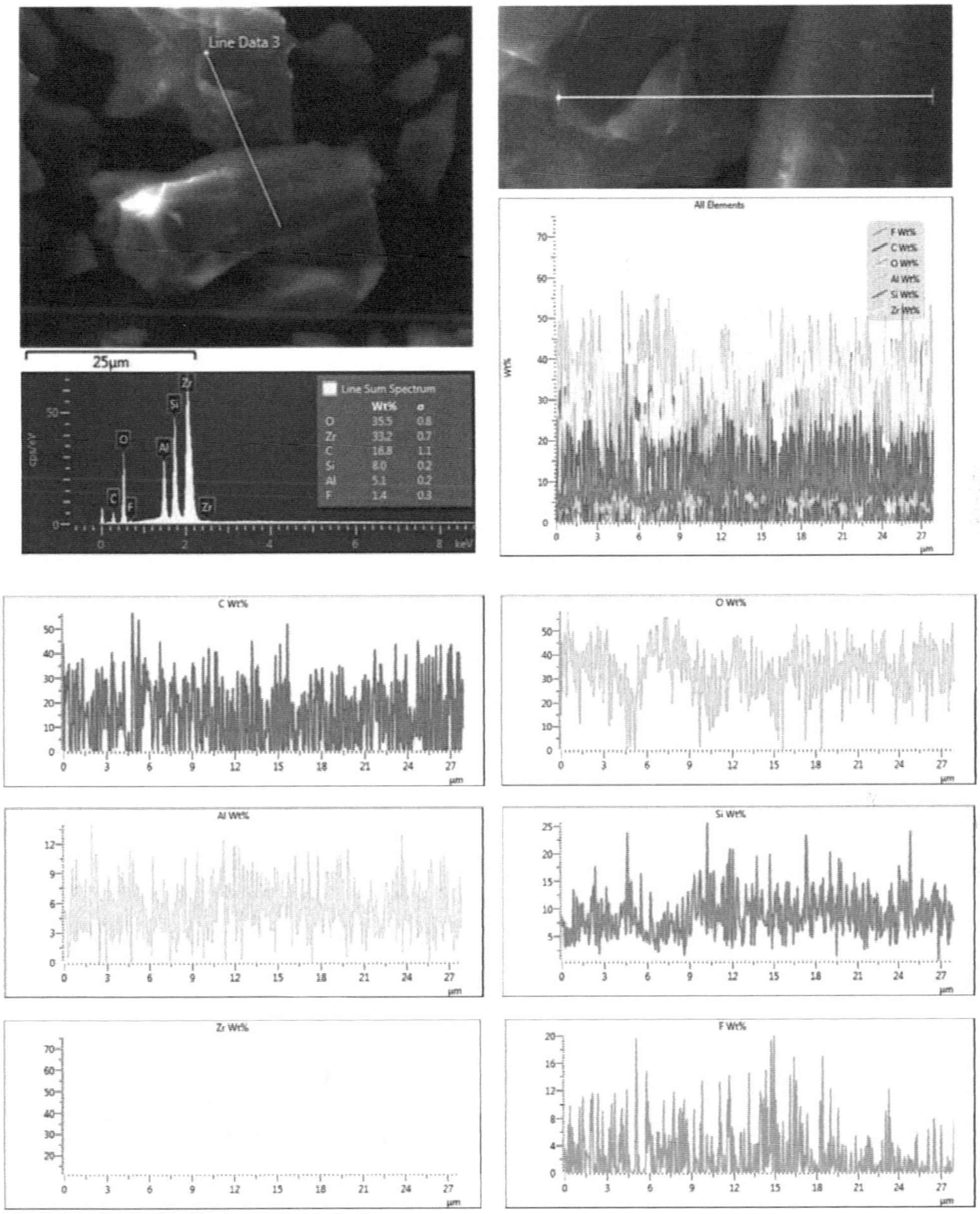

Figure 3. SEM image with EDS line scanning analysis of ZrO_2-Al_2O_3-SiO_2 particles.

Filler particles that incorporated to a resin matrix shall improve dental composite properties, which then provided well bond particles to a matrix. If not, the filler cannot provide reinforcement but will weaken the material [10].

Filler particles synthesized from zirconia as ceramic materials needs stabilization process to improve crack resistance as previously stated, so that the material can be applied to various kinds of prostheses. Zirconia can be stabilized as a whole (fully) or partially depending on the amount of stabilizer sufficient to form a particular crystalline phase. Partial stabilization involves the addition of stabilizers such as Al_2O_3, SiO_2, CaO, MgO, La_2O_3 and Y_2O_3 in concentrations lower than those required for full stabilization resulting in cubic forms [4, 5, 10].

When stress occurs on the zirconia surface, the energy accumulated at the crack will cause the tetragonal crystal transition to become monoclinic and increase in volume. This crystal modification is followed by an expansion that can close existing cracks. The crystal transformation produces compressive stress as a result of volume expansion, thereby slowing down the propagation of the crack. This process is known as transformation toughening mechanism, which can ultimately improve the mechanical properties [4-6].

4. Conclusions

The ZrO_2-Al_2O_3-SiO_2 particles as dental composite fillers were successfully synthesized. Analyses using SEM demonstrated a heterogeneous grain size distribution ranges from 0.5–150 μm. The average composition of Zr, Si and Al in the obtained particle is 34.65±7.09 wt%, 7.88±2.00 wt%, and 4.85±0.57 wt% respectively. Based on the EDS line scanning, the obtained particles have homogenous composition of Zr, Si and Al. These results suggest that the samples provide good alternatives for ceramic fillers of dental composite.

Acknowledgments

The authors were grateful to Directorate of Higher Education Ministry of Research, Technology and Higher Education, Republic of Indonesia through Universitas Padjadjaran for funded the research.

References

[1] Anusavice K J, Shen C and Rawls H R 2013 *Phillips' science of dental materials: Resin- based composites* 12th ed (Missouri: Elsevier Inc. Saunders) p 275
[2] Denry I and Kelly J R 2008 *Dental Materials* **24** 299
[3] Piconi C and Maccauro G 1999 *Biomaterials* **20** 1
[4] Manicone PF, Iometti PR and Raffaelli L 2007 *J of Dentistry* **35** 819
[5] Ozkurt Z, Iseri U and Kazazoglu E 2010 *Dent Mater J* **29** 233
[6] Kohal RJ, Att W, Bachle M and Butz F 2008 *Periodontology 2000* **47** 224
[7] Hasratiningsih Z, Cahyanto A, Takarini V, Karlina E, Djustiana N, Febrida R, Usri K, Faza Y, Hardiansyah A and Purwasasmita B S 2016 *Key Engineering Materials* **696** 93
[8] Hasratiningsih Z, Takarini V, Cahyanto A, Faza Y, Asri LATW, Hardiansyah A and Purwasasmita B S 2017 *IOP Conf. Ser.: Mater. Sci. Eng.* **172** 012067
[9] Olea-Mejia O, Contreras-Bulnes R, Zamudio-Ortega CM, Morales-Luckie RA, Olea-Cardoso O, and López-Castañares R 2014 *Microscopy: advances in scientific research and education (A. Méndez-Vilas, Ed.)* pp 70
[10] Anusavice K J 2003 *Phillips' science of dental materials: Mechanical properties of dental materials, dental composite, dental ceramics* 11th ed (Missouri: WB Saunders Company) pp 73-100, 401-437 and 655-714
[11] Saini M, Chandra S, Singh Y, Basu B and Tripathi A 2013 *Contemporary Clinical Dentistry* **1** 3

IOP Conf. Series: Journal of Physics: Conf. Series **1080** (2018) 012015 doi:10.1088/1742-6596/1080/1/012015

Influence of silica spray coating from rice husk ash on Pb, Cu, and Zn metals toward growth of *Staphylococcus Aureus* and *Escherchia Coli* bacterium

F A Prasetya[*], Y Zakaria, S E Santosa, S Sholihah, A J In'am, E R Hartiyana and R D P Amalia

Department of Chemical Engineering, Universitas Internasional Semen Indonesia, Gresik 61122, Indonesia

[*]fandi.prasetya@uisi.ac.id

Abstract. Synthesis of amorphous Silica from rice husk ash as a coatin°g on Pb, Cu, Zn metals to prevent growth of bacterium on it has been carried out by co-precipitation method. There are two variables of rice husk ash, first one was pure rice husk ash and the second one it was immersed in HCl. Resulted silica was characterized by X-ray diffraction (XRD) and X-ray fluorescence (XRF). Based on XRD data, both Silica of pure rice husk ash and immersed HCl have same pattern. They have peak (2θ) at 22.16°, indicates it formed amorphous Silica. XRF data clarified that percentage of Silica on both sample were 78.7% and 85.5% respectively. Silica powder was dissolved in Ethanol with ratio of Silica and ethanol were 1:10 and 1:100. Silica solution was sprayed on Pb, Cu, and Zn metals and *Staphylococus Aureus* and *Escherchia Coli* Bacterium was grown on it. Based on observation of bacterium growth showed that rate of bacterium growth decreases when Silica was coated on metal than without coating.

1. Introduction

Around 60 million tons rice per year was produced in Indonesia [1]. The increase in population leads to increase consumption of rice. Rice production has increased year by year based on statistics [2]. Rice husk is one of the most abundant agricultural wastes in Indonesia but its utilization is not yet optimal. Milling industry can produce 65% rice, 20% rice husk, and the rest is lost [3]. Until now, the utilization of rice husks commercially is still relatively low. This is due to the characteristics of rice husk which is rough, low nutritious, low density, and the content of ash is quite high [4]. Rice husk is commonly used as a cultivation medium, as an energy source in the form of charcoal husk briquettes, feed livestock, or destroyed by uncontrolled combustion [5, 6]. Rice husks have a lot of potential that can be developed, one of them is as a source of silica. Rice husk contains the most silica compared to other rice products. The ash content in rice husk is about 20% and more than 90% of the ash is silica [7]. The application of Silica (SiO_2) itself is currently as silica gel which is used to maintain the moisture of plant seeds. However, the application of the silica will have higher value when the purity is high. Co-precipitation method is one of the treatment to increase purity of silica from raw material with high silica content become higher purity amorphous silica which is near to nanoparticle [8, 9]. Surface modification by silica coating is one of the application alternative of silica. One of them was used to reduce the toxicity of ZnO Nanoparticle [10]. This study is aimed to extend utilization of rice husk ash silica as coating on Pb, Cu, and Zn metals to prevent growth of bacterium. Metal-based engineered nanomaterials (ENMs) are known to affect bacterial processes and metabolic activities [11], for example Pb, Cu, and Zn metals. Bacterium observed in this study is *Staphylococcus Aureus* and *Escherchia Coli. Staphylococcus Aureus* is major human pathogen that causes a wide range of clinical infections, many of which preferentially colonize the human body [12, 13]. *Escherichia coli* is a ubiquitous bacterial species commensal of humans and warm blooded animals [14].

2. Methods

For preparation, rice husk ash was preheated in an oven at 110°C for 24 hours to remove water content. Rice husk ash as raw material in this study was varied by immersing it in HCl and not. The purpose of immersion process is to reduce impurities in rice husk ash such as; Potassium, Magnesium, and Calcium by binding mechanism with HCl. Furthermore, the husk ash which has been immersed in HCl and not, both of them were treated by co-precipitation process to increase the purity of Silica and reduce the particle size. The alkali used in this process was NaOH 7 M with ratio of 1:15 between rice husk ash and NaOH. Then the titration process was carried out with HCl 37%. Co-precipitation results were characterized by X-Ray Diffraction to determine the formed phase and X-ray Fluorescence to determine the content of the elements therein. Furthermore, rice husk ash resulted by co-precipitation was mixed with ethanol with a ratio of 1:10 and 1:100. The use of ethanol was intended as a carrier medium of Silica for spray coating on metal. There were three kinds of metal used; Pb, Zn, and Cu. Furthermore, the bacteria were grown around the metal, to observe how the effect of spray coating on the metal on the growth of bacteria around the metal qualitatively. The bacteria used in this study were *Staphylococcus Aureus* and *Escherchia Coli.*

3. Result and Discussion

XRD data for rice husk ash immersed in HCl and not, showed in figure 1. The XRD pattern on both images was almost the same, which has a peak around 2θ; 22.16°, indicate that the main content was Silica which is in amorphous phase because of width of the peak is wide. XRD pattern formed below 2θ; 10° is only background. There are no another peaks after 2θ; 22.16°, indicate that Silica almost pure or other impurities still formed but they are not dominant.

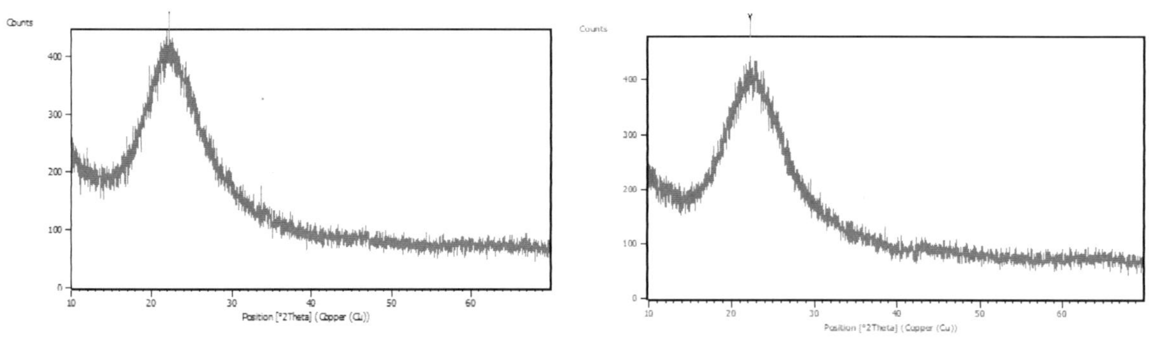

(a) (b)

Figure 1. XRD Pattern of Rice Husk Ash (a) immersed in HCl and (b) no immersed.

XRF data of rice husk ash resulted by co-precipitation showed in table 1. Immersing in HCl gives significant result in compositions. Some elements bind with Cl^- ion from HCl, such as Mg, K, Fe, and Na. Immersing in HCl make those elements decrease compared to rice husk ash without immersing, even Mg become 0%. This result gives evidence that immersing in HCl is able to decrease any impurities which can bind with HCl and increase purity of Silica.

Table 1. Composition of rice husk ash resulted by co-precipitation.

	% Weight	
Composition	No Immersed	Immersed HCl
Si	78.7	83.5
Mg	4.2	-
Al	0.814	0.854
Cl	0.174	0.175
K	0.106	0.0718
Mn	0.0711	0.0477
Fe	0.108	0.0806
Na	15.6	15.1
Ca	7.5	14.4

Table 2 shows the growth of bacteria around metal Pb, Cu, and Zn which is illustrated by distance of bacterium growth ratio to metals. The use of the code in the title of the image is done with the aim of simplifying the sentence. Code A denotes the *S.Aureus* bacteria while B for *E.Coli* bacteria. NC denotes No Coating. While 10 and 100 show the ratio between silica and ethanol 1:10 and 1: 100 respectively. Furthermore, the N code at the end indicates that rice husk ash is not immersed in HCl, whereas H code indicates a soaking process.

Tabel 2. Distance of bacterium growth ratio to metals.

	Bacterium distance to metals (cm)		
Code	Pb	Cu	Zn
ANC	0.167	0.25	1.14
A10N	0.667	0.667	0.01
A10H	0.667	2	0.1
A100N	0.667	0.5	0.5
A100H	1	0.625	0.0067
BNC	0	0.1	0.5
B10N	1	0	0.2
B10H	1	0.5	0.5
B100N	1	0	0.1
B100H	0.067	0.15	0

Growth is define as the addition of cells and structures can be used with size, followed by increasing number, increase in cell size, augmentation advisor or mass, and other parameters. In this study, growth of bacterium was calculated by ratio distance of bacterium to metals. Higher the value of those ratio, bacterium grow lower. Average value for Pb, Cu, and Zn metals are 0.6235, 0.4792 and 0.3056, respectively. It shows that Pb metal has higher resistance of bacterium growth than the others. This is related to the fact that characteristic of Pb metal resist to corrosion and acid. Based on the data in table 2, it showed that for each bacteria, most effective variable to inhibit bacterium growth are using 1:10 ethanol ratio and immersing in HCl. Average value for *S.Aureus* and *E.Coli* in those variables are 0.923 and 0.667 respectively. Growth of both kinds of bacteria around the metal Pb shows the regularity of the pattern. Uncoated Pb is seen to be overgrown with bacteria very quickly, while bacterial growth slows down in the Pb metal coated by silica. Similarly, the ratio of silica and ethanol equally slows the rate of bacterial growth with an unobtrusive difference. While on Cu metal that is not coated with silica bacteria grow rapidly on. The growth inhibition pattern on Cu metal coated with silica is slightly different from that of Pb. There are several types of layers where metals continue to grow rapidly around silica-coated Cu metals, which are B10N and B100N. Significant bacterium growth differences occur around Zn metal. Almost in all Zn metal, either coated with silica or not growing bacteria occurs quickly. Only in A10H, A100N, and B10H bacteria can inhibited by the presence of silica coating.

4. Conclusion

Based on the data and discussion, high purity silica produced from co-precipitation of rice husk ash whether immersed in HCl or not, it formed amorphous phase. Overall, it can be concluded that in general, the silica coating on the metal is slightly able to inhibit bacterial growth around the metal, except for most Zn metal. Development of this study allow for determining silica from another sources or metals coating with other methods.

Acknowledgement

Authors would like to thank LPPM Universitas Internasional Semen Indonesia who was fully funded the research through Applied Research Grant.

References

[1] Rusono N, Suanri A, Candradijaya A, Muharam A, Martino I, Tejaningsih, Hadi P U, Susilowati S H and Maulana M 2013 *Rencana Pembangunan Jangka Menengah Nasional (RPJMN) Bidang Pangan dan Pertanian 2015-2019* p 11
[2] Zhongkui H, Liu A, Li C and Xuesi C 2009 *Journal of Non-Crystalline Solids* **355** 368
[3] Ismunadji M 1988 *Padi Book I 1st ed.* (Bogor: Agricultural Research and Development)
[4] Houston D F 1972 *Rice Chemistry and Technology* (American Association of Cereal Chemist *Inc.*)
[5] Badar R and Qureshi S A 2014 *Journal of Botany* **427**648
[6] Arévalo J, Quispe G and Raymundo C 2017 *Energy Procedia* **141** 138
[7] Sapei L, Miryanti A and Widjaja L B 2012 *Proceedings SINTECH-1 The First Symposium in Industrial Technology* A-8
[8] Ramadhan N I, Munasir and Triwikantoro 2014 *JURNAL SAINS DAN SENI POMITS* **3** 2337
[9] Munasir Z, Sulton A, Triwikantoro T, Zainuri M and Darminto D 2013 *AIP Conf. Proc.* **1555** 28
[10] Chia S L and Leong D T 2016 *Heliyon* **2** e00177
[11] Zhai Y, Hunting E R, Wouterse M, Peijnenburg W J G M, Vijver M G 2017 *Ecotoxicology and Environmental Safety* **145** 349
[12] Harris L G, Foster S J and Richards R G 2002 *European Cells and Materials* **4** 39
[13] Tong S Y C, Davis J S, Eichenberger E, Holland T L, Fowler Jr V G 2015 *Clin Microbiol Rev.* **28** 603
[14] Kabiru L M, Bello M, Kabir J, Grande L and Morabito S 2015 *Int. J. Environ. Res. Public Health* **12** 679
[15] Triyono E B S and Syarip S 2017 *Indonesian Journal of Nuclear Science & Technology (JSTNI)* **18** 1

IOP Conf. Series: Journal of Physics: Conf. Series **1080** (2018) 012016 doi:10.1088/1742-6596/1080/1/012016

Effect of KOH concentration on characteristics of polydimethylsiloxane synthesized by ring opening polymerization method

Fitrilawati[1], A N Fauza[1], A Ardi[1], R M Novianti[1], N Syakir[1], A S Kartasasmita[2] and Risdiana[1*]

[1]Department of Physics, Faculty of Mathematics and Natural Sciences, Universitas Padjadjaran, Jl. Raya Bandung-Sumedang km 21 Jatinangor, Sumedang, West Java, Indonesia 45363

[2]Department of Ophthalmology, Faculty of Medicine, Universitas Padjadjaran , Jl. Raya Bandung-Sumedang km 21 Jatinangor, Sumedang, West Java, Indonesia 45363

[*]risdiana@phys.unpad.ac.id

Abstract. Polydimethylsiloxane (PDMS) that widely known as silicone oil is commonly used as vitreous humour substitutes in vitreoretinal surgery. PDMS is needed to be produced in Indonesia and should be provided with sufficient quantity and quality for domestic needs. We synthesized PDMS using ring-opening polymerization method from octamethylcyclotetra-siloxane as monomer, hexamethyldisiloxane as chain terminator, and KOH as catalyst. KOH concentrations are varied from 0.5 M to 2.5 M to study the effect of KOH concentration on characteristics of PDMS. PDMS We successfully synthesized PDMS with a yield in between 39.73% and 65.30% and various viscosities in the range between 0.58 Pa.s and 9.36 Pa.s. From FTIR spectroscopy, it is found that all synthesized samples have structure and functional groups similiar with that of commercial PDMS. From UV-Vis and refractometer measurements, all samples have transparency almost 100% and refractive indexes in the range of 1.3989 to 1.4039. The effect of KOH concentration on the synthesized PDMS gel is on its viscosity, the higher concentration of KOH, the higher viscosity of the obtained PDMS.

1. Introduction

Eyes are optical system, which is very important to support our daily activities. The largest part of human eyes is a liquid called vitreous humour. Disorders of some parts of human eyes in vitreous humour is commonly occurred. One of disorders in vitreous humour is a damage causes retinal detachment. In that case, the retinal neurosensory layer is separated from the epithelium pigment. Same factors that usually cause retinal detachment are acute posterior vitreous detachment (PVD), myopi, afakia, and trauma [1]. A treatment of the retinal detachment can be done by using artificial replacement fluid to replace the vitreous humour in the human eyes. The artificial replacement fluid is known as vitreous substitutes or tamponade agents. An ideal vitreous substitutes is a liquid that has high surface tension, optically clear, and biologically inert [1].

One of liquids that commonly used as vitreous substitutes in vitreoretinal surgery is polydimethylsiloxane (PDMS) known as silicone oil. The aims of replacement of vitreous humour are to restore a normal retinal position in the eyes, to restore the volume of vitreous cavity, and to assist surgeons in the membrance surgery [2]. The study of optical and structure properties of commercial PDMS before and after vitreoretinal surgery in Indonesia has been performed and found that some additional functional groups were observed after a few months of vitreoretinal surgery [3]. It causes commercial PDMS should be replaced several times. There are limitations in availability of commercial PDMS in Indonesia, PDMS is needed to be produced and should be provided with sufficient quantity and quality for domestic needs.

There are many polymerization methods in synthesis PDMS such as addition polymerization, condensation polymerization, and ring opening polymerization. We studied a synthesis of PDMS as a vitreous substitute for domestic production. Previous works reported synthesis of PDMS by ring-opening polymerization method [2, 4]. In this work, a cyclic chain of monomer is breaking into a linear chain by applying high temperature treatment. A catalyst needed in this polymerization technique. Some catalyst such as tetramethylammonium silanolate, tetramethylphosphonium-siloxanolate, tetramethylammoniumsiloxanolate, alkaline metal hydroxides are frequently used in the ring opening polymerization [2]. We used KOH as one of alkali metal hydroxides since this chemical is readily available.

2. Experiment

The polymerization process was initiated by mixing 7.8 ml octamethylcyclotetrasiloxane (D4) as monomer and 3 ml hexamethyldisiloxane (MM) as chain terminator. The mixture was stirred using a magnetic stirrer at 300 rpm in oil bath at a temperature 170°C. The oil bath was used to maintain temperature stability during polymerization process. Then, 0.06 ml KOH catalyst with a certain concentration was added slowly into the solution. The solution was further strirred for 45 minutes to form a thick and clear gel. The various concentrations of KOH were performed with value in between 0.5 M and 2.5 M.

In order to remove residual monomers and a residual KOH from the PDMS gel, the synthesized PDMS was purified by using solvent extraction method. The purification process was started by dissolving the synthesized PDMS with chloroform with volume ratio of 1:1. The solution was mixed until it became homogeneous. Mili-Q water was added into the PDMS solution with ratio of 1:2. The mixture was stirred and then waited for about 10 minutes until form a two phase mixture of the solution. The bottom phase is a PDMS solution in chloroform and the upper phase is mili-Q water containing KOH and residuals of monomer. The PDMS solution was separated from the aqueous phase and the pH of the aqueous was checked. If the pH of the aqueous phase is not 7, some milli-Q waters were added again into the PDMS solution. The process was repeated until the pH of the aqueous phase is 7. The PDMS solution in chloroform was stirred and heated at 50°C to remove the chloroform solvent to obtain a transparent PDMS gel.

All samples were characterized using viscometer to measure viscosity, FTIR spectroscopy to identify its functional groups, refractometer to determine refractive index and UV-Vis spectrosphotometer to measure transparency.

3. Results and Discussion

The synthesized PDMS gel obtained using ring opening polymerization method is thick and transparent. After purification, the synthesized gel seem more clear as compare to the condition before purification. A yield of PDMS calculating from different amount of monomer and resulted polymer is shown in table 1. The highest yield was obtained in sample with 1 M KOH and the smallest yield was obtained in sample using 0.5 M KOH. From yield value, it is found that optimal concentration of KOH was 1 M.

The highest yield obtained in this synthesis is 65 % which is still relatively small. It is might be due to the high polymerized temperature causing some monomers to be evaporated during the polymerization process.

Viscosity of the synthesized PDMS was displayed in table 1. The value of viscosity of PDMS obtained using 2.5 M KOH was 8.66 Pa.s, while the sample using 0.5 M KOH was 0.58 Pa.s. The viscosity of synthesized PDMS was significantly increased with increasing concentration of KOH from 0.75 M to 1 M KOH. The viscosity slightly decreased when KOH concentration increased larger than 2 M KOH. Liquids with short polymer chains (low molecular weight) have low viscosity, and liquids with long polymer chains (high molecular weight) have high viscosity. This suggests that increasing of KOH concentration affects propagation of the polymer chain and resulted increasing of molecular weight. The longer PDMS chain formed, the thicker viscosity and the higher PDMS molecular weight. Synthesized PDMS using 0.75 M KOH has viscosity that very close with the viscosity of commercial PDMS of 3.55 Pa.s.

Table 1. The yield of synthesized PDMS at varied concentrations of KOH.

Concentration of KOH (M)	Mass of Monomer (gr)	Mass of Polymer (gr)	*Yield* (%)	Viscosity (Pa.s)
2.50	10.28	5.35	52.04	8.66
2.00	10.33	5.80	56.20	8.36
1.00	10.46	6.83	65.30	9.36
0.75	10.13	4.59	45.29	3.65
0.50	10.15	4.03	39.73	0.58

In vitreoretinal surgery, when low-viscosity tamponade agents or liquid was used, it will be readily reabsorbed in short time. Conversely, PDMS that has high viscosity is expected to be durable as a tamponade agent.

Figure 1 shows FTIR spectra of all synthesized PDMS. For comparison, the spectrum of commercial PDMS was also displayed in figure 1(f). It is found that all samples had functional groups of PDMS. The FTIR spectra of the commercial PDMS have vibration of C – H bond in the ranges of 2905 cm^{-1} and 2962 cm^{-1}, Si – CH_3 bond in the range of 1262 cm^{-1}, Si – O – Si bond in the range of 1099 cm^{-1} and 1025 cm^{-1}, Si – C bond in the range of 800 cm^{-1}, and CH_3 bond in the range of 1412 cm^{-1}. FTIR spectra of all synthesized PDMS using various KOH concentrations show identical frequency of C-H bond in the range of 2897 – 2969 cm^{-1} dan 2897 – 2906 cm^{-1}, Si – CH_3 bond in the range of 1262 – 1264 cm^{-1}, CH_3 bond in the range of 1407 – 1418 cm^{-1}, Si – O – Si bond in the range of 1091 – 1098 cm^{-1} and 1019 – 1026 cm^{-1} and Si – C bond in the range of 799 – 805 cm^{-1}. This indicates that all synthesized PDMS samples have same structure which is similar with commercial PDMS.

Table 2 shows the value of refractive index for PDMS with various KOH concentrations. The refractive index measurements were carried out at a maintained room temperature of 20° C. The value refractive index was observed in the range of 1.3989 – 1.4036. The differences in refractive index is not significant but still in the range of the values.

Table.2. The refractive index of PDMS synthesized with various KOH concentrations at 20°C.

Concentration of KOH (M)	Refractive Index
2.50	1.4036
2.00	1.4034
1.00	1.4030
0.75	1.4008
0.50	1.3989

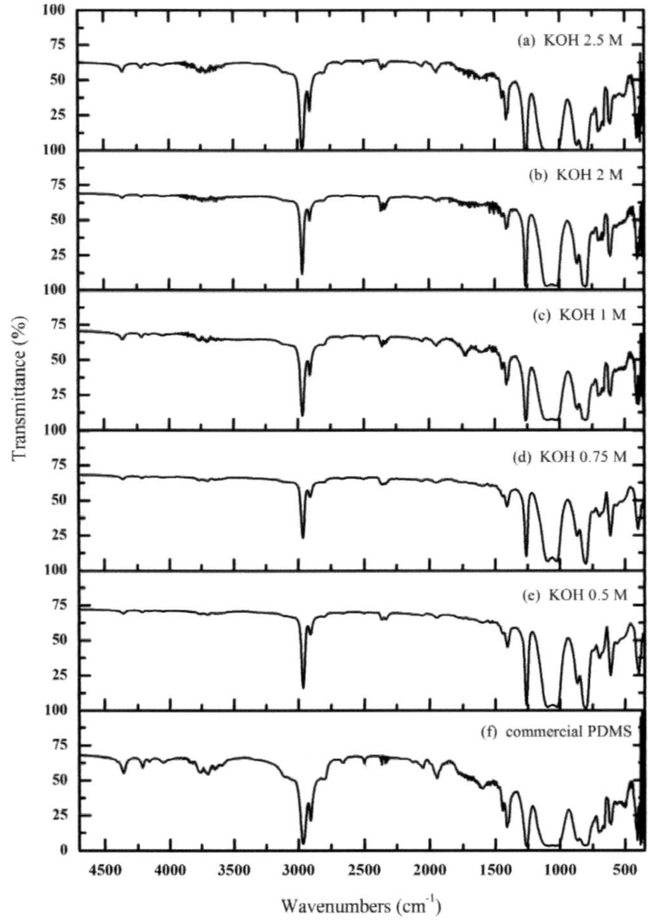

Figure 1. FTIR spectra of synthesized PDMS (a) KOH 2.5 M, (b) KOH 2 M, (c) KOH 1 M, (d) KOH 0.75 M, (e) KOH 0.5 M and (f) commercial PDMS.

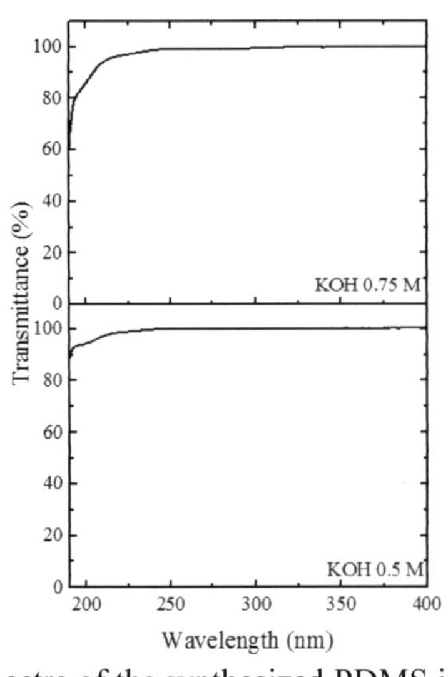

Figure 2. Transmittance spectra of the synthesized PDMS in the UV wavelength range.

IOP Conf. Series: Journal of Physics: Conf. Series **1080** (2018) 012016 doi:10.1088/1742-6596/1080/1/012016

The refractive index of PDMS is higher compared to the refractive index of actual vitreous humour which is in the range of 1.3345 to 1.3348 [5]. Significant differences of refractive index values can cause complications of myopia to vitreoretinal surgical patients. Changes in refractive index values result in the adhesion of silicone oil to intraocular lenses [6]. Since the refractive index of PDMS measured at outside the human body, there is a possibility of refractive index value of PDMS changed when PDMS gel is injected into the eye. It is because the value of refractive index is highly depend on the ambient temperature.

Transparency of synthesized PDMS was measured in transmittance mode. PDMS samples synthesized using KOH with concentration of 0.5 M and 0.75 M were measured in the range of 190 nm to 400 nm as representation of absorption of PDMS gel in the UV range. Through visual observation, PDMS gel with KOH concentration of 0.5 M is more clear than the one synthesized with KOH concentration of 0.75 M. The quantitative transmittance of both PDMS samples is shown in figure 2.

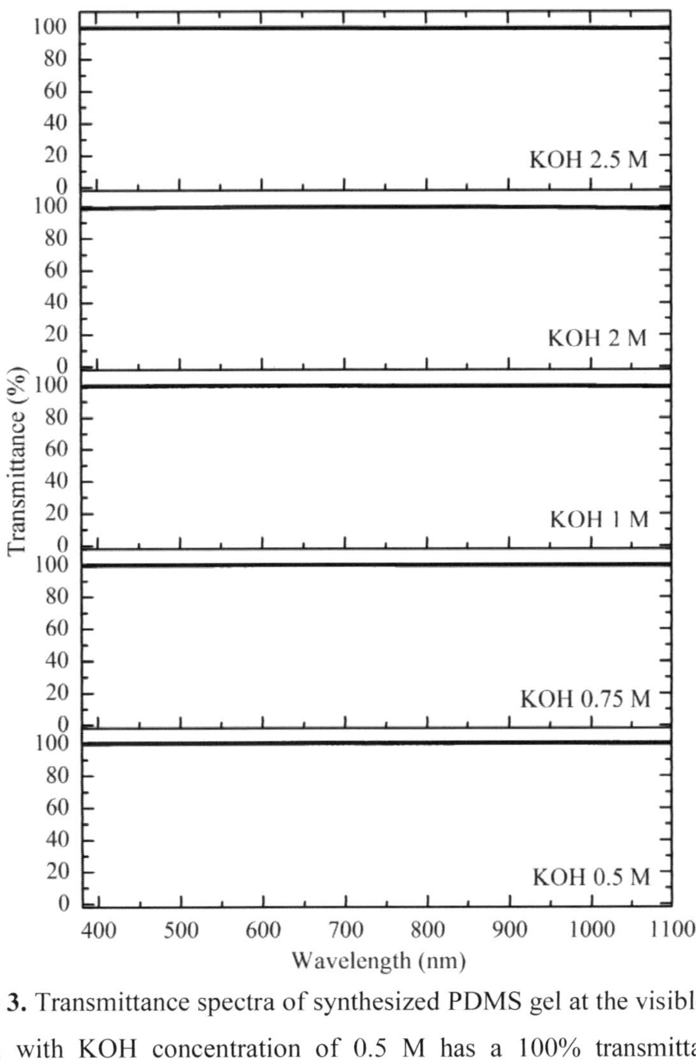

Figure 3. Transmittance spectra of synthesized PDMS gel at the visible range.

The PDMS gel with KOH concentration of 0.5 M has a 100% transmittance value and the transmittance decrease at 249.8 nm. The decreasing transmittance indicates that PDMS gel sample was being absorbed at UV wavelength. While the PDMS synthesized with KOH concentration of 0.75 M has a low transmittance in the UV range and start to decrease at of 249.9 nm. These data shows that PDMS was absorbed the UV wavelength.

The synthesized PDMS samples were measured in the visible wavelength in the range of 380 nm to 1100 nm are shown in figure 3. From the spectra, it can be seen that all transparencies of PDMS gel are almost 100%. PDMS synthesized using 0.5 M KOH is 100%. Other PDMS samples have a transmittance slightly below 100 %. The transparency corresponds to visual observation that PDMS gel synthesized with 0.5 M KOH seem more clear than PDMS synthesized with other concentration of 0.75 M, 1 M, 2 M, and 2.5 M.

4. Conclusion

We have succeeded to synthesis PDMS gel using ring-opening polymerization method using octamethylcyclotetrasiloxane (D4) as monomer, hexamethyldisiloxane (MM) as chain terminator at various concentrations of KOH. The synthesized PDMS gel are transparent, viscous and has yield in the range of 40 % to 65 0% with various viscosities in the range between 0.58 Pa.s and 9.36 Pa.s. The synthesized PDMS gel has similar structure as commercial silicone oil with refractive index value in the range of 1.3989 - 1.4036. All transparency properties of PDMS gel are almost 100%. The KOH concentration strongly affect the viscosity of the PDMS, the higher concentration of KOH, the higher viscosity of synthesized PDMS.

Acknowledgment

This works was supported by Penelitian Unggulan Perguruan Tinggi (PUPT) of Universitas Padjadjaran contract no.2519/ADD/UN6.3.1/LT/2017 and also was partially supported by Academic Leadership Grant of Universitas Padjadjaran no. 872/UN6.3.1/LT/2017.

References

[1] Khurana A K 2007 *Comprehensive Ophthalmology*. (New Age International (P) Ltd.)
[2] Giordano G G and Refojo M F 1998 *Prog. Polym. Sci.* **23** 509
[3] Nusa H S, Astuti W, Kartasasmita A S, Virgana R, Syakir N, Bahtiar A, Safriani L, Risdiana 2015 *Materials Science Forum* **827** 99
[4] Zhang Z, Zhou N, Xu C, Xie Z 2001 *Chinese Journal of Polymer Science* **19** 7
[5] Swindle K E, Ravi N 2007 *Expert Rev. Ophthalmol.* **2** 255
[6] Lee D H, Rah S H, Yoon I N 2009 *Korean J. Ophthalmol.* **23** 309

Synthesis and characterization of carboxymethyl cellulose (CMC) from salak-fruit seeds as anode binder for lithium-ion battery

S Hidayat[1*], P Ardiaksa[1], N Riveli[1] and I Rahayu[2]

[1]Department of Physics, Faculty of Mathematics and Natural Sciences, Universitas Padjadjaran, Jl. Raya Bandung-Sumedang km 21 Jatinangor 45363
[2]Department of Chemistry, Faculty of Mathematics and Natural Sciences, Universitas Padjadjaran, Jl. Raya Bandung-Sumedang km 21 Jatinangor 45363

*sahrul@unpad.ac.id

Abstract. Lithium-ion batteries are most popular battery for applications as portable electrical energy storage. Lithium-ion battery have some advantages compared to other battery such as have large specific energy capacity, light weight and long-life span. One of the components that contribute to the quality of Lithium-ion batteries is the binder material. Carboxymethyl Cellulose (CMC) is one of the most widely used binder materials other than PVDF. CMC can be synthesized from various organic materials such as empty bunches of palm, water hyacinth and salak-fruit seed. In this research, CMC is obtained from waste of salak-fruit seed (*Salacca zalacca*) and then used as anode binder in lithium-ion battery. The steps of CMC synthesis consist of cellulose isolation from salak-fruit seed, followed by alkalization process, carboxymethylation and finally the purification process. FTIR spectrum of CMC shows the successfully of CMC synthesis, the presence of strong absorption band at 1600 cm-1 and 1438 cm-1 is related to stretching vibration from the carboxyl group (COO-) and carboxyl group as it salts (COO-Na), respectively. The absorption in the region 1321–1400 cm^{-1} is due to symmetrical deformations of CH_2 groups while broad absorption at around 3400 cm^{-1} is due to the stretching of the hydroxyl groups (-OH). Application CMC as binder of graphite anode with 5wt% CMC shows good surface profile and evenly distributed cavities with conductivity of about 0.815 S/cm.

1. Introduction

Lithium-ion battery (Li-ion) has been known as the best electrochemical devices for electrical energy storage. Li-ion battery has high energy density, long life cycle (500-1000 cycles), light weight and environmentally friendly [1]. Energy density of the battery during charging or discharging process depends on its active material. The active material of Li-ion battery plays an important role on the intercalation of Li^+ ion from anode to cathode and vice versa. The active material needs a binder in order to keep the compactness and a high particle density of electrode. On the other hand, existence of binder material will affect the stability and loss capacity of the battery. If binder material is poor, then the loss energy of battery will increase and cause battery damaged [2].

Generally, polyvinylidene fluoride (PVDF) was used as binder in the cathode and anode of commercial Li-ion batteries. PVDF has good electrochemical stability and act as a good adhesive for the active material on the substrate. However, PVDF have some disadvantages, i.e.: expensive, not easy to recycle and involved the use of volatile organic compounds for its processing such as such N-methyl-2-pyrrolidone (NMP) which a harmful and toxic solvent [3]. The use of water-based binders has been done by Lestrie *et al.* 2007 [3]. Water-based binders also simplify the process of making the battery electrodes. One of the most used water-based binder materials is Carboxymethyl cellulose (CMC).

CMC is a cellulose derivative that is widely used in many industries such as food, pharmaceutical, detergent, textile and cosmetic products. CMC is mostly used as thickeners, emulsion or suspension stabilizers and binders. CMC can be made from plants therefore it can be readily available in large quantities as well as it is renewable. Initially CMC was mostly made of wood cellulose because of its sufficient cellulose content, which is about 42-47%. However, many CMC today is developed from non-wood materials such as stem and empty bunches of palm oil, bananas, water hyacinth plants and salak-fruit seeds.

Salak-fruit seeds are waste from the salak-fruit and they are abundant in Indonesia. The research of salak-fruit seeds showed that salak-fruit seeds contained about 35% of cellulose [4]. Cellulose can be processed into derivatives such as CMC (Carboxy Methyl Cellulose), MC (Methyl Cellulose), HPC (Hydroxy Propyl Cellulose), and HPMC (Hydroxy Propyl Methyl Cellulose). The CMC is a linear cellulose polymer with an anion compound which properties are biodegradable, colorless, odorless, non-toxic, soluble in water (powder form), pH range of 6.5 to 8.0, transparent and un-reacting with organic compounds. The chemical structure of CMC is showed in figure 1.

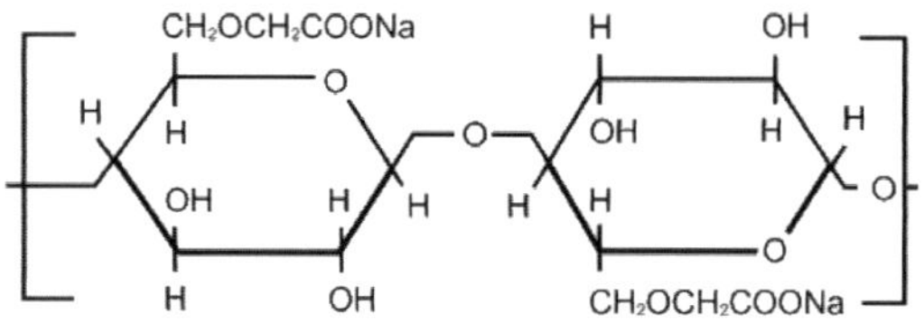

Figure 1. Chemical structure of CMC.

The solubility of CMC in water can be changed by controlling the amount of carboxymethyl sodium compounds that are substituted into the CMC chain. The amount of the hydroxyl group that substituted by the sodium carboxymethyl group is called as the substitution degree. Increasing the substitution degree will increase the solubility of CMC in the water [5]. High solubility of CMC will increase the electrode surface area and improves the efficiency of ion exchange between the electrode and electrolyte, furthermore will increasing capacity of battery.

2. Method

Several materials were used in obtaining CMC from salak-fruit seed, such as $NaClO_2$, NaOH, sodium monocholoacetic (NaMCA), CH_3COOH, ethanol and distilled water. Firstly, Salak seeds are separated from the fruit and then crushed into powder, and sieved through 100 mesh. Furthermore, the powder was heated at 110°C in the oven for about 2 hours. The second step is the isolation process of cellulose that consists of de-hemicellulose and bleaching. The step of de-hemicellulose was conducted by dissolving salak-seed powder in 17.5% NaOH for about 3 hours. The next step is removing lignin or bleaching process which conducted by dispersing powder precipitate from previous process in 1 wt% $NaClO_2$ [6].

Alkalization process started by adding the cellulose powder into solvent of isopropanol and ethanol, with volume ratio of 80:20, and stirred using magnetic stirrer for about 15 minutes. Then, 30% NaOH (w/v) was added dropwise into the mixture and was stirred for 1 hour at 30°C. The final step is carboxymethylation process which was done by adding sodium monochloroacetate (NaMCA) dropwise. Prior to use, NaMCA was dissolved in water and isopropanol. The carboxymethylation process was carried out at 55°C for about 4 hours. The solution was then cooled to room temperature and the slurry was filtered. The slurry was filtered and then dissolved in 70% of ethanol. The solution was neutralized using 98% acetic acid. After it was filtered and washed using 70% ethanol to remove undesirable reactants, the product was dried at 70°C for 5 hours [6]. The final product of CMC was in the powder form.

The produced CMC was analysed using FTIR to characterize its molecular structure and test its purity. Moreover, the characterization was also carried out to anode layer of CMC-graphite-C on the glass substrate. The characterization consists of conductivity measurement and surface profile imaging using Scanning Electron Microscopy (SEM). These characterizations are carried out to investigate the quality of CMC in its application as binder in the battery electrode.

3. Results and Discussion

Figure 2 shows the FTIR spectrum of CMC powder. The peaks of absorption are related to vibration frequency of the chemical bonds in the CMC compound. The presence of the absorption peak at 1600 cm^{-1} corresponds to stretching vibration of the carboxyl compound (COO-) while the absorption peak at 1438 cm^{-1} corresponds to streaking vibration of the carboxyl compound in the form of salt (COO-Na). In the fingerprint region, strong absorption occurs in the wide range of 950-1250 cm^{-1}, this shows the presence of ether bonds in the CMC compound. In addition, the absorption bands at 1321 cm^{-1} corresponds to symmetrical deformation of CH_2 molecule, and presence of broad bands absorption at 3400 cm^{-1} corresponds to stretching frequency of hydroxyl molecules (-OH).

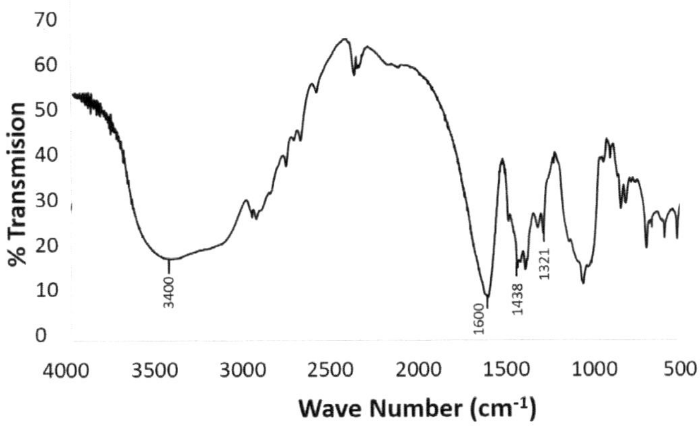

Figure 2. FTIR spectrums of CMC from salak-fruit seed.

Purity test of CMC was carried out by titrating the CMC solution using $AgNO_3$ and K_2CrO_4 as indicator. The purity of CMC can be calculated using equation (1) and (2) [7].

$$\text{Purity of CMC} = 100\% - \%\,\text{NaCl} \quad (1)$$

$$\%\,NaCl = \frac{V\times N\times 58{,}45\times 100\%}{A\times(100-B)} \quad (2)$$

where V is Volume of $AgNO_3$ (ml), N: Normality of $AgNO_3$ (0,1N), A: mass of CMC (500 mg) and B: water content of CMC (%). Based on the purity test, within three samples tested (CMC mesh-100, CMC mesh-60 and CMC, mesh-40), the results are 94%, 93% and 93%, respectively. These results prove that CMS has actually been formed with high enough purity.

Figure 3 shows the SEM images of CMC-graphite-C composites with different concentration of CMC. All samples show some slab which are connected to each other. Morphology of CMC-graphite-C layer is observed as flaky texture reflecting the graphite particle wrapped by CMC layer as shown in figure 3 for all concentration at magnification 1000x. The larger interspaces of the layer and the thinner layer edges of CMC-graphite-C can be clearly seen in figure 3 with magnification 5000x. The material composites show many cavities, which may be formed during the stirring in aqueous media and drying process. The composition of 1% CMC shows too many cavities and uneven distribution of CMC. Furthermore, the composition of 5% CMC shows good surface profile with evenly distributed cavities, while the composition of 10% CMC shows homogenous surface profile but too much CMC that will effect in too low conductivity.

Figure 3. SEM images of CMC-graphite-C layer (a) 1wt% CMC, (b) 5wt% CMC, (c) 10wt% CMC.

Figure 4 shows the conductivity measurement which was carried out by four-line probe method. The measurement was performed for different concentration of CMC started from 2% to 10%. Generally, the addition of CMC powder into graphite-C layer can increase the mechanical properties, but reduce its conductivity. The optimal composition of CMC in the CMC-graphite-C composite is 5% CMC which has the highest conductivity of 0.815 S/cm. This result is correlated with the SEM image where the composition of 5% CMC has good surface profile with evenly cavities distribution.

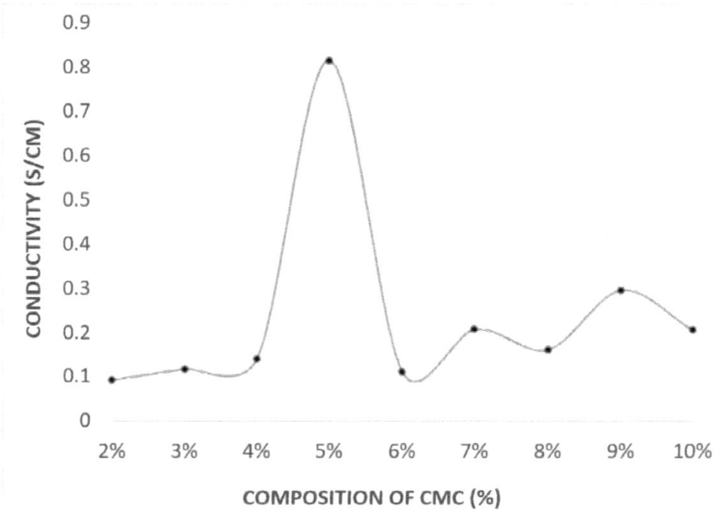

Figure 4. Conductivity of CMC-graphite-C layer with various composition of CMC.

4. Conclusions

Carboxymethyl cellulose (CMC) can be extracted from salak-fruit seed with high purity up to 94%. The analysis results of FTIR spectrum shows that molecular structure of CMC synthesized from salak-fruit seed has been formed. Based on SEM image, the CMC-graphite-C layer shows some slab which are connected to each other with cavities distribution. The addition of CMC into graphite layer can increase the mechanical properties, but reduce its conductivity. The optimum composition of CMC-graphite-C is 5:85:10, which has good surface profile with evenly distributed cavities, and has the highest conductivity 0.815 S/cm.

Acknowledgments
The authors would like to thank to Riset Kompetensi Dosen UNPAD (RKDU) Project 2017 for financial support.

References

[1] Hamid N A, Wennig S, Hardt S, Heinzel A, Schulz C and Wiggers H 2012 *J. Power Sources* **216** 76
[2] Nitta N, Wu F, Lee J T and Yushin G 2015 *Mater. Today* **18** 252
[3] Lestrie B, Bahri S, Sandu, Roué L and Guyomard D 2007 *Electrochem. Commun.* **9** 2801
[4] Anggrahini S, Marseno D. W, Setiyoko A and Wahyuningtyas A 2017 *Indonesian Food and Nutrition Progress* **14** 101
[5] Haleem N, Arshad M, Shahid M and Tahir M A 2014 *Carbohydr. Polym.* **113** 249
[6] Hidayat S, Susanty, Riveli N, Suroto B, Rahayu I 2018 *AIP Conf. Proc.* **1917** 030023-1
[7] Saputra A H, Qadhayna L and Pitaloka A B 2014 *Int. J. Chem. Eng. Appl.* **5** 36

IOP Conf. Series: Journal of Physics: Conf. Series **1080** (2018) 012018 doi:10.1088/1742-6596/1080/1/012018

Synthesis and conductivities of the Ti-doped apatite-type phases $La_{9.33}Si_{6-x}Ti_xO_{26}$

A R Noviyanti[1*], D R Eddy[1], F Lastiyanti[1], I Rahayu[1], D G Syarif[2], T Saragi[3] and Risdiana[3]

[1]Department of Chemistry, Faculty of Mathematic and Natural Sciences, Padjadjaran University, Jl. Raya Bandung Sumedang km 21 Jatinangor, West Java, Indonesia 45363
[2]PTNBR – BATAN, Jl. Taman Sari 71 Bandung 40132 Indonesia
[3]Department of Physics, Faculty of Mathematics and Natural Sciences, Padjadjaran University, Jl. Raya Bandung Sumedang km 21 Jatinangor, West Java, Indonesia 45363

[*]atiek.noviyanti@unpad.ac.id

Abstract. Solid oxide fuel cell is one of the alternative energy sources with high efficiency and low emissions. Solid oxide electrolyte is an important component of this fuel cell. A new electrolyte that suitable at medium temperature is needed to be developed. Apatite type lanthanum silicate is one of the electrolytes, which can operate at medium temperature. Increasing performance of apatite lanthanum silicate such as ionic conductivity can be achieved by doping. The purpose of the present study was to synthesize the electrolyte of Lanthanum silicates oxide with hydrothermal method and to investigate the effect of Ti^{4+} dopant on apatite electrolyte characteristics such as crystal lattice parameter and conductivity. The results showed that samples of $La_{9.33}Si_{6-x}Ti_xO_{26}$ with x = 0.1, 0.2, 0.3 were successfully synthesized. The value of *c* on the crystal lattice parameter increases with increasing the doping content of Ti^{4+}. Increasing *c* value on the crystal lattice parameter is directly proportional to the conductivity. Apatite $La_{9.33}Si_{5.7}Ti_{0.3}O_{26}$ showed the highest conductivity of 2.4×10^{-4} S/cm at 700°C.

1. Introduction

Solid oxide fuel cell is one of the alternative energy sources with high efficiency and low emissions. Solid oxide electrolyte is an important component of this fuel cell. Yttria stabilized zirconia is commonly used electrolyte material in solid oxide fuel cells. Unfortunately, operational temperatures of this material were in high temperatures in between 800°C and 1000°C. In order to overcome this problem, new electrolytes that suitable at medium temperature has been developed. Apatite type of lanthanum silicate is one of the electrolytes, which can operate at medium temperature in between 500°C and 700°C.

Lanthanum silicates apatite oxide structure is based on isolated silicate tetrahedral (SiO_4) that share oxygen with La polyhedral to form rigid network. The La cations have two kinds of sites. First site is located at seven-coordinated sites named *6h* and the second site is located at nine-coordinated sites named *4f*. In these compounds, oxide ions, occupying the centre of one dimensional channels running along the *c*-axis, are responsible for ionic conduction [1]. The open structure of oxy-apatite suggests that this material should be appropriate for the electrolyte applications for intermediate temperature SOFC (solid oxide fuel cell) [2].

Structure of apatite makes it possible to doped various dopants in both La [3] and Si positions [4]. Structures with stoichiometric compositions for example, $La_8Sr_2Si_6O_{26}$ have lower conductivity and higher activation energies than cation vacancy structures ($La_{9.33}Si_6O_{26}$) or excess oxygen ($La_9Sr_1Si_6O_{26.5}$). The highest conductivity found in excess oxygen-containing samples indicates that the oxygen ions are the result of conduction through an oxygen ion interstitial mechanism [5].

The doping of the lanthanum silicate apatite is well known increasing its conductivity [6-10]. Doping of cations with a relatively low valence for Si site is effective to increase the conductivity of lanthanum silicates apatite when compared to doping for La position [11, 12]. Kendrick *et al.* [13] reported that a significant number of oxide ions displaced into interstitial positions and significant local structural changes are the reason why the conductivity is increased by doping with Zn^{2+}. In addition, doping of trivalent cations such as $Al^{3+}$and Ga^{3+} for the Si^{4+} site has also a beneficial effect on conductivity [14].

Tolchard *et al.* [15] reported that calculation of atomic simulations indicates that the doping of M^{4+} at the La position is strongly not suitable. That is, Titanium, one of the dopants at the Si position is more suitable because of isovalent M^{4+} [16]. In addition, the titanium ionic radius (0.42 Å) is in the range of dopant ionic radii (0.01-0.60 Å) proven for doping the apatite of lanthanum silicate apatite at Si position [17, 18].

Doping of titanium to lanthanum silicate apatites for series of $La_{9.33}Si_{6-x}Ti_xO_{26}$ decrease its conductivity significantly when the dopant content of x > 1 [19, 20]. This is due to the trapped oxygen ions with the higher dopant content. We attempted to prepare Titanium (Ti) doping in lower concentration. Conductivity properties of $La_{9.33}Si_{6-x}Ti_xO_{26}$ with dopant content x = 0.1; 0.2; 0.3 have not been reported in the previous study yet. Here, we reported the effect of Ti^{4+} dopant on apatite electrolyte characteristics, namely, crystal structure and conductivity, in lanthanum silicate apatite with dopant content of x = 0.1; 0.2; 0.3.

2. Experiment

High-purity La_2O_3 (Aldrich, 99.999%), Na_2SiO_3 (Sigma, 97%), TiO_2 (Sigma Aldrich 99%) were used to prepare apatite-type phase $La_{9.33}Si_6O_{26}$ (standard), $La_{9.33}Si_{5.9}Ti_{0.1}O_{26}$, $La_{9.33}Si_{5.8}Ti_{0.2}O_{26}$ and $La_{9.33}Si_{5.7}Ti_{0.3}O_{26}$. Sample compositions were chosen to examine the effect of Ti content on the conductivity. A NaOH 3 M solution was used in the preparation of these apatites as mineralizer. The La_2O_3 was pre-calcined at 1100°C for 12 h in order to achieve decarbonation. In a typical experiment for preparing $La_{9.33}Si_6O_{26}$, and $La_{9.33}Si_{6-x}Ti_xO_{26}$ (x = 0.1, 0.2, 0.3), The dried starting materials were ground in the appropriate ratios thoroughly mixed and added to the NaOH solution, after which the mixtures were transferred into a stainless-steel autoclave (CIT-HTC230-V100 Co1-Int 125 mL) until 60 % volume of the autoclave was occupied. After heated at 240°C for 3 days, the precipitate was collected by filtration and washed with demineralized water. The resulting powder was dried at 120°C. Phase purity was then examined using powder X-ray diffraction using Cu-K_α radiation. Composition of apatite was measure by using EDAX spectrometer, ionic conductivity was measured using an LCR (GW Instek 61056) at 600°C–700°C and a frequency range of 20 Hz – 5 MHz on the pellet sample that was obtained by applying 6.000 kg cm^{-2} pressure, followed by sintering at 1500°C for 3 hours. The value of resistivity (ρ) was calculated by equation (1).

$$\rho = R\frac{A}{l} \quad (1)$$

where R is the total resistance, A is surface area and *l* is thickness, respectively. After the value of ρ obtained, the value of conductivity (σ) was calculated by equation (2).

$$\sigma = \frac{1}{\rho} \quad (2)$$

The activation energy (E_a) was calculated by equation (3).

$$k = Ae^{\frac{-Ea}{RT}} \quad (3)$$

where k is the rate constant, T is absolute temperature (in Kelvin), *A* is the pre-exponential factor, and *R* is the universal gas constant.

3. Result and Discussion`

Figure 1 shows the X-ray diffraction patterns of $La_{9.33}Si_{6-x}Ti_xO_{26}$ with x = 0.1, 0.2, 0.3 specimens. All specimens were almost the same as listed in powder diffraction file (JCPDS No. 010760340) reported by Nakayama and Sakamoto [21]. In addition, there are small diffraction peak at 2θ ~ 15° and 44° that could be assigned to a secondary phase, these peaks ware observed in all the Ti-doped specimens with different intensities.

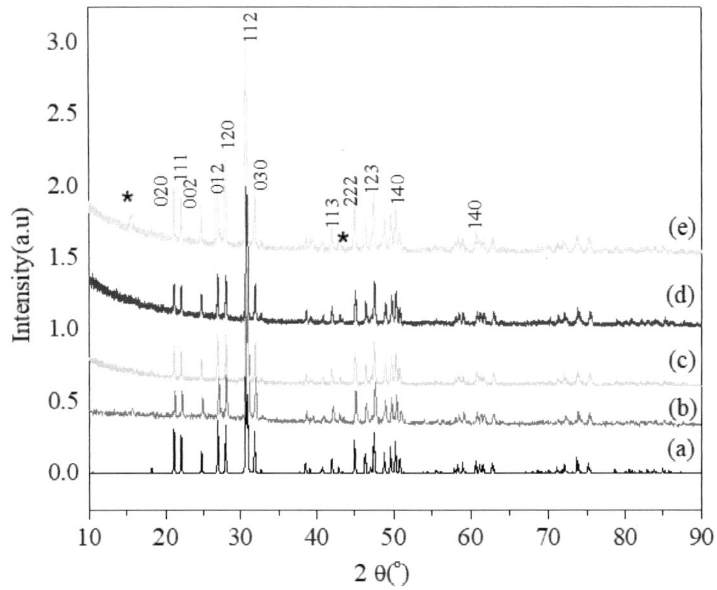

Figure 1. X-Ray diffraction patterns of (a) $La_{9.33}Si_6O_{26}$ (JCPDS No 010760340), (b) $La_{9.33}Si_6O_{26}$, (c) $La_{9.33}Si_{5.9}Ti_{0.1}O_{26}$, (d) $La_{9.33}Si_{5.8}Ti_{0.2}O_{26}$ and (e) $La_{9.33}Si_{5.7}Ti_{0.3}O_{26}$.

Table 1 shows lattice parameters for $La_{9.33}Si_{6-x}Ti_xO_{26}$ with x = 0.1, 0.2, 0.3. The *c* parameter increases with increasing Ti content as expected from cation size effects. The largest *c* parameter was found in the sample of $La_{9.33}Si_{5.7}Ti_{0.3}O_{26}$.

Table 1. Lattice parameters (hexagonal phase) for Ti doped $La_{9.33}Si_{6-x}Ti_xO_{26}$ with x = 0.1, 0.2, 0.3.

Specimens	*a* = *b* / Å	*c* / Å	Cell volume / Å^3
$La_{9.33}Si_6O_{26}$	9.7231(1)	7.1899(1)	588.6563(1)
$La_{9.33}Si_{5.9}Ti_{0.1}O_{26}$	9.6294(1)	7.1895(1)	583.8781(1)
$La_{9.33}Si_{5.8}Ti_{0.2}O_{26}$	9.6928(1)	7.1920(1)	585.1637(1)
$La_{9.33}Si_{5.7}Ti_{0.3}O_{26}$	9.7140(1)	7.2074(1)	588.9865(1)

Slightly different from the literature [22-25], Ti substitution for Si shows the larger *c* parameter for larger Ti content samples, but have smaller *a* and *b* parameter. This fact proves that the substituted Ti (0,42 Å) occurs at the Si (0,26 Å) position, which confirms that the substitute at Si position is very sensitive to change the *c* parameter [19].

EDX analysis of composition ratio for $La_{9.33}Si_{6-x}Ti_xO_{26}$ with x = 0.1, 0.2, 0.3 using experimental and theoretical data was shown in table 2.

Table 2. EDX analysis for $La_{9.33}Si_{6-x}Ti_xO_{26}$ with x = 0.1, 0.2, 0.3.

Specimens	Experimental ratio of La:Si:Ti	Theoretical ratio of La:Si:Ti
$La_{9.33}Si_{5.9}Ti_{0.1}O_{26}$	9.7231(1)	7.1899(1)
$La_{9.33}Si_{5.8}Ti_{0.2}O_{26}$	9.6294(1)	7.1895(1)
$La_{9.33}Si_{5.7}Ti_{0.3}O_{26}$	9.6928(1)	7.1920(1)

Overall, the experimental ratio is lower than the theoretical one, which may result from the small unidentified impurities and homogeneity of specimens. These ratios are reasonable for lanthanide apatites. The presence of impurities is indicated as peak in the XRD pattern at 2θ of 15° and 44°.

The Nyquist diagrams at 650°C exhibited single semicircle as shown in Figure 2. As they were difficult to accurately separate grain and grain boundary contributions on the Nyquist diagrams, the reported conductivities were total conductivities. Nyquist diagrams obtained were fitted using the Zview software [26].

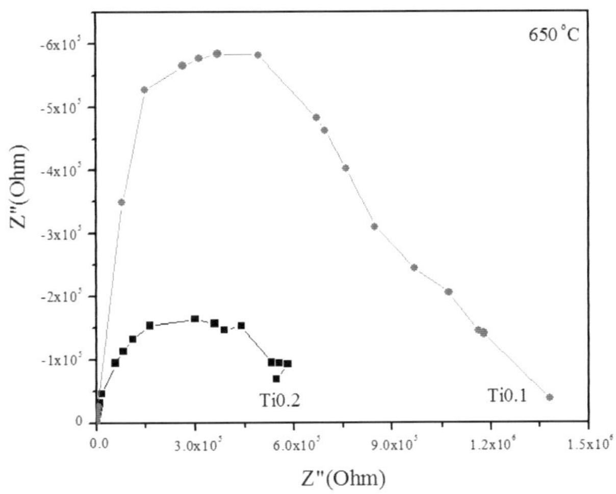

Figure 2. Nyquist diagram of $La_{9.33}Si_{5.9}Ti_{0.1}O_{26}$ (red circle) and $La_{9.33}Si_{5.8}Ti_{0.2}O_{26}$ (black rectangular) specimens sintered at 1500°C.

The conductivity values of $La_{9.33}Si_{6-x}Ti_xO_{26}$ with x = 0.1; 0.2; 0.3 at 600°C, 650°C, and 700°C are presented in table 3. The conductivity increased with increasing of Ti content. Apatite phase of $La_{9.33}Si_{5.7}Ti_{0.3}O_{26}$ showed the highest conductivity of 2.4 × 10^{-4} S/cm at 700°C.

Table 3. Conductivity of $La_{9.33}Si_{6-x}Ti_xO_{26}$ with x = 0.1, 0.2, 0.3 at variation operation temperature.

Specimens	Conductivity /(10^{-5}S/cm)		
	600 °C	650 °C	700 °C
$La_{9.33}Si_{5.9}Ti_{0.1}O_{26}$	0.23	0.81	0.95
$La_{9.33}Si_{5.8}Ti_{0.2}O_{26}$	0.60	1.48	2.57
$La_{9.33}Si_{5.7}Ti_{0.3}O_{26}$	9.07	17.80	24.00

Conductivity is still lower than that of standard requirement for solid electrolyte of 10^{-3}-10 S/cm [27, 28]. It is probably due to low sintering temperature of specimens, resulted lower densification of the oxyapatite [29, 30].

Generally, conductivity values of Ti doped apatite were higher than that of related phases without Ti doping. The order of conductivity value from small to high value is $La_{9.33}Si_6O_{26}$, $La_{9.33}Si_{5.8}Ti_{0.1}O_{26}$, $La_{9.33}Si_{5.8}Ti_{0.2}O_{26}$, and $La_{9.33}Si_{5.7}Ti_{0.3}O_{26}$, respectively [16]. Effect of Ti content to *c* parameter and conductivity is displayed in Figure 3.

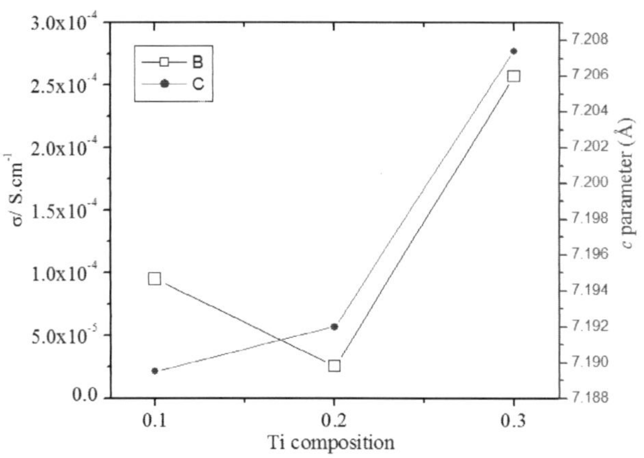

Figure 3. Effect of Ti content to c parameter and conductivity of $La_{9.33}Si_{6-x}Ti_xO_{26}$ with x=0.1, 0.2, 0.3.

An interesting feature was the fact that the highest conductivity was observed at higher Ti content and *c* parameter. In previous studies, generally that the highest conductivities of apatite systems observed at lower Ti [31]. This phenomenon can be probably explained by the fact that the expansion of lattice size is effectively increasing the mobility of the oxide ion. The magnitude of the *c* parameter in the crystal lattice is suspected to extend the oxygen ion channel, and affected the oxygen ion motion that increasing the conductivity.

Figure 4 shows the total conductivity of $La_{9.33}Si_{6-x}Ti_xO_{26}$ with x = 0.1, 0.2, 0.3 specimen with different Ti contents. Diffusion process of oxide-ions is thermally activated, which is well fitted to the Arrhenius equation.

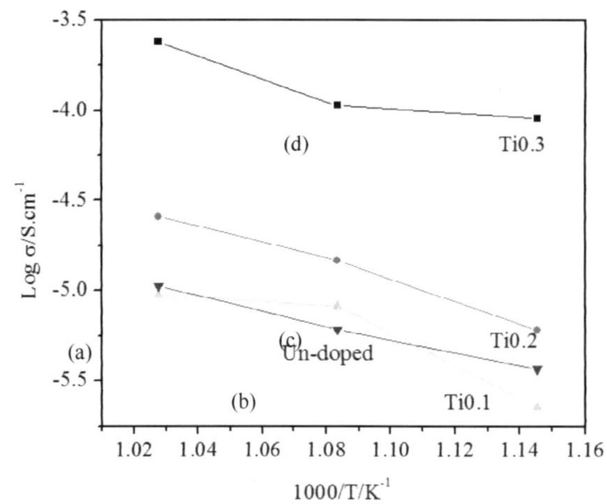

Figure 4. Arrhenius plot of un-doped (a) $La_{9.33}Si_6O_{26}$, (b) $La_{9.33}Si_{5.9}Ti_{0.1}O_{26}$, (c) $La_{9.33}Si_{5.8}Ti_{0.2}O_{26}$ and (d) $La_{9.33}Si_{5.7}Ti_{0.3}O_{26}$.

The activation energy (E_a) was calculated from the slope and the intercept of the linear fits in the logarithmic form of the Arrhenius equation for all specimens. The value of E_a, for $La_{9.33}Si_{6-x}Ti_xO_{26}$ with x = 0.1, 0.2, 0.3 are shown in table 4.

Table 4. The activation energy (E_a) of $La_{9.33}Si_{6-x}Ti_xO_{26}$ with x = 0.1, 0.2, 0.3.

Specimens	Activation Energy /eV
$La_{9.33}Si_{5.9}Ti_{0.1}O_{26}$	1.062
$La_{9.33}Si_{5.8}Ti_{0.2}O_{26}$	1.061
$La_{9.33}Si_{5.7}Ti_{0.3}O_{26}$	0.716
$La_{9.33}Si_6O_{26}$	0.734

Activation energy related to the ion migration mechanism in apatite structure. It was found that the value of activation energy for all specimens was smaller than 1.1 eV indicating the conduction via interstitial mechanism [13, 32].

4. Conclusions

The apatite-type oxide ion conductors of $La_{9.33}Si_{6-x}Ti_xO_{26}$ with x = 0.1, 0.2, 0.3 have been successfully synthesized in order to prepared new electrolytes that suitable at medium temperature. It was found that the conductivities of the Ti doped specimens are generally higher than that of undoped samples. From the crystal structure, it was found that the value of *c* parameter increased with increasing of Ti content and conductivity. Apatite $La_{9.33}Si_{5.7}Ti_{0.3}O_{26}$ showed the highest conductivity of 2.4×10^{-4} S/cm at 700°C. From activation energy, it was found that the value of E_a was smaller than 1.1 eV indicating the conduction ion via interstitial mechanism.

Acknowledgement

Authors are grateful to Directorate General for Higher Education, Ministry of Ristek Dikti through PUPT no 718/UN6.3.1/PL/2017, Universitas Padjadjaran for research Grant of Academic Leader Grant No.872/UN6.3.1/PL/2017 as a financial support. A part of this research also supported by PKM No. 547/UN6.1.1/KM/2017.

References

[1] Gasparyan H, Neophytides S, Niakolas D and Stathopoulos. V 2011 *Solid State Ionics* **192** 158
[2] Cao X G and Jiang S P 2014 *J Maters ChemiA* **2** 20739
[3] Ma Y, Fenineche N, Elkedim O, Moliere M, Liao H, and Briois P 2016 *Int J Hydrogen Energy* **41** 9993
[4] Ding X, Hua G, Ding D, Zhu W and Wang H 2016 *J Power Sources* **306** 630
[5] Kioupis D and Kakali G 2015 *J Ceram Int.* **42** 9333
[6] Najib A and Sansom J E H 2004 *Dalton T* **3106** 3106
[7] Vincent A, Savignat S B and Gervais F 2007 *J Eur Ceram Soc* **27** 1187
[8] Kharlamova T, Pavlova S, Sadykov a V, Lapina O and Khabibulin D 2008 *Solid State Ionics* **179** 1018
[9] Xiang J, Jia-Hu O, and Zhan-Guo G 2015 *J Power Sources* **284** 49
[10] Noviyanti A R, Ismunandar, Prijamboedi B and Marsih I N 2012 *ITB J Sci* **44** 193
[11] Qingle S and Hua Z 2012 *J Rare Earths* **30**
[12] Marrero-Lópeza D, Martín-Sedeno M C, Pena-Martínez J, Ruiz-Morales J C and Núnez P 2010 *J Power Sources* **195**
[13] Kendrick E, Islam M S and Slater P R 2007 *J MaterChem* **17** 3104
[14] Kendrick E, Islam M S and Slater P R 2007 *Solid State Ionics* **177** 3411
[15] Tolchard J R 2007 *Adv FuncMater* **17** 2564
[16] Sansom J E H, Sermon P A and Slater P R 2005 *Solid State Ionics* **176** 1765

[17] Kendrick E, Sansom J E H, Tolchard J R, Islam M S and Slater P R 2007 *Faraday Discuss* **134** 181
[18] Shannon R D 1976 *Acta Cryst A* **32**
[19] Al-Yasari A, Jones A, Orera A, Apperley D C and Driscoll D 2009 *J Mater Chem.* **19** 5003
[20] Sansom J E H, Sermon P A and Slater P R 2005 *Solid State Ionics* **176** 1765
[21] Nakayama S, Aono H and Sadaoka Y 1995 *Chem. Lett.* **24** 431
[22] Tolchard J, Sansom J, Islam M and Slater P 2004 *Ionics* **10** 353
[23] Sansom J E H, Rudge-Pickard H A, Smith G, Slater P R and Islam M S 2004 *Solid State Ionics* **175** 99
[24] Sansom J E H, Najib A and Slater P R 2004 *Solid State Ionics* **175** 353
[25] Noviyanti A R, Ismunandar, Prijamboedi B, Marsih I N and Mukti R R 2011 (*ICICI-BME* Bandung Indonesia)
[26] Bonhomme C, Beaudet-Savignat S, Chartier T, Maître A, Sauvet A L and Soulestin B 2009 *Solid State Ionics* **180** 1593
[27] Goodenough J B 2003 *Annl Rev Mater* **33** 91
[28] Noviyanti A R, Hastiawan I, Irwansyah F S, Hidayat S, Hardian A, Syarif D G and Yuliyati Y B 2016 *AIP Conf Proceedings* Bandung Indonesia, 050002-1--6.
[29] Xiang J, Liu Z G, Ouyang J H and Yan F Y 2014 *Jornal of Power Sources* **251** 305-10
[30] Rahayu I, Hidayat S, Noviyanti A R, Rakhmawaty D and Ernawati E 2017 *IOP Conf Series: J Phys: Conf. Series* **812** 1
[31] León-Reina L, Porras-Vázquez J M, Losilla E R and Aranda M A G 2006 *Solid State Ionics* **177** 1307
[32] Noviyanti A R, Hastiawan I, Yuliyati Y B, Rahayu I, Rosyani D and Gustaman S D 2017 *AIP Conf. Proceedings* 0400011

IOP Conf. Series: Journal of Physics: Conf. Series **1080** (2018) 012019 doi:10.1088/1742-6596/1080/1/012019

The effect of pH and sintering treatment on magnetic nanoparticles ferrite based synthesized by coprecipitation method

T Saragi[1*], B Permana[1], M Saputri[1], B L Depi[1], S W Butarbutar[1], L Safriani[1], I Rahayu[2] and Risdiana[1]

[1]Department of Physics, Faculty of Mathematics and Natural Sciences, Universitas Padjadjaran, Jl. Raya Bandung-Sumedang km 21 Jatinangor, West Java, Indonesia 45363

[2]Department of Chemistry, Faculty of Mathematics and Natural Sciences, Universitas Padjadjaran, Jl. Raya Bandung-Sumedang km 21 Jatinangor, West Java, Indonesia 45363

[*]t.saragi@phys.unpad.ac.id

Abstract. Ferrite nanoparticles synthesized by coprecipitation method depend mostly on parameters synthesis such as reaction temperature, pH of the suspension, and initial molar concentration. Here, we reported the physical properties of nanoparticles magnetite (Fe_3O_4) synthesized by coprecipitation method with variation of pH in the range between 8 and 11, in order to investigated the effect of pH treatment to its optical properties, particles size, and phase changes after sintering treatment. The band gap parameters of magnetite nanoparticles and crystal structure of magnetite powder are determined. From UV-Vis measurement, it is found that the band gap of magnetite nanoparticles was in the range of 1.3 eV to 1.5 eV with the smallest size of nanoparticles was 9.69 nm obtained at condition of pH=10. From XRD characterization, it was found that all samples sintered at 100°C for 4h have magnetite phase of Fe_3O_4 with cubic structure, while all samples sintered at 1100°C for 4 h have hematite phase of α-Fe_2O_3 with hexagonal structure.

1. Introduction

The size of magnetic particles ferrite-based is very important for biological application. In many in-vivo biological applications, the nano-sized of materials is highly beneficial because of it properties that can penetrate cells without any agglomeration in bloodstream. In ferrite-based materials, there is a critical length showing the sized limit in which the materials have two different magnetic properties. For samples that has smaller sized than that of critical length, superparamagnetic state appears showing a strong magnetic properties under an applied magnetic field. In the case of particles size larger than critical length in the few tens of nm, the thermal agitation dominates the ferromagnetic order [1-2]. Magnetite (Fe_3O_4) with spherical nanoparticles and diameters less than 20 nm, for example, has advantages properties such as chemical stability, chemical and biological compatibility, non-toxicity, high saturation magnetization, M_s (90 emu/gr), environmental stability, simple preparation and can be used for biomedical applications due to their ability to affect the relaxation rate of water proton [3-4].

IOP Conf. Series: Journal of Physics: Conf. Series **1080** (2018) 012019 doi:10.1088/1742-6596/1080/1/012019

Controlling the size of nanoparticles in the order of critical length, several methods by using of chemical methods have been done, such as precipitation method [5], sol-gel process [6-8], and hydrothermal synthesis [9-10]. All the methods performed various experimental parameters such as the pH of the solution, ionic strength, capping agent, reaction temperature, composition and pressure. Recently, we have been synthesized iron oxide by using hydrothermal method. However the size of nanoparticles was in hundreds nm [9], so that, the effect of particles size to its phase behavior could not be observed. This nanoparticles was still in large size, due to the solubility of precursor, which is limited to the pressure in reactor [11].

In this paper, we reported the synthesis of iron oxide prepared by co-precipitation method in order to investigate the effect of pH on optical and morphology of nanoparticles and explored the sintering effect on phase change of iron oxide at sintering temperature of 100°C and 1100°C. This method is a simple, reliable and cheap process.

2. Experiment

Magnetite (Fe_3O_4) nanoparticles were synthesized by co-precipitation method. The ratio of Fe^{2+}:Fe^{3+} was 1:2, with 25% of $NH_3.H_2O$ solution. 0.033 mol $FeCl_3.6H_2O$ and 0.016 mol $FeCl_2.4H_2O$ were dissolved in 100 ml DI-water. The $NH_3.H_2O$ solution was then drop wisely to the Fe^{2+}/Fe^{3+} solution until pH condition reach appropriate value. In this study, we synthesized the magnetite nanoparticles at room temperature (25°C) and pH condition was varied in the range of 8-11 to investigate the effect of pH solution on optical properties, size, and morphology of synthesized magnetite nanoparticles. After the reaction, precipitates of magnetite were washed two times by using of n-hexane and ethanol, and magnetic decantation to remove all the soluble substances. Magnetite nanoparticles was re-dispersed in ethanol and characterized by UV-Vis and TEM measurement. The magnetite nanoparticles was then dried at 80°C, sintered at 100°C for 4 hours and 1100°C for 4 hours, and characterized by XRD measurement to investigate the phase change of Fe_3O_4.

3. Result and Discussion

After synthesized by co-precipitation method, the precipitates nanoparticles were black color which indicates that the particles formed are magnetite [3]. The absorption spectra of (UV-Vis) of as-prepared nanoparticles magnetite synthesized at various pH condition is shown in figure 1. The normalized UV-Vis spectrum is shown in the left side, and the normalized tauc-plot curve (band gap dependent of $(\alpha h\nu)^{1/2}$) of magnetite nanoparticles is shown in the right side. It can be seen that all samples exhibited strong absorption bands in the visible region. It was calculated that the band-gap are in the range of 1.3 eV – 1.5 eV. The lowest band-gap observed at sample prepared at pH condition of 11, while the largest band-gap observed at sample prepared at pH condition of 10. All the band-gap values were less than 2 eV and it is indicated the band-gap is in direct transition [12].

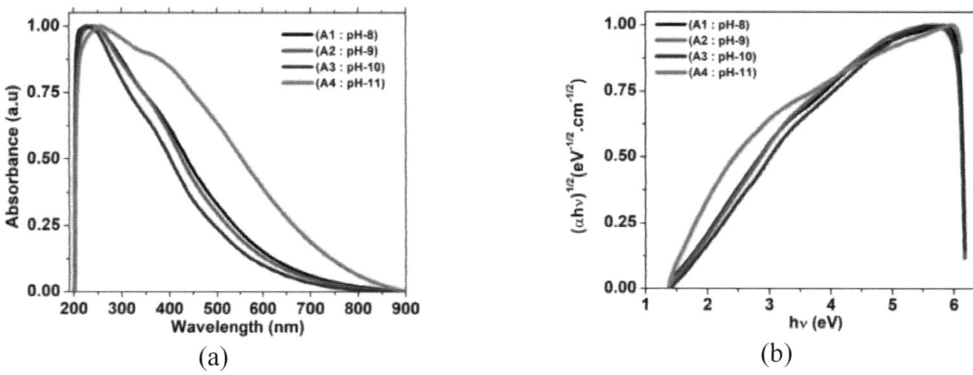

Figure 1. (a) The normalized UV-Vis spectrum and (b) tauc-plot curve of magnetite nanoparticles synthesized by co-precipitation method at room temperature and variation of pH in the range of pH 8, 9, 10 and 11.

IOP Conf. Series: Journal of Physics: Conf. Series **1080** (2018) 012019 doi:10.1088/1742-6596/1080/1/012019

The value of band-gap and the average diameter of magnetite (Fe_3O_4) nanoparticles growth at room temperature with various pH conditions is shown in table 1. The morphology and size of particles were determined by TEM. The mean particles and distribution were determined by size average of some transparent particles which are randomly selected on the TEM image. Figure 2 shows TEM image of magnetite nanoparticles synthesized at room temperature with various of pH condition of 8 (A1), 9 (A2), 10 (A3), and 11 (A4) with scale bar of 20 nm. It is found that samples are mainly composed of quasi-spherical particles. The smallest average size of magnetite nanoparticles is 9.69 ± 1.06 nm observed at pH condition of 10 and the largest size of magnetite nanoparticles is 12.48 ± 0.85 nm observed at pH condition of 9.

Table 1. The band-gap properties and average size of Fe_3O_4 nanoparticles growth at room temperature with various pH condition.

Sample	pH	Band-gap (eV)	Average of Diameter (nm)
A1	8	1.4	11.03 ± 1.06
A2	9	1.4	12.48 ± 0.85
A3	10	1.5	9.69 ± 1.06
A4	11	1.3	13.24 ± 1.62

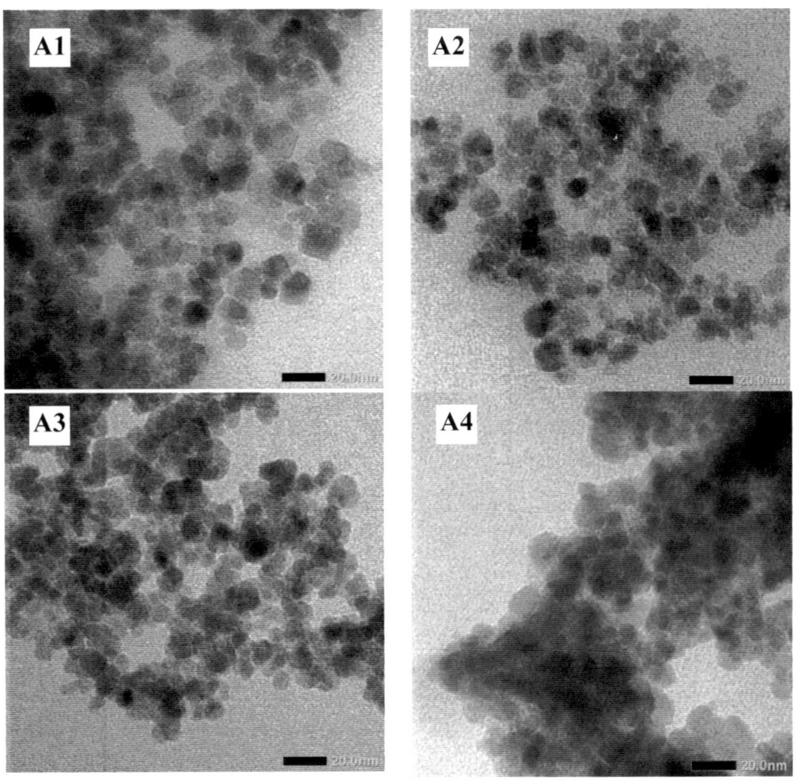

Figure 2. The TEM image of magnetite nanoparticles (Fe_3O_4) with scale bar of 20 nm growth at room temperature with various pH condition of (A1) pH = 8, (A2) pH=9, (A3) pH=10 and (A4) pH=11.

The structural analysis of a dried powder of magnetic particles was conducted by using of XRD with radiation source of Cu-Kα. Figure 3(a) illustrates the XRD pattern of iron oxide growth at room temperature with various pH condition and sintered at 100°C for 4 hours (a) and 1100°C for 4 hours, respectively. From XRD measurement, it was found that all samples sintered at 100°C for 4h have a low quality of crystal characteristics, while all samples sintered at 1100°C for 4 hours have a high quality of crystal characteristics. The XRD patterns of sample sintered at 100°C for 4 hours have all characteristic peaks of magnetite Fe_3O_4 with cubic phase and consistent with JCPDS data #11-614, and the XRD patterns of sample sintered at 1100°C for 4 hours had all characteristic peak of hematite α-Fe_2O_3 without impurity peak with hexagonal phase and consistent with JCPDS data #1309-37-1. This experiment shows that the sample sintered at 1100°C for 4 hours is hexagonal phase [9]. It was found that samples sintered at high temperature of 1100°C transformed from cubic phase of Fe_3O_4 to hexagonal phase of α-Fe_2O_3 without any influence of pH value. This transformation is probably due to the arrangement of Fe and O ions in unit cells. Compared with another report, it is found that samples sintered at 100°C have similar crystal structure of cubic phase [3], while samples sintered at 1100°C in this report, has different crystal structure with samples sintered at 1000°C [3]. It is found that crystal structure changes with increasing sintering temperature from rhombohedral α-Fe_2O_3 at sintering temperature of 1000°C to hexagonal α-Fe_2O_3 at sintering temperature of 1100°C.

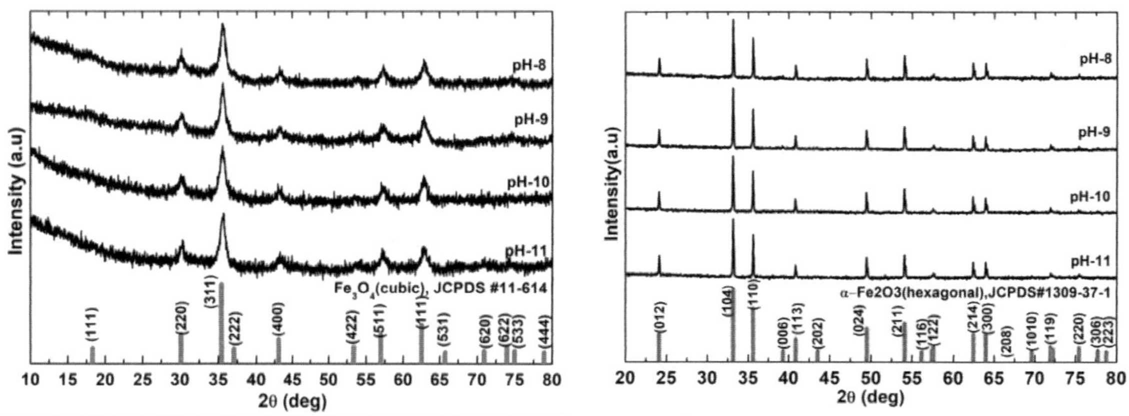

(a) (b)

Figure 3. The XRD pattern of iron oxide growth at room temperature with various of pH condition and (a) sintered at 100°C for 4 hours and (b) 1100°C for 4 hours respectively.

The average crystallite size in the sample was calculated using the Scherrer equation [13] and tabulated in table 2. The largest size of iron oxide was 81.02 nm. It is indicated that the sample growth at pH of 10 has the best condition.

Table 2. The crystal size of iron oxide sintering at 1100 °C for 4h calculated by using of Debye-Scherrer equation.

Sample	pH	Average of diameter (nm)
A1	8	64.80
A2	9	54.00
A3	10	81.02
A4	11	54.00

4. Conclusion

The synthesis of magnetite (Fe_3O_4) has been successfully prepared by co-precipitation method. The obtained nanoparticles have been synthesized in different pH condition of 8,9,10 and 11, and different sintering temperature of 100°C and 1100°C. All samples were mainly composed of quasi-spherical particles with the smallest average size of magnetite nanoparticles of 9.69 nm observed at pH condition of 10 and the largest size of magnetite nanoparticles was 12.48 nm observed at pH condition of 9. It is found that samples sintered at 100°C have crystal structure of magnetite Fe_3O_4 with cubic phase, while sample sintered at 1100°C have crystal structure of hematite α-Fe_2O_3. Samples sintered at 1100°C transformed from cubic phase of Fe_3O_4 to hexagonal phase of α-Fe_2O_3 without any influence of pH value. Crystal structure changes with increasing sintering temperature from rhombohedral α-Fe_2O_3 at sintering temperature of 1000°C to hexagonal α-Fe_2O_3 at sintering temperature of 1100°C.

Acknowledgment

This work was supported by Riset Fundamental Unversitas Padjadjaran (RFU) 2017, No. 855/UN6.3.1/PL/2017, and partially supported by Academic Leadership Grant of Universitas Padjadjaran 2017, No. 872/UN6.3.1/LT/2017.

References

[1] Pankhurst Q A, Connolly J, Jones S K and Dobson J 2003 *J. Phys. D: Appl. Phys.* **36** 167
[2] Mornet S, Vasseur S, Grasset F dan Duguet E 2004 *J. Mater. Chem.* **14** 2161
[3] Kazeminezhad I and Mosivand S 2014 *Acta Physica Polonica A.* **125** 1210
[4] Revia R A and Zhang M 2016 *Materials Today* **19** 157
[5] Ghandoor H E, Zidan H M, Khalil M M H and Ismail M I M 2012 *Int. J. Electrochem. Sci.* **7** 5734
[6] Saragi T, Nurjannah S, Novia R, Syakir N, Simanjuntak E, Safriani L, Risdiana and Bahtiar A 2015 *Materials Science Forum* **827** 219
[7] Saragi T, Syakir N, Nainggolan T H, Alboin C and Risdiana 2013 *AIP Conf. Proc.* **1554** 123
[8] Shirsath S E, Mane M L, Yasukawa Y, Liu X and Akimitsu M 2013 *J Nanopart Res.* **15** 1976
[9] Saragi T, Santika A S, Permana B, Syakir N, Kartawidjaja M and Risdiana 2017 *IOP Conf. Series: Materials Science and Engineering* **196** 012025
[10] Hayashi H and Hakuta Y 2010 *Materials* **3** 3794
[11] Yanyan Y, Lifen X, Yanqiang Z and Fengyun W 2008 *Int. J. Electrochem. Sci.* **3** 67
[12] Ghandoor H E, Zidan H, Khalil M M and Ismail M I 2012 *Int. J. Electrochem* **7** 5734
[13] Callister J, Rethwisch W D and D G 2009 *Materials Science and Engineering An Introduction 8th Ed.* (John Wiley & Sons, Inc.)

IOP Conf. Series: Journal of Physics: Conf. Series **1080** (2018) 012020 doi:10.1088/1742-6596/1080/1/012020

The study of wettability in reduced graphene oxide film on copper substrate using electrostatic spray deposition technique

B J Suroto, H F Sembiring, M T Fathurrahman, Risdiana, N Riveli, S Hidayat and N Syakir*

Department of Physics, Faculty of Mathematics and Natural Sciences, Universitas Padjadjaran, Jl. Raya Bandung-Sumedang km 21 Jatinangor, Sumedang, West Java, Indonesia 45363

*n.sjakir@phys.unpad.ac.id

Abstract. We report the preliminary study of wettability properties of reduced-graphene oxide (rGO) film prepared by electrostatic spray deposition technique. Wettability is an important property in surface material that can drive the direction of development for various applications. The experiments were performed using copper plate as substrate and graphene oxide (GO) solution of 4 mg/ml as coating material. We examined and investigated the effect of various high voltages to surface morphology and wettability of rGO. The voltages are varied at 1.0 and 1.5 kV and substrate temperature was fixed at 150°C. The rGO film was obtained through thermal treatment of GO film at 200°C for 1 hour. The surface morphology of rGO films were examined using optical microscope and scanning electron microscope. The wettability was examined using static contact angle method to determine water contact angle. The results show that the surface roughness and water contact angle of rGO film was affected by deposition voltage. The highest water contact angle of rGO film of 123.98° that indicated good hydrophobic properties was obtained for deposition condition on 1.5 kV.

1. Introduction

Graphene sheet is the thinnest material in the world. It is known as 2D organic material and has many superior properties such as highest thermal conductivity [1], highest transparency [2], highest electric conductivity [3] and highest mechanical properties [4]. Therefore, graphene is a promising material for many applications including boiling heat transfer. The previous study reported that mixed-wettability of hydrophobic-dot and superhydrophilic material are able to enhance heat transfer performance [5], and shown an unique bubble behavior on subcooling condition [6]. Recent studies show that reduced graphene oxide (rGO) as heater and working fluid increases the value of heat transfer coeficient (HTC) and critical heat flux (CHF) [4]. The same result was also found by An *et al.* [7] by using supersonic spraying technique of rGO films. The surface imperfection is characterized by rGO flakes structure as a factor affecting bubble or bubble nucleation [5]. Several studies have shown that the surface morphology of graphene coating influence the surface wettability that is indicated by changing of water contact angle (WCA). Number of graphene layers have shifted the wettability [8]. Thermal treatment also affects the value of wettability [6]. With increasing thermal annealing of GO/TiO_2 coating, the value of WCA is increased. Uniquely, the shift in wettability properties from

IOP Conf. Series: Journal of Physics: Conf. Series **1080** (2018) 012020 doi:10.1088/1742-6596/1080/1/012020

superhydrophobicity having the value of WCA larger than 150° to superhydrophilicity that the maximum value of WCA is 5° may occur due to the effects of UV irradiation and wettability values can be reversible by the heat treatment process. Environmental factors also affected the value of wetttability of graphene coating. When graphene coating binds water molecules from the environment then the wettability value shifts from hydrophobic (WCA > 90°) to hydrophilic (WCA < 90°) [9]. Providing high quality of rGO with good hydrophobic properties become one of big challenges. Some researchers reported preparation of rGO by spraying technique with various improvement such as supersonic spraying [7], substrate temperature assisted [10], superhydrophlic assisted [11], and electrostatic [11, 12]. Here, we reported the synthesis of rGO thin film by electrostatic spray deposition and studied the effect of electrostatic high voltage condition to its morphology and wettability properties in rGO thin film.

2. Experiment Method

A commercial graphene oxide (*Graphenea*) was used as GO water solution sources with concentration of 4 mg/ml. The GO solution was stirred using magnetic stirrer for 30 minutes then sonicated for 30 minutes at room temperature using ultrasonic bath. A subsection samples substrate was made from pure copper plate (*Nilaco*: 113519) with size 1 cm x 1 cm without polished treatment. The copper surface was cleaned by acetone prior to deposition of GO films. The deposition of GO film was prepared by electrostatic spray deposition (ESD). The schematic ESD that utilized in this experiment is shown in figure 1. We use a copper wire ring with diameter (d) of 5 cm as a cathode and the nozzle of sprayer as an anode. The distance between electrodes was 7.5 cm and the distance of an anode to substrate in hot plate (H) was 10 cm. GO solution is injected and controlled using micrometer staging that fixed to the spray gun (*Meiji* F100) and connected to the compressor. GO aerosols are generated using high pressure of 8 bar and adjusted the nozzle sprayer to get the desired aerosol size. The copper substrates are cleaned using standard procedure prior to deposition of GO film. High voltage source can be adjusted from 0 to 2 kV. In this report, ESD was adjusted to be 1.0 kV and 1.5 kV. Hot plate temperature was fixed to 150°C as temperature of substrate.

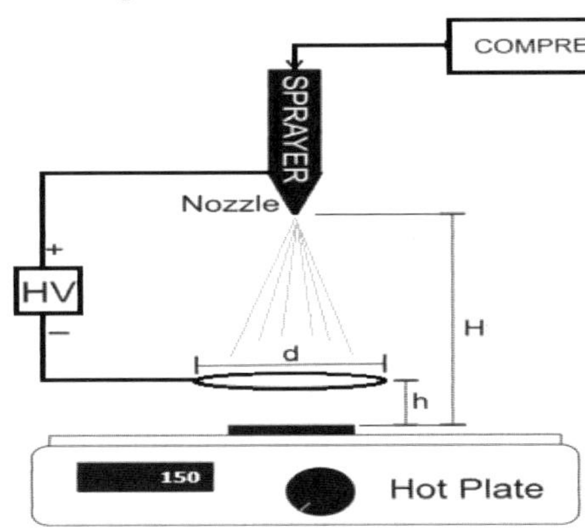

Figure 1. Schematic of modified electrostatic spray deposition (ESD).

The thermal treatment was used to obtain rGO films from GO films on Cu substrate as a result of electrostatic spray deposition. Samples of GO films on Cu substrate were heated at 200°C for 1 hour.

All rGO films were characterized by optical microscope (*Zeiss Axio* imager 2) to define the homogeneity of rGO films in microscopic scale, static contact angle measurement to define the value of WCA and scanning electron microscope (SEM) to measure nanostructure of rGO films. Additionally, the energy dispersive spectroscopy (EDS) measurement was also performed to obtain the ratio of carbon and oxygen in rGO films. Static contact angle measurement is based on drop test of deionized (DI) water on the surface of rGO film at room temperature.

3. Results and Discussion

Surface images of rGO films on cupper substrate were performed macroscopically using an optical microscope with 100 times magnification as shown in figure 2(a) and 2(b) with electrostatic high voltage of ESD of 1.0 kV and 1.5 kV, respectively. The surface area of each image is 880 μm × 660 μm. Both images have different contrasts and gray level distribution indicated the presence of different surfaces homogeneity and roughness. It can also be seen that size of rGO films are unequally distributed but rather form a spotted pattern indicated the surface rGO films more rough. Figure 2(b) has a brighter contrast indicating that the sample deposition using high voltage of 1.5 kV is thicker than that of deposited using a voltage of 1.0 kV.

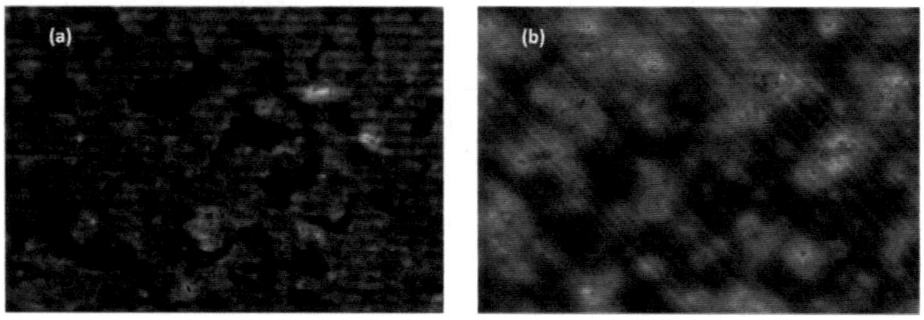

Figure 2. Surface images of rGO films on Cu substrates deposited using electrostatic high voltage condition of (a) 1.0 kV and (b) 1.5 kV.

Further evaluation to figure 2 can be based on 2.5D surface profile as shown in figure 3. Figure 3(a) and 3(b) represent surface profile of sample film deposited at high voltage condition of 1.0 kV and 1.5 kV, respectively. For comparison, surface profile of bare copper was also displayed in figure 3(c). From figure 3(c), it can be observed that the bare substrate has a homogeneous surface profile indicated by no contrast difference. The bare substrate itself indicated to have black contrast. Figure 3(a) for sample deposited with 1.0 kV has a coarser surface profile, which is indicated by contrast differences and has patterns forming steep peaks indicating by a spotted pattern small and tight spot. Figure 3(b) shows surface profile with large spot. It is also found that the intensity of the top of image in figure 3(a) has higher intensity than that of the top in figure 3(b). Both of images are caused greater stresses resulting in cone-shaped when spraying becomes sharper with the same droplet volume. The coverage area will be larger and thicker addressed by the spraying result with the larger voltage having a larger spot pattern and higher intensity [13].

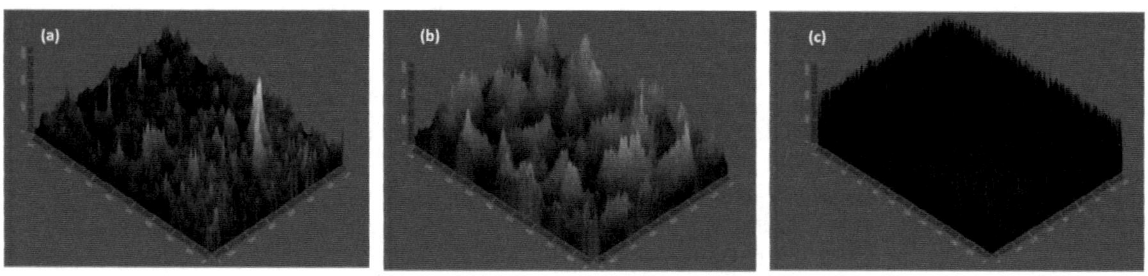

Figure 3. 2.5D surface profile of rGO films on Cu substrates deposited using electrostatic high voltage condition of (a) 1.0 kV, (b) 1.5 kV and (c) Cu bare surface.

The WCA testing was performed by droplet dripping DI water with a volume of 12 μL on the sample surface of rGO film deposited using high voltage condition of 1.0 kV, 1.5 kV and Cu bare as seen in this figure 4(a), (b) and (c), respectively. There is evolution of contact angle from 77.88^{o} of WCA for Cu bare to 123.98^{o} of WCA for 1.5 kV rGO film. The value of WCA has good correlation with the roughness on the sample surface. Increasing the WCA value will increase roughness on the sample surface. Samples deposited with 1.0 kV has greater water contact angle of 103.05° than that of the Cu substrate bare of 77.88° because of the contribution of surface coarseness. However, an increasing in WCA of samples deposited with 1.0 kV and 1.5 kV stresses occurs due to the morphological differences of the repositioned rGO layer. The rGO film deposited with a voltage of 1.5 kV has a greater roughness. This result can be explained using a dynamic molecular simulation where the height of roughness gives the dominant effect of the roughness of a surface [8]. From WCA values of rGO films, it is found that both samples of rGO films deposited using electrostatic high voltage of 1.0 kV and 1.5 kV have hydrophobic properties.

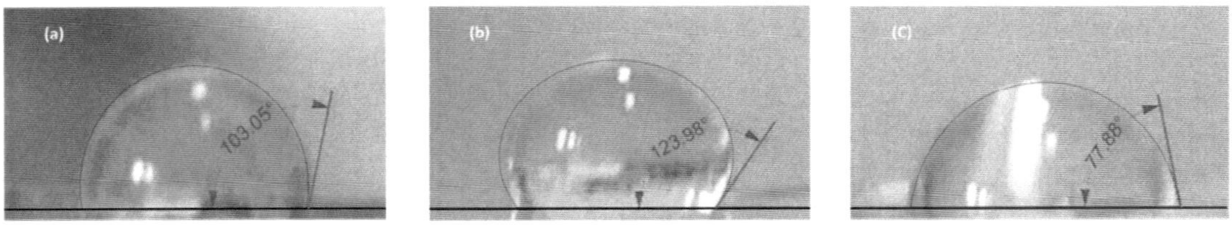

Figure 4. DI water drop test contact angle of rGO films on Cu substrates for electrostatic high voltage condition of (a) 1.0 kV, (b) 1.5 kV and (c) Cu bare surface.

The SEM imaging results of rGO samples deposited using 1.0 kV and 1.5 kV are shown in figure 5. The rGO flake is identified as wrinkled pattern on the image which is pointed by the bold black arrow. Figure 5 shows two magnifications for each sample that are magnification of 5000 (figure 5(a) and 5(b)) and 30000 (figure 5(c) and 5(d)). It is clearly visible in figure 5(a) a collection of rGO flakes deposited above substrate but not covering the entire surface of Cu substrate formed a pattern spotted. With larger magnification of the square dot line area, it is clearer that rGO flake has not deposited flatly. Some parts are making up its own cluster and contribute to its surface roughness. In contrast to samples deposited with 1.5 kV, SEM imaging of 5000 times magnification for this sample has not visible substrate surface indicating the deposition covers all areas. There is a cluster forming section as seen in the enlarged area with square dot line of figure 5(c).

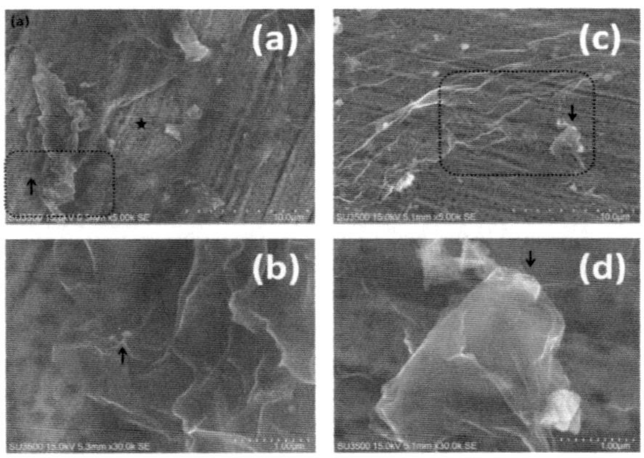

Figure 5. SEM images of rGO films on Cu substrates deposited using high voltage condition of 1.0 kV with (a) 5000 times magnification, (b) 30000 times magnification, (c) rGO film with 1.5 kV with 5000 times magnification and (d) 30000 times magnification.

The height of the cluster formed in this sample is due to the effect of increasing electrostatic voltage during deposition as has been discussed previously in the 2.5D surface profiles. The same position pointed by an arrow in the square dot line was analyzed using EDS to justify the existence of rGO flakes. Sample tested was sample deposited using 1.0 kV. It is found that the film composition was consist of Carbon (C) of 39.72%, Oxygen (O) of 10.34% and Copper (Cu) of 49.95%. Since the value of 49.95% of Cu element is belong to the Cu substrate, normalized composition of rGO film found the percentage value of C to be 79.35% and O to be 20.65%. The composition of commercial GO, the presentation value of C was 49-56% and O was 41-50%. It is clear that in our sample, the ratio of C/O was 3.84, while for commercial GO, the ratio C/O was 1.15. In addition to the morphology, this deposited material type will also affect the WCA value. GO material has a smaller surface tension than rGO material. This result is in good agreement with modeling study by Perrozia *et al.* [14].

4. Conclusions

Deposition of rGO film by electrostatic spraying technique has successfully prepared with electrostatic high voltage condition of 1.0 kV and 1.5 kV. The rGO film deposited using electrostatic high voltage of 1.5 kV, has highest value of WCA and roughness. It is found that increasing in WCA value of samples deposited with 1.0 kV and 1.5 kV was due to the morphological differences of the repositioned rGO layer. WCA values of rGO films deposited using electrostatic high voltage of 1.0 kV and 1.5 kV were 103.05^{o} and 123.98^{o}, respectively, indicating that both samples of rGO films have hydrophobic properties.

Acknowledgement

Thanks to Universitas Padjadjaran for supporting research funding of HIU (Riset Kompetensi Dosen UNPAD, PI: Dr.Eng. Bambang Joko Suroto). This project was also partially supported by ALG (Academic Leadership Grant, PI: Prof. Dr. Risdiana, M.Eng.).

References

[1] Balandin A A, Ghosh S, Bao W, Calizo I, Teweldebrhan D, Miao F and Lau C N 2008 *Nano Letter* **8** 902
[2] Novoselov K S, Geim A K, Morozov S V, Jiang D, Katsnelson M I, Grigorieva I V, Dubonos S V and Firsov A A 2005 *Nature* **438** 197
[3] Novoselov K S, Geim A K, Morozov S V, Jiang D, Zhang Y, Dubonos S V, Grigorieva I V and Firsov A A 2004 *Science* **306** 666
[4] Lee C, Wei X D, Kysar J W and Hone J 2008 *Science* **321** 385
[5] Suroto B J, Tashiro M, Hirabayashi S, Hidaka S, Kohno M, Takata Y 2013 *Journal Thermal Science and Technology* **8** 294
[6] Shen B, Suroto B J, Hirabayashi S, Yamada M, Hidaka S, Kohno M, Takahashi K and Takata Y 2015 *Applied Thermal Engineering* **88** 230
[7] An S, Kim D Y, Lee J G, Jo H S, Kim M W, Al-Deyab S S, Choi J, Yoon S S 2016 *International Journal of Heat and Mass Transfer* **98** 124
[8] Wei N, Lv C and Xu Z 2014 *Langmuir* **30** 3572
[9] Shen B, Yamada M, Orejon D, Suroto B J, Hidaka S, Kohno M, Takahashi K and Takata Y 2015 *Formation of air-rich bubbles on a superhipohilic surface in sub cooled boiling* (9th International Conference on Boiling and Condense Heat Transfer)
[10] Woo J S, Lee G W, Park S Y and Han J T 2017 *Thin Solid Films* **638** 367
[11] Wang L J, Li L, Yu J, Wu Y, He H, Ouyang X, Zhao X, Yen Y C and Lee L J 2014 *Carbon* **79** 294
[12] Wilhelm O 2005 *Electro hydrodynamic spraying–transport, mass and heat transfer of charged droplets and their application to the deposition of thin functional films* (Zurich: Dissertation at Swiss Federal Institute of Technology)
[13] Wang Z H, Wang F P, Fan J R, Gao Q J and Wang J Q 2016 *Journal of Engineering and Technology Research* **8** 31
[14] Perrozzia F, Crocea S, Treossi E, Palermo V, Santuccia S, Fioravantia G and Ottaviano L 2014 *Carbon* **77** 473

Neutronic analysis of subcritical assembly for ^{99}Mo isotope production based on thorium nitrate fuel

F Yassar[1], A W Harto[1] and Syarip[2*]

[1]Department of Nuclear Engineering and Engineering Physics, Jl. Grafika No. 2, Kampus UGM, Yogyakarta 55284
[2]Center for Accelerator Science & Technology, National Nuclear Energy Agency (BATAN) Jl. Babarsari, Yogyakarta 55281

[*]syarip@batan.go.id

Abstract. The paper discusses a neutronic analysis of the SAMOP reactor based on thorium nitrate as fuel and target materials. The reactor core model consists of an annular cylindrical tube with a tube in the center, filled with thorium nitrate as fuel. The analysis is done for the variation of fuel enrichment, the first case use thorium nitrate with 16% ^{233}U enrichment both on the annulus and the center of the core, and the second use 10% ^{233}U enrichment in the annulus and 14% enrichment in the center. The methods used for this analysis is a modelling calculation using MCNPX computer code. The analysis results show that the neutron multiplication factor (k_{ef}) is 0,97706 ± 0,00250 with conversion ratio (CR) of 0.1175 and k_{eff} = 0,98113 ± 0,00216 with CR = 0,1704, for the first and second case respectively. The need of ^{233}U for the first case and the second case are 1.448 kg and 4.383 kg, respectively. The results show that the sub-criticality requirement (k_{eff} < 1) for both cases was achieved, but from the fuel utilization, the first case is more efficient. Based on the CR values the SAMOP cannot be categorized as a breeder reactor.

1. Introduction

The Subcritical Assembly for Molybdenum-99 Production (SAMOP) is a subcritical reactor device to produces ^{99}Mo. ^{99}Mo is a parent radioisotope used to produces ^{99m}Tc which is used as medical diagnostics and cancer therapy. Generally, ^{99}Mo production to ^{235}U target (over 90%) and nuclear reactors with high neutron flux. The use of highly enriched uranium is at risk of being misused for nuclear weapons. In order to prevent such abuse, the international community is coordinating a Reduced Enrichment for Research and Test Reactor (RERTR) program, which aims to replace highly enriched uranium targets into low-enriched uranium targets or with non-uranium targets [1].

The purpose of this research are to find the amount the neutron multiplication factor (k_{eff}) and the Conversion Ratio (CR). K_{eff} is the multiplication factor that takes leakage into account which is defined as the ratio of the neutrons produced by fission in one generation to the number of neutrons lost through absorption and leakage in the preceding generation. CR is defined as the ratio of fissile material created to fissile material consumed either by fission or absorption. The requirements of k_{eff} and CR for SAMOP are k_{ef} = 0.97-0.99 and CR > 1. To achieve these two conditions, variations of SAMOP-related parameters include of SAMOP geometry, fuel composition, and number of UZrH fuels in the ring is varied for optimization.

The previous researches in relation to SAMOP neutronic analysis was done [2-4]. In the fulfillment of its safety aspect, the previous SAMOP design was adjusted for parameters related to the criticality that is height and the diameter of the core is varied but the ratio should not be the same (H / D ≠ 1) [4]. A related literature review of thorium fuel information is mainly taken from IAEA-TECDOC-1450 [5]. Thorium is 3 to 4 times more abundant than uranium and is widely distributed in nature as an easily exploitable resource in many countries. Unlike natural uranium, which contains ~0.7% fissile ^{235}U isotope, natural thorium does not contain any fissile material and is made up of the fertile ^{232}Th isotope only. Hence, thorium and thorium-based fuel as metal, oxide or carbide, has been utilized in combination with fissile ^{235}U or ^{239}Pu in nuclear research and power reactors for conversion to fissile ^{233}U, thereby enlarging the fissile material resources [5].

The world's demand for uranium-based fuels per year is expected to grow annually. It is estimated that the uranium demand will increase to 82,000 tonnes per year by 2025. By 2025 it is also estimated that secondary sources such as military stocks and uranium re-enrichment will decline and are expected to meet only 4-6% of the deficiencies. At that time, the introduction of the thorium fuel cycle will play a complementary role and ensure the availability of easy fuel for nuclear fission energy [5, 6]. The thorium deposit in Indonesia was investigated by Ngadenin [7]. Before undertaking large-scale utilization of thorium for nuclear power fuel, special treatment is required, i.e. to change fertile ^{232}Th to the fissile ^{233}U. There are two ways that change thorium to ^{233}U ,i.e. first, is an open fuel cycle, which is based on ^{232}Th irradiation that produces fissile ^{233}U, without involving chemical separation of ^{233}U, and the second is a closed fuel cycle. The closed fuel cycle is based on chemical processing of irradiated thorium and recycled ^{233}U fuel [5].

2. Theoretical Background

The National Nuclear Energy Agency (BATAN) has designed a reactor that could produce ^{99}Mo but did not categorized it use as critical reactor and did not use highly-enriched uranium, known as SAMOP or Subcritical Assembly for ^{99}Mo Production [1, 3]. The system is simple and safer if it compared to critical reactor because it is operated at a subcritical condition, but it need a continuous external neutron source along it operation.

2.1. Nuclear Criticality

Criticality is the ratio of the rate of fission reactions in one generation to the rate of catch and the rate of neutrons leaked in the previous generation. In general, this reaction can be described in equation (1).

$$^{235}_{92}U + ^{1}_{0}n \rightarrow X1 + X2 + (2-3)\,^{1}_{0}n + E \quad (1)$$

$X1$ and $X2$ are fission fragments followed by 2 or 3 neutrons (n) and E is energy generated in a fission. In a nuclear reactor, for a stable fission reaction the fission production rate must be kept constant. Figure 1 is illustrating the process of fission chain reaction [8].

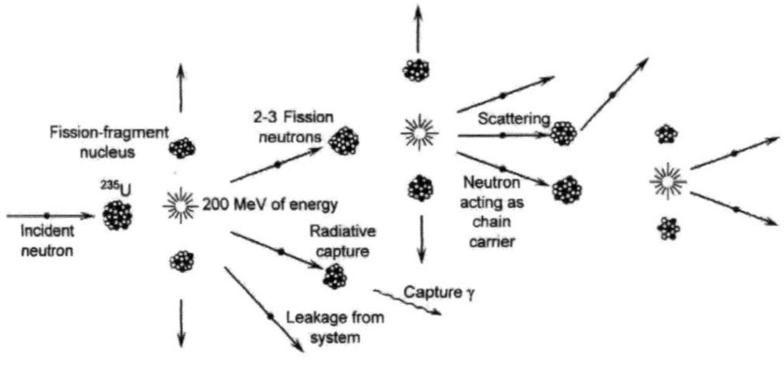

Figure 1. Chain fission scheme.

Neutrons will be generated from fission reaction, then the neutron will be absorbed or emitted out of the reactor. Comparison of the number of neutron populations in the new generation with the neutron population in the previous generation is called neutron multiplication factor (k_{eff}). The reactor criticality can be stated by the k_{eff}, when the number of neutron populations in one generation equals with the number of neutron populations in the previous generation or $k_{eff} = 1$, the reactor can be said to be critical or stable. In this critical condition the fission chain reaction will depend on time. When the number of populations in one generation is smaller than the neutron population in the previous generation (k_{eff} <1), the reactor condition is in subcritical condition. Under the subcritical conditions, the number of neutron populations will decrease as the fission reaction runs until the fission reaction is completely stopped. When the number of neutron populations in one generation is greater than the number of previous neutron populations (k_{eff} > 1), the reactor is said to be in supercritical condition.

2.2. Conversion Ratio
A quantity that characterizes this conversion of fertile into fissile material is known as conversion ratio (CR) or conversion factor. The CR is defined as the ratio of fissile material created to fissile material consumed either by fission or absorption [9]. If the ratio is greater than one, it is often referred to as the breeding ratio, the reactor is creating more fissile material than it is consuming. In this case, the CR is the ratio between the thorium capture reaction rate and the uranium uptake reaction rate in the center and the annulus such as described in equation (2).

$$CR = \frac{neutron\ absorbtion\ reaction\ rate\ in\ fertile\ material}{neutron\ fission\ reaction\ rate\ in\ fissile\ material} \quad (2)$$

The reactor may be said as a breeder if CR > 1 i.e. the neutron capture rate in the thorium is greater than the uranium absorption neutron reaction rate. The calculations performed to find the magnitude of CR can be expressed in equation (3).

$$CR = \frac{\phi_A.\sigma_{c,Th}.N_{Th^{232}}.V_A + \phi_{Triga}.\sigma_{c,U^{238}}.N_{U^{238}}.V_{Triga}}{\phi_C.\sigma_{f,U^{233}}.N_{U^{233}}.V_C + \phi_{Triga}.\sigma_{c,U^{235}}.N_{U^{235}}.V_{Triga}} \quad (3)$$

where: V_A , V_C and V_{Triga} are volume (cm^3) of annulus, center tube, and Triga fuel respectively, ϕ_A , ϕ_C and ϕ_{Triga} are neutron flux (n/cm^2s) at the annulus, center, and Triga fuel, $\sigma_{c,Th}$ and $\sigma_{a,U}$ are neutron capture and absorption cross section (barn) of thorium and uranium respectively. N_{Th} and N_U are atomic density of thorium and uranium ($atom/cm^3$). In equation (3), atomic density (N) is defined as the number of nuclides or atoms per unit volume. In this calculation the density of this atom can be approximated by modifying Avogadro's law which can be seen in equations (4) [10, 11].

$$N = \frac{g}{M} A_V, [atom] = \frac{\rho}{M} A_V, \left[\frac{atom}{cm^3}\right] = \frac{\rho_{eff}}{M_x} A_V \left[\frac{atom}{cm^3}\right] \quad (4)$$

where: β_{eff}, M_x and A_v are effective density of material (g/cm^3), mass number, and Avogadro number.

3. Methods
The reference design used for this research is the latest SAMOP BATAN design which was designed using uranil nitrate fuel. The SAMOP BATAN design will be re-modeled using MCNPX with the fuel replaced by thorium nitrate. Thorium nitrate is mixed with ^{233}U fissile material, and then the calculation and analyses is done for various geometry model configurations to find the value of k_{eff} and conversion ratio (CR).

The parameter varied to calculate the k_{eff} for the SAMOP BATAN design is ^{233}U mass fraction parameters on thorium nitrate fuel. If the k_{eff} obtained from SAMOP BATAN has not reached the expected k_{eff} range, then another parameter variation should be performed. In this study the geometry variation is done by adding the number of TRIGA fuel (UZrH) and changing the size of the fuel geometry of the solution. This research will also calculate the value of CR. The flow diagram used in the study can be seen in figure 2.

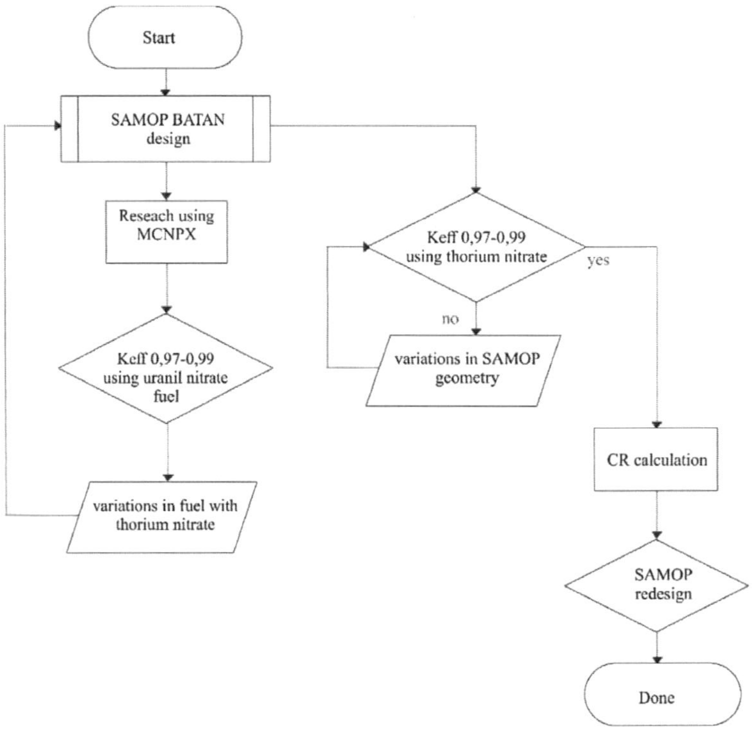

Figure 2. Flowchart for SAMOP neutronic analysis.

4. Result & Discussion

Redesigning of the SAMOP BATAN design, resulting in two new designs ie SAMOP Mod-1 and SAMOP Mod-2. The design has some supporting features in the form of addition (loading) of TRIGA UZrH fuels suround the reactor core. The function of TRIGA UZrH fuels to increase neutron multiplication, the characteristic of this fuel can be found in reference [12]. Loading of UZrH fuels continued until the SAMOP core reaches the optimum subcritical criterion value. The technical specification of SAMOP BATAN can be seen in figure 3, whiles, SAMOP Mod-1 and SAMOP Mod-2 are shown in figure 4. The technical specification of the SAMOP Mod-1 and SAMOP Mod-2 are described in table 1.

Table 1. Specification of SAMOP Mod-1 and SAMOP Mod-2.

Parameter	Material/ Dimension	
	SAMOP Mod-1	**SAMOP Mod-2**
Fuel	Thorium nitrate	Thorium nitrate
Diameter of Center Tube	5 cm	19 cm
Diameter of Annulus	31,4 cm	69.4 cm
Height of Center Tube	40 cm	40 cm
Height of Annulus	46 cm	46 cm
No. of TRIGA Fuel Rod	12	12
Fuel Diameter (TRIGA)	3,6 cm	3,6 cm
Reflector Thickness	20 cm	20 cm
Cladding Thickness	0,1 cm	0,1 cm
Reflector	Graphite	Graphite
Coolant	Water (H_2O)	Water (H_2O)

IOP Conf. Series: Journal of Physics: Conf. Series **1080** (2018) 012021 doi:10.1088/1742-6596/1080/1/012021

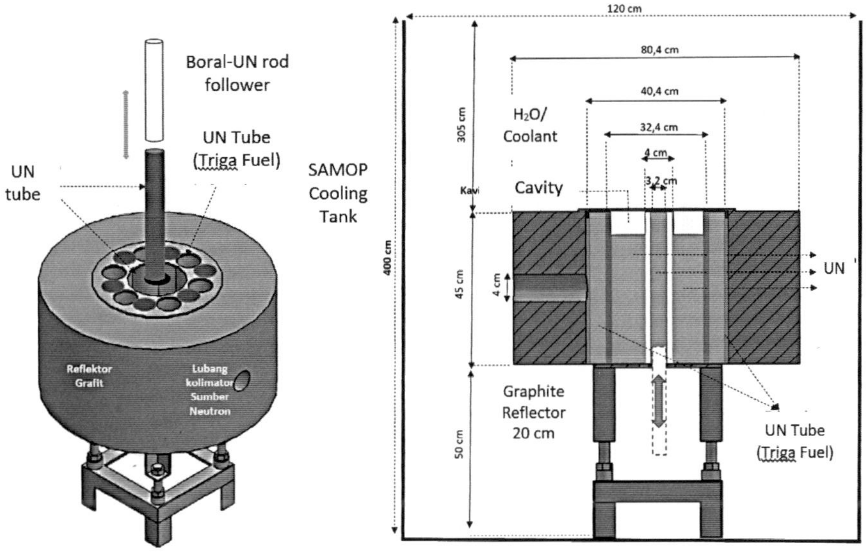

Figure 3. SAMOP BATAN Specification.

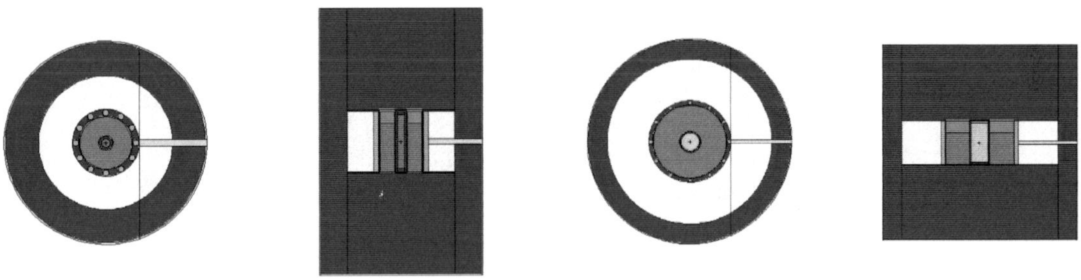

(a) (b)

Figure 4. Model of (a) SAMOP Mod-1 and (b) SAMOP Mod-2.

The calculation and analysis result of k_{eff} values as function of mass fraction of ^{233}U in thorium nitrate for the SAMOP model reactor based on the patent and SAMOP Mod-1 is described in figure 5. The variations of ^{233}U fraction in the thorium nitrate solution ranging from 10%, 12%, 14%, 16%, 18% and 20%. From figure 5 can be seen that for SAMOP design based on BATAN patent, at the maximum ^{233}U fraction of 20%, the maximum k_{eff} can be reached is 0,89787 ± 0,00269. Whiles at SAMOP Mod-1, at the maximum fraction of ^{233}U (20%), the reactor to be in critical condition (k_{eff} > 1). The optimum subcritical condition for SAMOP Mod-1 can be achieved i.e. k_{eff} = 0.97706 ± 0.00250 at the fraction of ^{233}U mass of 16%, with the CR = 0.1175.

In SAMOP Mod-2, a variation of fuel composition was performed. At the center, filled with a solution of thorium nitrate mixed with ^{233}U nitrate by 14% and in the annulus section filled by thorium nitrate solution with the addition of a mass fraction of ^{233}U of 10%. The calculation and analysis results on SAMOP Mod-2 shown that the k_{eff} = 0.98113 ± 0.00216, but this value is too close to the critical value. The k_{eff} of SAMOP Mod-2 is greater compared to SAMOP Mod-1 and SAMOP based on the patent. This is because the enlarging of the fuel column diameter for the aqueous solution thus increasing the chance for fission reactions to occur. The increase in k_{eff} is also followed by an increase in CR of 0.1704. The increasing of CR is caused by the fact that ^{233}U mass fraction required to reach the k_{eff} is smaller, so that the fertile material on the fuel of SAMOP Mod-2 much more than the SAMOP Mod-1. The more fertile material in the annulus and center increases the chance of neutron fission reaction from ^{235}U fissile material in UZrH fuels and ^{233}U in thorium nitrate solution so that resulting in CR of greater than SAMOP BATAN (patented design) and SAMOP Mod-1. The k_{eff} and

CR values for SAMOP BATAN, SAMOP Mod-1 and SAMOP Mod-2 is shown in table 2, where this result is in accordance with similar calculation reported by Syarip [13].

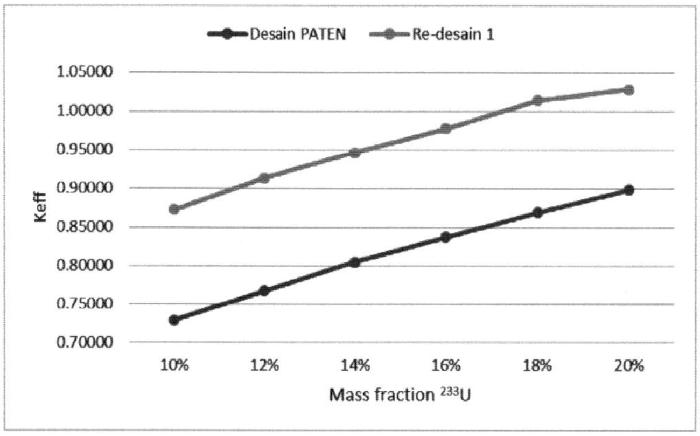

Figure 5. The k_{eff} values as function of mass fraction of ^{233}U in thorium nitrate for the model of SAMOP BATAN (patented) and SAMOP Mod-1.

Table 2. Neutron multiplication (k_{eff}) and CR of SAMOP BATAN, SAMOP Mod-1 and SAMOP Mod-2.

Model	k_{eff}	CR
SAMOP BATAN	0.89787 ± 0.00269	0.0871
SAMOP Mod-1	0.97706 ± 0.00250	0.1175
SAMOP Mod-2	0.98113 ± 0.00216	0.1704

5. Conclusion

The SAMOP reactor model based on BATAN patent can not be fueled using thorium nitrate because of resulting in smaller neutron multiplication factor (k_{eff}). The k_{eff} obtained for SAMOP re-desain i.e. SAMOP Mod-1 is 0.97706 ± 0.00287 with the CR of 0.1175. The redesign result of SAMOP Mod-2 give a better k_{eff} CR values i.e. k_{eff} = 0.98113 ± 0.00216 and CR = 0.1704 but it is not recommended to be implemented because it is too close to the critical value. The re-design of SAMOP still need to be developed further in order to achieve a breeder reactor with operating under subcritical condition.

Acknowledgement

This work was part of design assessment activities, related to the neutronic aspect of the SAMOP (Subcritical Assembly for Mo^{99} Production) system, INSINAS Project No.: RT-2016-0151 titled "Production technology development and application of ^{99}Tc for medical diagnoses", funded by Ministry of Research, Technology& Higher Education (Ristekdikti). The author would like to thank to the director of PSTA-BATAN Dr. Susilo Widodo and his staffs for supporting the implementation of the project.

References

[1] BATAN 2005 *Perangkat Reaktor Subkritik untuk Memproduksi ^{99}Mo* (Jakarta: PATEN. Paten No. P-00200500760. DJKI)
[2] Hermawan D P, Rionaldy and Syarip S 2017 *Neutronic Analysis of SAMOP Reactor Experimental Facility Using SCALE Code System* (Presented at the International Conference on Computation in Science and Engineering, ITB)
[3] Syarip S, Sutondo T, Trijono EBS and Susiantini E 2018 *Proceedings of the Pakistan Academy of Sciences: A. Physical and Computational Sciences* **55** 21
[4] Sutondo T, Syarip S and Santoso S 2009 *Safety design limits of main components of the*

proposed SAMOP system (Bandung: Proceedings of the 3rd Asian Physics Symposium (APS 2009))
[5] INTERNATIONAL ATOMIC ENERGY AGENCY 2005 *Thorium fuel cycle—potential benefits and challenges* (Google Scholar : Tech. Rep. IAEA-TECDOC-1450)
[6] Baldova D *et al.* 2011 J Radioanal Nucl Chem **288** 37
[7] Ngadenin, Syaeful H, Widana K S dan Nurdin M 2014 *Potensi Thorium dan Uranium di Kabupaten Bangka Barat* (BATAN: Magazine of Eksplorium)
[8] Duderstadt J J, Hamilton L J 1976 *Nuclear Reactor Analysis* (New York: John Wiley & Sons, Inc.)
[9] Permana S, Takaki N and Sekimoto H 2011 *Annals of Nuclear Energy* **38** 337
[10] Zsolt V, Nicholl A, Wallenius M and Mayer K 2012 *Analytica Chimica Acta* **718** 25
[11] Kokubu Y S *et al.* 2012 *International Journal of Mass Spectrometry* **310** 52
[12] Sutondo T and Syarip 2014 *Ganendra Majalah Iptek Nuklir* **17** 83
[13] Syarip *et al.* 2018 *J. Phys.: Conf. Ser.* **978** 012072

The effect of heating treatment in electron doped superconductor $Eu_{1.85}Ce_{0.15}CuO_{4+\alpha-\delta}$

Y Maryati[1], A I Hanifah[1], M A B Subardhi[1], E A Rahayu[1], Y R Tayubi[2], M Manawan[1], T Saragi[1] and Risdiana[1*]

[1]Department of Physics, Faculty of Mathematics and Natural Sciences, Universitas Padjadjaran, Jl. Raya Bandung-Sumedang km 21 Jatinangor, Sumedang, Indonesia
[2]Department of Chemistry, Faculty of Mathematics and Natural Sciences, Universitas Padjadjaran, Jl. Raya Bandung-Sumedang km 21 Jatinangor, Sumedang, Indonesia

[*]risdiana@phys.unpad.ac.id

Abstract. Electron-doped superconductor $Eu_{1.85}Ce_{0.15}CuO_{4+\alpha-\delta}$ has been synthesized using solid state reaction method. The purpose of this research is to study the effect of heating process with and without annealing treatment to its superconductivity. The sample without annealing treatment was covered by CuO to prevent the excess of oxygen in sample during heating process. All samples were characterized by XRD, resistivity and susceptibility measurement to study structure, electrical and magnetic properties of $Eu_{1.85}Ce_{0.15}CuO_{4+\alpha-\delta}$ samples with different heating process. From XRD measurement, it is found that all samples have tetragonal crystal with T' structure. From magnetic susceptibility and resistivity measurement, superconductivity was observed only in the sample with annealing treatment. The absence of superconductivity in sample without annealing was probably due to the existence of excess oxygen.

1. Introduction

Superconductivity is a phenomenon when the resistivity became zero at low temperature below its critical temperature called T_c. This T_c depends on the structure and composition of the material. First discovery of High-T_c Superconducting Cuprates (HTSC) is cuprates compound called 214 cuprates system. There are two types of 214 cuprates systems, hole-doped and electron-doped superconductor. For hole-doped system, some researchers were performed and reported various properties such as impurity effects, ground state, thermal conductivity and magnetic properties [1-7]. For electron-doped, in other side, only a few reports were published [8-11]. One of possible reason is some difficulties in preparing high quality samples.

The formula of 214 electron-doped cuprates is $Ln_{2-x}Ce_xCuO_4$ (Ln = La, Pr, Nd, Sm, Eu). Compared with other compound of electron-doped cuprates, $Eu_{2-x}Ce_xCuO_4$ (ECCO) has the narrowest range of doping concentration. However, the study of ECCO is very important because nonmagnetic properties of Eu give some advantages to explore magnetic properties in CuO_2 plane without any disturbing from rare earth magnetic moment [8, 9].

In process of synthesis, ECCO formed as-grown sample after sintering process. Usually, in as-grown sample, the superconducting phase has not formed yet even though the value of doping concentration was in optimum-doped condition. The oxygen content became one of important

IOP Conf. Series: Journal of Physics: Conf. Series **1080** (2018) 012022 doi:10.1088/1742-6596/1080/1/012022

parameters to form superconducting phase [9]. Controlling oxygen content can be done in several ways, such as, by flowing argon gas (Ar) with a certain time and temperature, called annealing process. In addition to the treatment of annealing, oxygen content can also be controlled by covering the sample with CuO material during the second sintering process. Covering by CuO is expected to prevent the addition of oxygen content during the heating process.

In this research, we synthesized electron-doped superconductor ECCO to study the effect of heating process with and without annealing treatment to its superconductivity.

2. Experiments

The samples of ECCO were prepared by the solid-state reaction method with $x = 0.15$. The raw materials of Eu_2O_3, CeO_2, and CuO were grinded finely. After that, the samples were pre-fired in the furnace at 900°C for 20 hours. After pre-fired, samples were grinded again finely and were sintered at 1050°C for 16 hours. The heating treatment was performed in different ways for each sample. Some samples were covered by CuO in order to prevent the addition of oxygen content during sintering, and other samples were annealed in argon gas (Ar) flow at 900°C for 10 hours. The samples were characterized by X-ray difraction (XRD), resistivity by four point probe method and magnetic susceptibility by superconducting quantum interference device (SQUID) measurement.

3. Result and Discussion

The first grinding process of ECCO was visually produced a grey sample, indicating that the mixture of the materials has been homogeneous as shown in figure 1(a). After pre-fired process, the colour of samples changed to be black as shown in figure 1(b). The first sintering process was produced a fairly hard and black sample, shown in figure 1(c). The sample is formed into a 10 mm of pellet by using a hydraulic press as shown in figure 1(d).

Figure 1. (a) Mixed raw materials (b) After pre-fired (c) After sintering and (d) as-grown sample.

After the sample was formed into a pellet, one of the samples was subjected to annealing treatment at 900°C for 10 hours as shown in figure 2(a) and other samples were subjected to cover by CuO material as shown in figure 2(b).

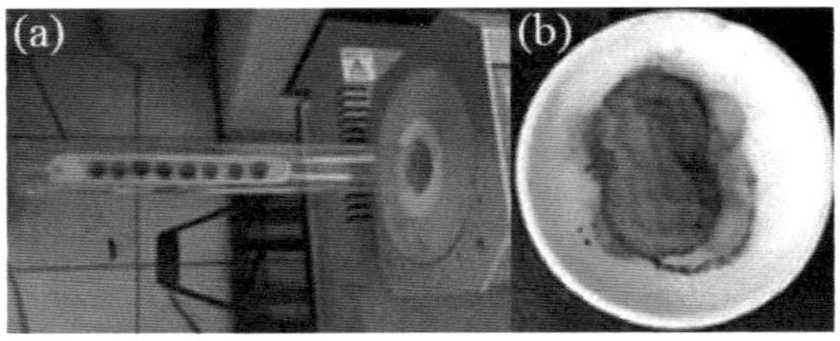

Figure 2. (a) Samples with annealing treatment in tube furnace and (b) Samples with covering by CuO without annealing treatment.

XRD pattern of sample $Eu_{1.85}Ce_{0.15}CuO_{4+\alpha-\delta}$ are shown in figure 3. Figure 3(a) shows samples with annealing treatment (δ = 0.04), and figure 3(b) sample covered by CuO without annealing treatment.

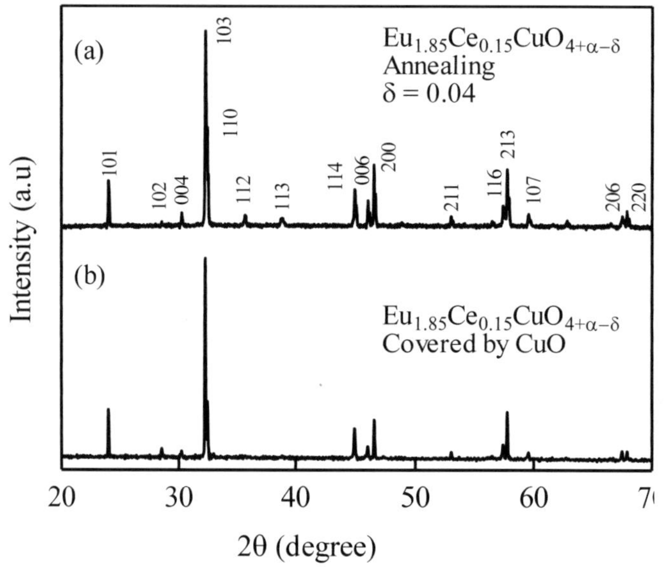

Figure 3. XRD pattern for ECCO samples: (a) with annealing treatment and (b) covered by CuO without annealing treatment.

All samples have shown tetragonal crystal with T' structure of common electron-doped system. The analysis results show the values of lattice parameters (*a*, *c*) and unit cell volumes (*V*) as shown in table 1.

Table 1. Lattice Parameter of $Eu_{1.85}Ce_{0.15}CuO_{4+\alpha-\delta}$.

Sample	*a*(Å)	*c*(Å)	*V*(Å^3)
Covered by CuO	3.90868	11.84645	180.9877
Annealing treatment	3.90914	11.85363	181.1395

The values of lattice parameters *a* and *c* for samples without annealing had smaller than that of the samples with annealing treatment.

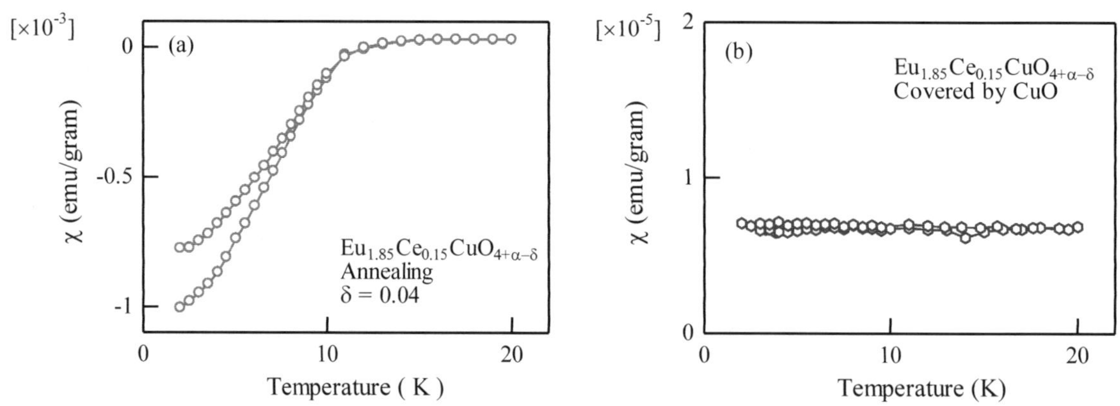

Figure 4. Graph χ-T for ECCO samples (a) with annealing treatment and (b) covered by CuO without annealing treatment.

Figure 4 shows magnetic susceptibility measurements from 2 K to 20 K for samples $Eu_{1.85}Ce_{0.15}CuO_{4+\alpha-\delta}$ with δ = 0.04 (figure 4(a)) and samples covered by CuO without annealing treatment (figure 4(b)). The result of magnetic susceptibility measurement for $Eu_{1.85}Ce_{0.15}CuO_{4+\alpha-\delta}$ with δ = 0.04 [9] shown T_c at around 12 K. However, for samples covered by CuO, superconductivity did not appear down to 2 K.

Figure 5 shows resistivity measurements for ECCO sample with annealing treatment with δ = 0.04 (figure 5(a)) [9], and sample with δ = 0 or covered by CuO (figure 5(b)).

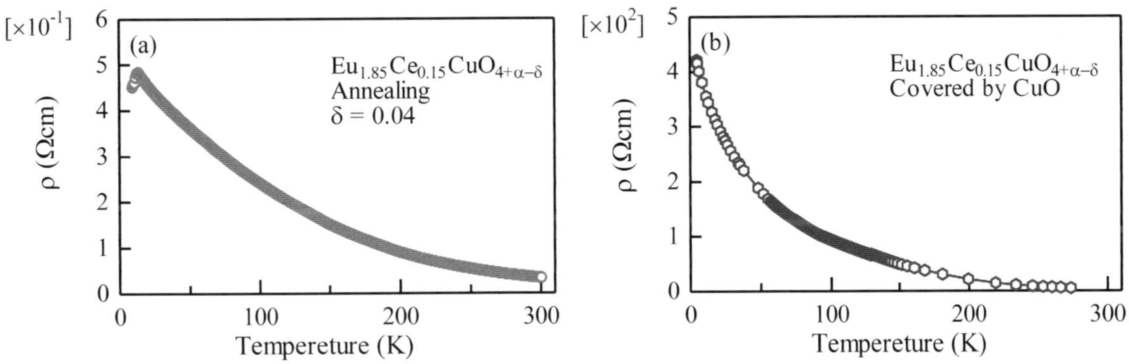

Figure 5. Graph ρ-T for ECCO samples (a) with annealing treatment and (b) covered by CuO without annealing treatment.

For samples with annealing treatment, the trace of superconductivity observed at 12 K. For samples covering by CuO, resistivity increases with decreasing temperature without any trace of superconductivity. This result indicated that the value of oxygen content still cannot be controlled by covering the samples during sintering process.

4. Conclutions

Electron doped superconductor of ECCO has been synthesized with different heating treatment, namely, with and without annealing process. From XRD measurements, it is found that all samples have tetragonal crystal with T' structure. From susceptibility and resistivity measurement, it is found that superconductivity was observed in sample with annealing process with T_c around 12 K. However, no trace of superconductivity was observed in samples without annealing process, which is probably due to the excess of oxygen still exist even though samples were covered by CuO during sintering process.

Acknowledgment

We would like to thank Professor Yoji Koike and Koike Laboratory members for assisting in measuring of XRD characterization, resistivity and susceptibility measurement at Department of Applied Physics Graduate School of Engineering at Tohoku University. These works were supported by Academic Leadership Grant Universitas Padjadjaran 2017, No: 872/UN6.3.1/LT/2017 and International Research Collaboration and International Publication Grant No. 718/UN6.3.1/PL/2017.

References

[1] Tanabe Y, Adachi T, Suzuki K, Kawamata T, Risdiana, Suzuki T, Watanabe I and Koike Y 2011 *Physical Review B* **83** 144521
[2] Suzuki K, Adachi T, Tanabe Y, Koike Y, Kawamata T, Risdiana, Suzuki T and Watanabe I 2010 *Physical Review B* **82** 054519
[3] Tanabe Y, Adachi T, Risdiana, Kawamata T, Suzuki T, Watanabe I and Koike Y 2009 *Physica B* **404** 717
[4] Tanabe Y, Suzuki K, Adachi T, Koike Y, Kawamata T, Risdiana, Suzuki T and Watanabe I

2010 *J. Phys. Soc.. Jpn.* **79** 023706
[5] Risdiana *et al.* 2008 *Phys. Rev. B* **77** 054516
[6] Adachi T, Oki N, Risdiana, Yairi S, Koike Y and Watanabe I 2008 *Phys. Rev. B* **78** 134515
[7] Koike Y *et al.* 2005 *Physica C* **426-431** 189
[8] Risdiana, Safriani L, Somantri W A, Saragi T, Adachi T, Kawasaki I, Watanabe I and Koike Y 2014 *Advanced Materials Research* **896** 354
[9] Risdiana, Pratiwi S, Suhendar D, Somantri W A, Safriani L and Saragi T 2016 *AIP Conf. Proc.* **1712** 050020
[10] Risdiana, Adachi T, Koike Y, Watanabe I, and Nagamine K 2005 *Physica C* **426-431** 355
[11] Risdiana, Adachi T, Oki N, Koike Y, Suzuki T and Watanabe I 2008 *Phys. Rev. B* **82** 014506

IOP Conf. Series: Journal of Physics: Conf. Series **1080** (2018) 012023 doi:10.1088/1742-6596/1080/1/012023

Microwave absorbing properties of $Nd_2Fe_{14}B/\alpha Fe$ nanocomposite magnet

E Handoko[1*], I Sugihartono[1], S Budi[2], M Randa[3], Z Jalil[4] and C Kurniawan[5]

[1]Department of Physics, Universitas Negeri Jakarta, Jl. Rawamangun Muka, Jakarta Indonesia 13220
[2]Department of Chemistry, Universitas Negeri Jakarta, Jl. Rawamangun Muka, Jakarta Indonesia 13220
[3]Balitbang Kementerian Pertahanan Republik Indonesia, Jakarta, Indonesia
[4]Department of Physics, Syiah Kuala University, Banda Aceh, Indonesia
[5]Research Center of Physics, Indonesian Institute of Sciences (LIPI) Puspitek, Tangerang, Indonesia

*erfan@unj.ac.id

Abstract. We have investigated microwave absorbing properties of $Nd_2Fe_{14}B/\alpha Fe$ nanocomposite prepared by melt spinning method. Structures, magnetic characteristics and microwave absorbing properties were characterized using XRD, PERMAGRAPH and Vector Network Analyzer (VNA), respectively. The complex permeability and permittivity, as well as reflection loss (RL) values of the $Nd_2Fe_{14}B/\alpha Fe$ nanocomposites were calculated in the range of 8.2–12.4 GHz. Magnetic hysteresis loop of $Nd_2Fe_{14}B/\alpha Fe$ nanocomposite with a weight ratio of 52.2% of $Nd_2Fe_{14}B$ and 47.8% of αFe resulted the magnetic saturation (Ms) 0.8 Tesla and coercivity (Hc) 300 kA/m. The RL value of $Nd_2Fe_{14}B/\alpha Fe$ is less than -10 dB and tends to decrease up to -20 dB at 11.5 GHz. As a consequency, it has absorbed microwave minimum 70% in 8.2 - 12.4 GHz.

1. Introduction

Nowadays, the development of the use of communication and electronic equipment continues to increase [1]. It is also followed by the use of high frequencies of electromagnetic waves on the gigahertz scale. Instead of the advantages, the application in high frequency tools also causes many problems, for example the occurrence of wave interference, electromagnetic irradiation and electromagnetic (EM) pollution [2, 3]. The basic requirement for electromagnetic absorption materials are lightweight, thin, wide wave band and have strong absorptions ability [3–5]. To overcome the problems, various kinds of microwave absorption (MA) materials have been synthesized such as carbon black (CB), nano size silicon carbide (SiC) [6], conductive polymers [7], and magnetic materials [8–11]. Among of them, $Nd_2Fe_{14}B/\alpha Fe$ nanocomposite as a hard and soft magnetic phase and a permanent magnet with high saturation (M_S), coercive force (H_C) and remanent (M_r) magnetization is considered to be a promising material for electromagnetic wave absorber [12–17]. Many efforts to modify composition, structure, and method of manufacture of the $Nd_2Fe_{14}B/\alpha Fe$ nanocomposite to obtain optimum magnetic properties have been reported [18–20]. To improve the value of remanent and high magnetic saturation, $Nd_2Fe_{14}B/\alpha Fe$ as nano-composite has been developed using $Nd_2Fe_{14}B$ as hard magnets and αFe as soft magnets [21–23]. Due to the uniqueness of magnetic

IOP Conf. Series: Journal of Physics: Conf. Series **1080** (2018) 012023 doi:10.1088/1742-6596/1080/1/012023

properties (higher M_S, larger H_C and M_r) of the conventional ferrites, large resistivity, and high permeability value, so $Nd_2Fe_{14}B/\alpha Fe$ nano-composite is a potential candidate as a microwave absorbing material in the high frequency range [17].

In this paper, we have studied a crystal structure and magnetic properties of the $Nd_2Fe_{14}B/\alpha Fe$ nano-composite which was produced by melt spinning method. The microwave absorbing properties of $Nd_2Fe_{14}B/\alpha Fe$ nano-composite were investigated in the 8–13 GHz ranges.

2. Experimental method

$Nd_2Fe_{14}B/\alpha Fe$ nano-composite was prepared in the form of small ribbons by melt spinning method from Magnet Quench Ltd. A fine composite magnetic powder was obtained by using a mortar in toluene to avoid oxidation. Then, a powder was characterized using x-ray diffraction (XRD) (Phillips Co-Kα with 40 kV and 30 mA) to analyze crystal structure. The magnetic properties of $Nd_2Fe_{14}B/\alpha Fe$ nano-composite were obtained from the measurements on the maximum external field 1 Tesla with PERMAGRAPH type Electromagnet EP 3 from magnet physik Dr. Steingroever GmbH. $Nd_2Fe_{14}B/\alpha Fe$ nanocomposite magnet powder mixed with polyvinyl alcohol as adhesive is prepared in a 5 mm thick sample holder and then measured using vector network analyzer (VNA) Rohde-Schwarz ZVL13 to analyze microwave absorbing properties in 8–13 GHz frequency range. The values of complex permeability (μ) and permittivity (ε) were determined from the reflected signal data S11 and transmission (S21) which calculated using Nicholson Ross Weir (NRW) formula [24] and then the reflection loss (RL) value of $Nd_2Fe_{14}B/\alpha Fe$ nanocomposite can be obtained based on the measured complex permittivity and permeability using the transmit line theory [25–27].

3. Results and discussion

3.1. Crystal structure

The powder was characterized using XRD with 2θ interval 20°-105° for analyzing its structure. Figure 1(a) shows XRD patterns of $Nd_2Fe_{14}B/\alpha Fe$ magnet powder after matching results with the Powder Diffraction File (PDF) reference patterns number 00-039-0473. Based on the qualitative analysis of the observed peaks, the XRD pattern indicated that the magnetic bands have two phases of $Nd_2Fe_{14}B$ and αFe. Each phase has formed a polycrystalline of magnetic composite with a weight ratio of 52.2% of Nd_2Fe_{14} B and αFe of 47.8% by weight. This indicates that the composite magnetic material synthesized by the melt spinning method has formed a crystalline state. FWHM and crystallites size from the XRD data are shown in table 1. By using Scherrer formula [8], the crystallite size was measured and shown in figure 1(b). The crystallites size of $Nd_2Fe_{14}B$ phase in the range of 30–40 nm and αFe phase at 40–50 nm. This result confirmed that $Nd_2Fe_{14}B/\alpha Fe$ ribbon is a magnetic nano-composite. This sample also indicated the change of lattice parameter values from the standard which refer to PDF number 00-039-0473 (a = b = 8.8020 Å and c = 12.1790 Å) to a = b = 8.82139 Å and c = 12.19574 Å for $Nd_2Fe_{14}B$ ribbon and a = b = c = 2.86842 Å for αFe phase (ICSD ref no. 98-018-0969, a = b = c = 2.8690 Å).

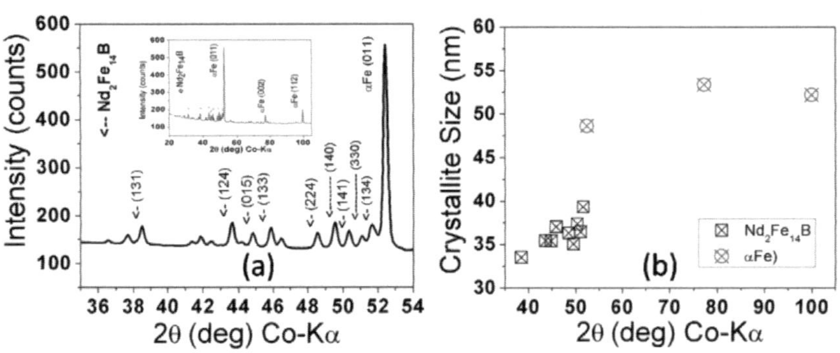

Figure 1. (a) XRD pattern and (b) distribution of crystallite size of $Nd_2Fe_{14}B/\alpha Fe$ nanocomposite magnet.

Table 1. Peak position and crystallite size of $Nd_2Fe_{14}B/\alpha Fe$ nanocomposite magnet.

Position 2θ (deg)	FWHM 2θ (deg)	h	k	l	Crystallite Size (nm)	Phase
38.4793	0.2714	1	3	1	33.51	$Nd_2Fe_{14}B$
43.5989	0.2754	1	2	4	35.42	$Nd_2Fe_{14}B$
44.7982	0.2764	0	1	5	35.39	$Nd_2Fe_{14}B$
45.8418	0.2773	1	3	3	37.02	$Nd_2Fe_{14}B$
48.5154	0.2798	2	2	4	36.29	$Nd_2Fe_{14}B$
49.4984	0.2807	1	4	0	35.04	$Nd_2Fe_{14}B$
50.3051	0.2815	1	4	1	37.35	$Nd_2Fe_{14}B$
51.0322	0.2823	3	3	0	36.43	$Nd_2Fe_{14}B$
51.5928	0.2829	1	3	4	39.32	$Nd_2Fe_{14}B$
52.4078	0.2285	0	1	1	48.61	αFe
77.2428	0.2562	0	0	2	53.36	αFe
99.9777	0.3036	1	1	2	52.17	αFe

3.2. Magnetic properties

Figure 2 shows a magnetic hysteresis loop of $Nd_2Fe_{14}B/\alpha Fe$ nanocomposite with the magnetic saturation (M_S) and coercivity (H_C) of 0.8 tesla and 300 kA /m, respectively. However, the H_C value confirmed the sample as hard magnet eventhough it is lower than H_C of $Nd_2Fe_{14}B$ nanocrystalline, where the M_S value is greater than 1 tesla thus increasing the remanent magnetization (M_r) value. The decrease of the values of saturation magnetization (M_S) and coercivity (H_C) can be caused by grinding process. Ribbon sizes after grinding and non-solid samples also reduced magnetic saturation (M_S) and remanent (M_r) values. On the other hand, a fine loop shape also shows that there has been an interaction between the hard magnet of $Nd_2Fe_{14}B$ and the soft magnet αFe and that also caused in the decrease of the coercivity (H_C) value. It can be explained by the exchange interaction between hard and soft magnets [15, 21].

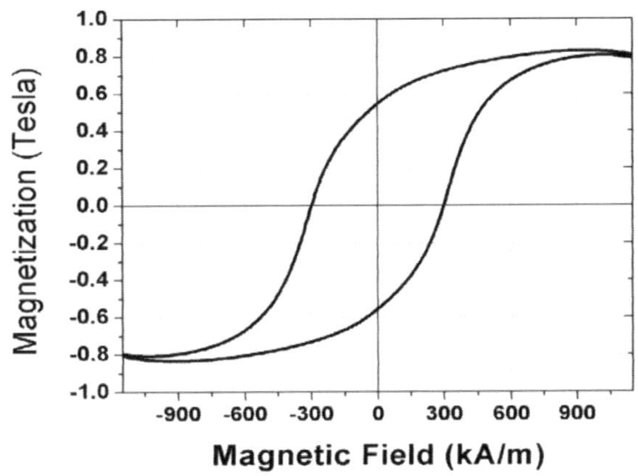

Figure 2. Room temperature magnetic hysteresis loop of $Nd_2Fe_{14}B/\alpha Fe$ nano-composite magnet.

3.3. Microwave absorbing properties

Based on the measured signal of S_{11} and S_{21} parameters of $Nd_2Fe_{14}B/\alpha Fe$ nanocomposite magnet, the complex permeability (μ) and permittivity (ε) were calculated by using NRW formula. As we can see in figure 3, the real (μ') and imaginary part (μ'') of the permeability tend to decrease in the range of 8–13 GHz. Similar conditions is also observed in the real (ε') and imaginary part (ε'') of the permittivity. Meanwhile, the complex relative permittivity tends to constant.

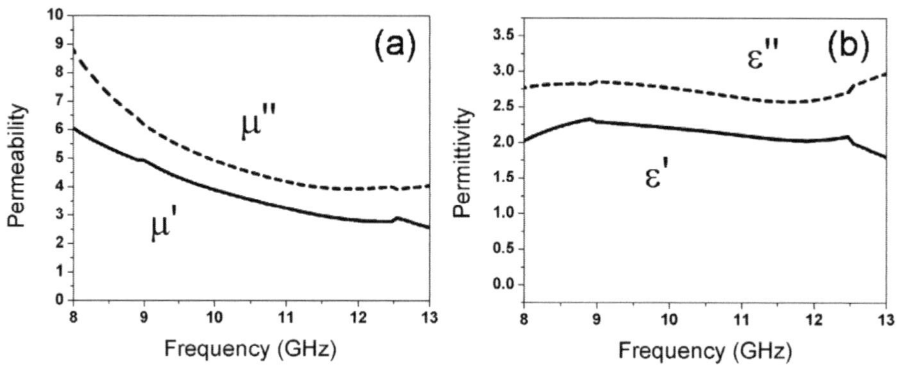

Figure 3. (a) Complex relative permeability and (b) permittivity of $Nd_2Fe_{14}B/\alpha Fe$ nanocomposite magnet.

Figure 4 shows the results reflection loss (RL) of $Nd_2Fe_{14}B/\alpha Fe$ nanocomposite magnet at 8–13 GHz in frequencies range is calculated based on the measured complex permeability ($\mu = \mu' - j\mu''$) and permittivity ($\varepsilon = \varepsilon' - j\varepsilon''$) (figure 3(a) and 3(b)) according to the transmittance line theory, which can be described by the following equations:

$$RL\ (dB) = -\ 20 \log \left|\frac{(Z-1)}{(Z+1)}\right| \tag{1}$$

$$Z = \sqrt{\mu/\varepsilon}\ \tanh\left[\left(-j2\pi f d/c\right)\sqrt{\mu.\varepsilon}\right] \tag{2}$$

where $\mu = \mu' - j\mu''$ and $\varepsilon = \varepsilon' - j\varepsilon''$, f is frequency and d is the thickness of the sample.

The RL of $Nd_2Fe_{14}B/\alpha Fe$ nano-composite magnet in the range of 8–13 GHz is less than -10 dB and tends to decrease to -20 dB at 11.5 GHz. However, $Nd_2Fe_{14}B/\alpha Fe$ nano-composite magnet has absorbed minimum 70 % in 8 – 13 GHz.

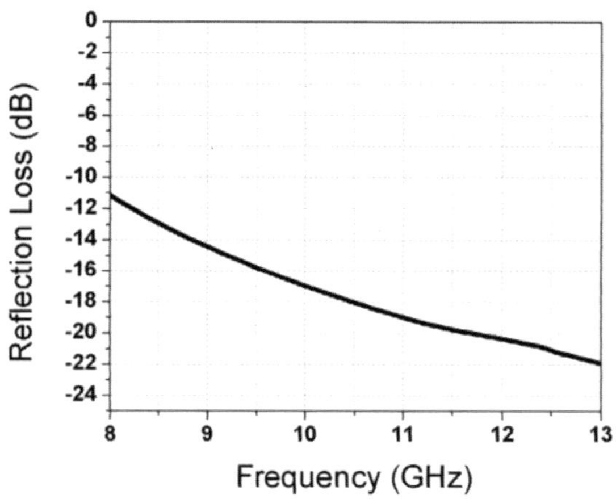

Figure 4. Absorption characteristics of $Nd_2Fe_{14}B/\alpha Fe$ nanocomposite magnet.

4. Conclusions

$Nd_2Fe_{14}B/\alpha Fe$ nanocomposite magnet ribbon that was prepared by melt spinning method has two phases and polycrystalline form. Their crystallite sizes are in the order of nanometer resulted a nanocomposite magnet with a weight ratio of 52.2% $Nd_2Fe_{14}B$ and αFe 47.8%. Magnetic hysteresis loop of $Nd_2Fe_{14}B/\alpha Fe$ nanocomposite resulted from the saturation magnetization (M_S) 0.8 Tesla and coercivity (H_C) 300 kA/m. The complex permeability tends to decrease in range 8–13 GHz and the complex permittivity tends to be constant. The reflection loss values (RL) of $Nd_2Fe_{14}B/\alpha Fe$ nanocomposite magnet is less than -10 dB and tends to decrease to -20 dB at 11.5 GHz.

References

[1] Tan K H, Ahmad R and Johan M R 2013 *Mater. Chem. Phys.* **139** 66
[2] Deng J, Li S, Zhou Y, Liang L, Zhao B, Zhang X and Zhang R 2017 *J. Colloid Interface Sci.* doi: 10.1016/j.jcis.2017.09.029
[3] Chen X G, Cheng J P, Lv S S, Zhang P P, Liu S T and Ye Y 2012 *Compos. Sci. Technol.* **72** 908
[4] Liu G, Wang L, Chen G, Hua S, Ge C, Zhang H and Wu R 2012 *J. Alloys Compd.* **514** 183
[5] Ni Q Q, Melvin G J H and Natsuki T 2015 *Ceram. Int.*, vol. **41** 9885
[6] Liu X, Zhang Z and Wu Y 2011 *Compos. Part B Eng.* **42** 326
[7] Akman O, Kavas H, Baykal A, Toprak M S, Oruh A C and Akta B 2013 *J. Magn. Magn. Mater.* **327** 151
[8] Handoko E *et al.* 2018 *Mater. Res. Express* in press https://doi.org/10.1088/2053-1591/aac4d7
[9] Wang T, Pan C, Wang Y, Tan G, Wang H, Li R and Li F 2014 *J. Magn. Magn. Mater.* **354** 12
[10] Gairola S P, Verma V, Singh A, Purohit L P and Kotnala R K 2010 *Solid State Commun.* **150** 147
[11] Zhou J, Ma H, Zhong M, Xu G, Yue Z and He Z 2006 *J. Magn. Magn. Mater.* **305** 467
[12] Harland C L, Lewis L H, Chen Z and Ma B 2004 *J. Magn. Magn. Mater.* **271** 53
[13] Hashino H, Tazaki Y, Ino H and Ohkubo T 2004 *J. Magn. Magn. Mater.* **278** 68
[14] Costa I, Oliveira M C L, De Melo H G and Faria R N 2004 *J. Magn. Magn. Mater.* **278** 348
[15] Xie G, Xu J, Lin P, Gu B and Lu M 2004 *J. Magn. Magn. Mater.* **278** 179
[16] Jurczyk M 2000 Journal of Alloys and Compounds **299** 283
[17] Lian L X, Deng L J, Han M, Tang W and Feng S D 2007 *Journal of Applied Physics* **101** 9
[18] Chen Z, Wu Y Q, Kramer M J, Smith B R and Ma B 2004 *J. Magn. Magn. Mater.* **268** 105
[19] Tang Y L, Kramer M J, Dennis K W, Mccallum R W and Lewis L H 2003 *J. Magn. Magn. Mater.* **267** 307
[20] Davies H A and Wang Z C 2004 *Materials Science and Engineering: A* **375–377** 78
[21] Sen Y, Shandong L, Xiansong L, Xiaoping S, Benxi G and Youwei D 2002 *J. Alloys Comp.* **343** 217
[22] Shindo M, Ishizone M, Kato H, Miyazaki T and Sakuma A 1996 *J. Magn. Magn. Mater.* **161** 1
[23] Jakubowicz J 2001 *Journal of Alloys and Compounds* **314** 305
[24] Handoko E, Sugihartono I and Mangasi A M *2017 IEEE Broadband Communication, Wireless Sensors and Powering (BCWSP)* p. 1
[25] Zhao B, Zhao W, Shao G, Fan B and Zhang R 2015 *ACS applied materials & interfaces* **7** 12951
[26] Handoko E, Mangasi A M and Iwan S 2016 *IEEE Radar, Antenna, Microwave, Electronics, and Telecommunications* p. 28
[27] Zhao B, Guo X, Zhao W, Deng J, Fan B, Shao G and Bai Z 2017 *Nano Research* **10** 331

Study of thermal efficiency of biomass carbonizing by direct method

O Nurhilal*, S Suryaningsih and I Indrana

[1]Department of Physics, Faculty of Mathematics and Natural Sciences, Universitas Padjadjaran, Jl. Raya Bandung Sumedang km 21 Jatinangor Sumedang West Jawa 45363 Indonesia

*otong.nurhilal@phys.unpad.ac.id

Abstract. In this study, we have designed a carbonization furnace comprising a combustion and a carbonization chamber. To know the performance of carbonization furnace, we calculated the thermal efficiency using direct method. Biomass that were used as a test material are rice husk and corn cob. We found that the highest temperature for 1.0 kg of wood fuel of combustion chamber and carbonization chamber are 648°C and 459°C, respectively. Each of the biomass has been carbonized three times at different average carbonization temperatures. It is found that the highest thermal efficiency was 2.82% and 4.43% for rice huks and corn cob, with carbonization temperature of 405.24°C and 399.80°C, respectively.

1. Introduction

Biomass is material forming from organic compound that using for energy production. The organic compound is coming from living species such as plants and animals [1]. There are two techniques for converting biomass into energy production or fuel. One of these techniques is thermochemistry. There are three thermochemical techniques, namely carbonization, phyrolysis, and gasification. Carbonization is a technique to obtain charcoal as the main product by heating biomass at 400°C-600°C in the almost or complete absence of air or oxygen. The main objective of carbonization is to increase the calorific value of charcoal [2]. The carbonization process can be carried out using a furnace having a fuel chamber as the heat generator and a carbonization chamber as the heat receiver.
Generally, the heat generator is producted from hot gas or electrical heater to get high temperatures. However, the use of hot gas or electrical heater required a high energy comsumption. To get a furnace with low energy consumption, it is necessary to design a simple furnace for carbonizing biomass sample. In this research, we designed simple furnace and performed carbonization of biomass sample using biomass waste as a fuel heat generator. The biomass to be carbonized was rice husks and corncobs, while the fuel for combustion was biomass waste from woods. Here, we studied thermal efficiency of heat transfer in the simple furnace for rice husks and corncobs samples in various temperatures.

2. Methodology

This research begins by designing a simple carbonization furnace. The simple furnace consists of two chambers that serve as a space for fuel and space for carbonization. The dimension of each chamber was 27 cm in diematers and 30 cm in hight, as shown in figure 1. In the carbonization chamber, a good heat distribution is required to make carbonization homogeniously. The cylindrical shape is chosen so that the heat spread homogeniously in all area of chamber. To reduce the heat loss, the carbonization chamber is coated on the outer wall by insulators form. For temperature control, K-type thermocouples are used. Furnace used for carbonization are designed to be simple with fuel from wood waste and small electrical energy for blowers.

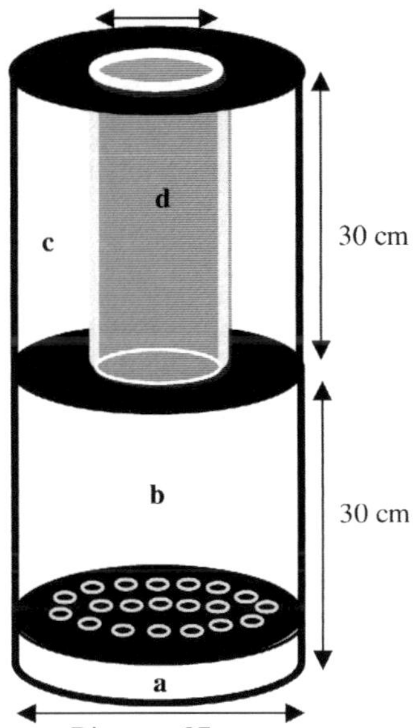

Figure 1. Design of simple furnace included (a) space of ash, (b) combustion chamber, (c) carbonization chamber and (d) chimney.

To know the temperature generated in the carbonization chamber, one kilogram of wood is burned in the combustion chamber. The amount of temperature produced is measured by the thermocouple. To determine the performance of the carbonization furnace, 250 grams of biomass each from rice husk and corncob has been carbonized on three different mean temperature variations.

Based on data of mean carbonization temperature measurement, carbonization time and other parameters then calculated thermal efficiency by direct method. This method was chosen because of simple calculation involved only two parameters, namely output heat transferred to the biomass (Q_{out}) and the total heat of the fuel input (Q_{in}). The equation to find the value of thermal efficiency was shown in equation (1) [4].

$$\eta = \frac{Q_{out}}{Q_{in}} \times 100\% \quad (1)$$

where η is thermal efficiency in percentage. The value of Q_{out} can be obtained from equation (2).

$$Q_{out} = m_B \times c_B \times \Delta T \quad (2)$$

where m_B is mass of biomass (kg), c_B is specific heat capacity of biomass (kJ/kg°C), and ΔT is surrounding and carbonization temperature difference (°C). The value of Q_{in} can be obtained from equation (3).

$$Q_{in} = Q_f + Q_e \quad (3)$$

where Q_f is input energy from fuel (kJ), Q_e is electrical energy of blower (kJ). Both values can be obtained from equation (4) and (5).

$$Q_f = m_f \times HHV_f \tag{4}$$

$$Q_e = V \times I \times t_b \tag{5}$$

where m_f is mass of fuel (kg), HHV_f is hight heating value of fuel, the value of HHV_f of woods is 8352.67 kJ/kg, V is voltage of blower (Volt), I is current of blower (Ampere), and t_b is time of blower (second).

3. Result and Discussion

3.1 The test of combustion and carbonization chamber temperature without biomass

The first test of the simple furnace was measuring the temperature distibution in combustion chamber using waste biomass of 1.0 kg of firewood without biomass in carbonization chamber. To increase temperature in combustion chamber, the air suppliers is needed using the blower with a voltage of 12 Volt and current of 1.2 Ampere. The result of temperature distribution in combustion chamber and carbonization chamber is shown in figure 2.

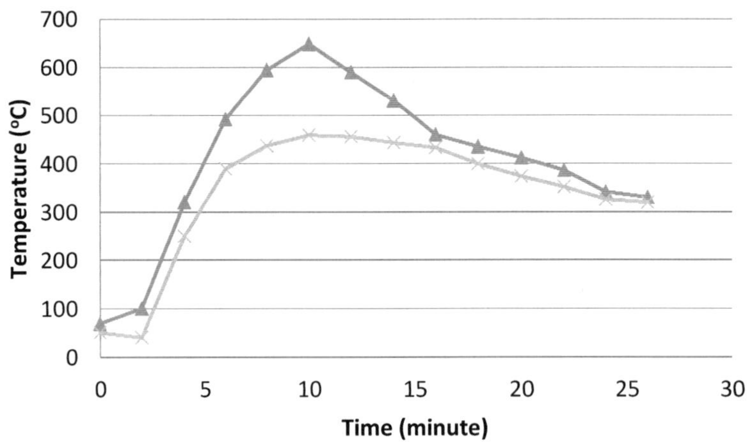

Figure 2. The graph of temperature distribution for every minutes in combustion chamber (▲) and carbonization chamber (×).

The temperature difference in the combustion and carbonization chamber indicated the evidence of heat loss in the carbonization chamber. The heat loss is due to convection and radiation processes in the carbonization furnace. The value of temperatures in carbonization chamber was reaching to 459°C, which qualified for carbonization processes [3].

3.2 Measurement of mean biomass carbonization temperature

The process of carbonizing rice husks and corncobs is done three times at different temperatures. The mean temperature of carbonization for rice husk and corncob during the time of carbonization are shown in figure 3, 4 and 5, respectively.

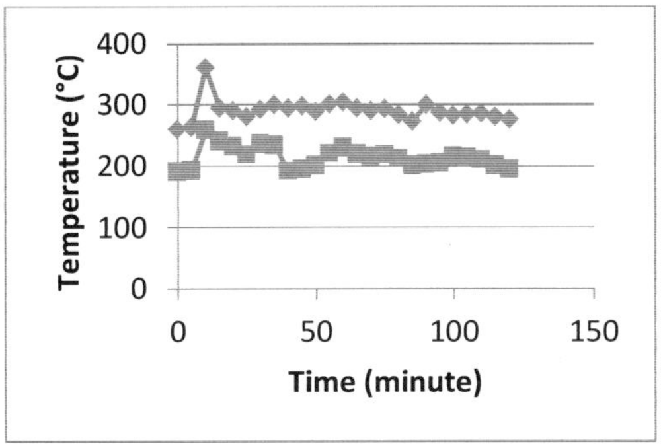

(a)

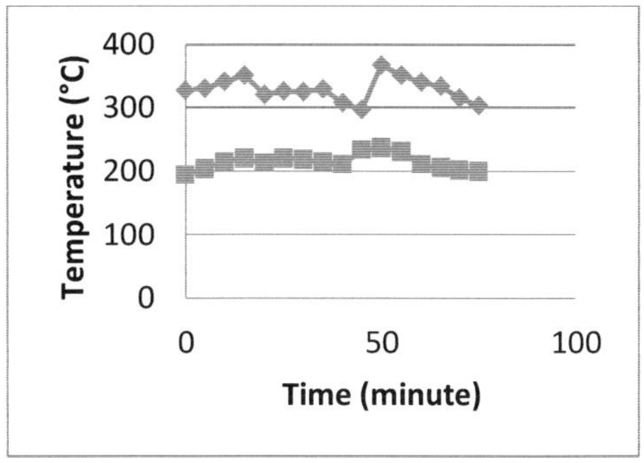

(b)

Figure 3. Temperature of carbonization (■) and combustion chamber (♦) (a) rice husk and (b) corncob at setting temperature 200°C.

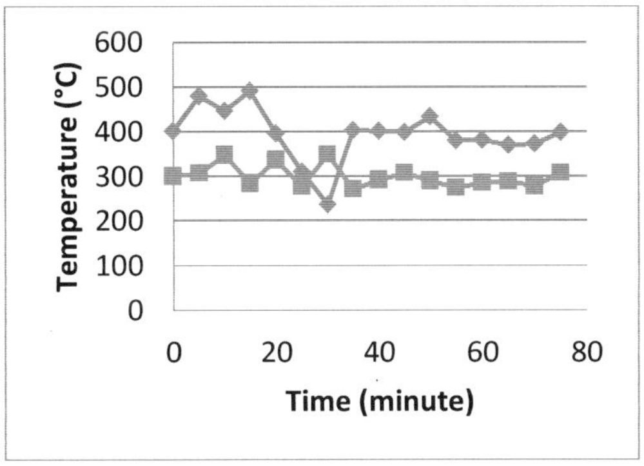

(a)

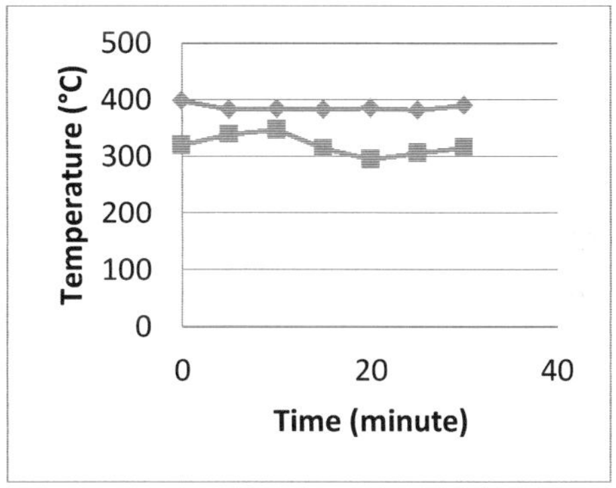

(b)

Figure 4. Temperature of carbonization (■) and combustion chamber (♦) (a) rice husk and (b) corncob at setting temperature 300°C.

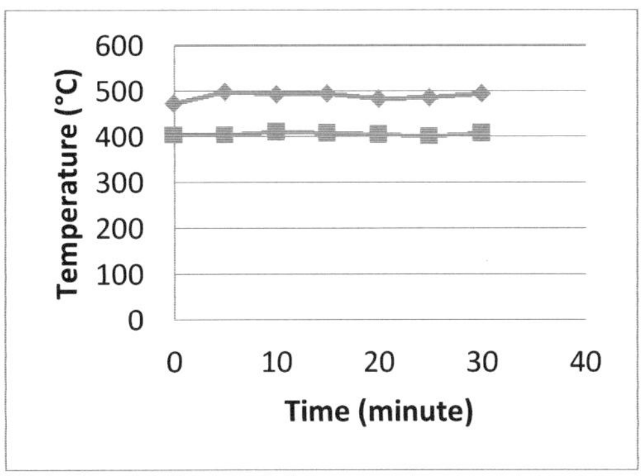

(a)

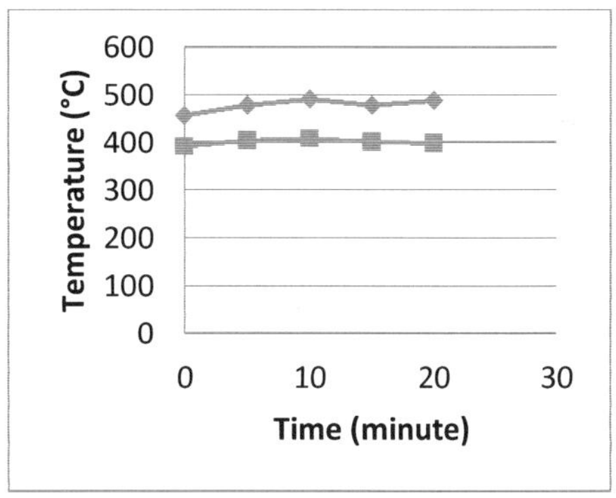

(b)

Figure 5. Temperature of carbonization (■) and combustion chamber (♦) (a) rice husk and (b) corncob at setting temperature 400°C.

The end of carbonization process is determined based on direct observation of carbonized charcoal. If all the biomass samples have turned into charcoal in a homogeneous fashion, then the carbonization process is stopped. How long it takes for the carbonization process depends on the carbonization temperature and the biomass type. The carbonization time (t_c), the surrounding temperature (T_s) before carbonization and the value of average temperature of carbonization ($\overline{T}_c$) are listed in table 1.

Table 1. The value of carbonisasi time, surrounding and average temperature of carbonization.

Biomass	t_c (min)	T_s (°C)	$\overline{T}_c$ (°C)
	120	25	213.88
Ricehusk	75	24	294.81
	35	24	405.24
	75	25	213.44
Corncob	30	25	319.14
	20	24	399.80

Based on table 1 it is seen that the greater the average temperature of carbonization, the time required during the carbonization process is smaller. It also appears that the process of carbonizing corncobs is faster than rice husks. In the process of carbonization, the average temperature increase is determined by the supply of fuel and air delivery into the combustion chamber through the blower. The air provided will improve the fuel combustion process. Thus, the heating energy generated by the combustion chamber comes from the fuel energy and electrical energy of the blower. The amount of electrical energy provided depends on the voltage, current and time of giving air (t_b) during the process of carbonization takes place.

3.3 The thermal efficiency

The thermal efficiency and various parameters values of rice husks and corncobs calculated by equations (1) to (5) are listed in table 2. The spesific calorie for rice husk and corncob is 2.772 kJ/kg°C [5] and 1.189 kJ/kg°C [6], respectively.

Table 2. Various parameters and **t**he thermal efficiency value of rice husk and corncob.

Biomass	ΔT (°C)	m_f (gr)	m_B (gr)	t_b (sec)	Q_e (kJ)	Q_f (kJ)	Q_{in} (kJ)	Q_{out} (kJ)	η (%)
	188.9	1000	250	7200	103.68	8352.67	8456.35	131.53	1.56
Rice husk	270.8	1050	250	4500	64.80	8770.00	8835.10	187.66	2.12
	381.2	1118	250	2100	35.28	9338.30	9373.57	264.10	2.82
	188.4	400	250	4500	64.80	3341.10	3405.87	56.20	1.65
Corncob	294.1	500	250	1800	25.92	4176.34	4202.26	87.57	2.08
	395.8	300	250	1200	17.28	2497.70	2514.98	111.52	4.43

The relationship between thermal efficiency and average temperature of carbonization has been illustrated in figure 6. Based on the curve in figure 6 it shows that the greater the average temperature of carbonization, the value of the thermal efficiency increases. This shows that at high carbonization temperature will cause the absorption of energy by carbonization samples getting better. Based on the time review, the higher the carbonization temperature then the time required for the carbonization process is getting shorter. The value of thermal efficiency of rice husk is various in between 1.56% and 2.82%. The highest value of η for rice husk was resulted with average carbonization temperature of 405.1°C. For corncob, the value of η is in the range between 1.65% and 4.43% with the highest value of η was resulted with average carbonization temperature of 399.8°C. The value of thermal efficiency of rice husk less than corncob. It is due to the calorivic value of rice husk less than corncob. The calorivic value of rice husk and corncob is 14.310 kJ/kg [5] and 17.620 kJ/kg [6], respectively.

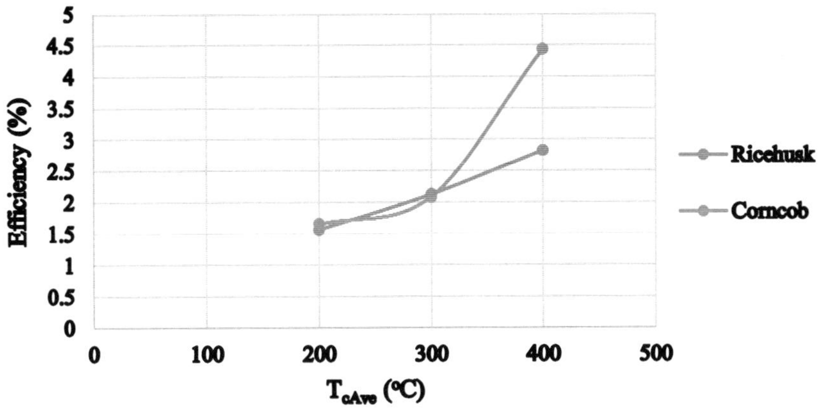

Figure 6. The curve of the thermal efficiency value as a function of the average temperature of carbonization.

4. Conclusion

Based on the results and discussion it is concluded that the designed carbonization furnace can be used for carbon biomass process with carbonization temperature reaching 459°C. Fuel used as a source of heat in the form of waste wood easily obtained. To increase the temperature in the combustion chamber is assisted with a blower with low electrical energy. It is found that the highest thermal efficiency was 2.82% and 4.43% for rice huks and corncob with carbonization temperature of 405.24°C and 399.80°C, respectively.

References

[1] Basu P 2010 *Biomass gasification and pyrolysis : practical design and theory* (United Kingdom: Elsevier, The Boulevard, Langford Lane Kidlington, Oxford, OX5 1GB)
[2] Yokohama S 2008 *The Asian Biomass Handbook* (The Japan Institute of Energy) p 107
[3] Speight J G 1994 *The Chemistry and Technology of Coal 2nd Edition* (New York: Macel Dekker)
[4] Patel C T, Patel B K and Patel V K 2013 *International Journal of Innovative Research in Science, Engineering and Technology* **2** 1518
[5] Mishra P *et al.* 1987 *Journal of Materials Science* **21** 2129
[6] Fang Z 1993 *Journal Fuel Science and Technology International* **11** 1037

IOP Conf. Series: Journal of Physics: Conf. Series **1080** (2018) 012025 doi:10.1088/1742-6596/1080/1/012025

Drop-coated reduced-GO thin film as counter electrode in DSSC

I Hanifah, F Yuliasari, Fitrilawati, N Syakir, S Hidayat, L Safriani, A Bahtiar, A Aprilia* and R E Siregar

Department of Physics, Faculty of Mathematics and Natural Sciences, Universitas Padjadjaran, Jl. Raya Jatinangor km 21, Sumedang, West Java, Indonesia 45363

*a.aprilia@unpad.ac.id

Abstract. In the last decade, many researchers have been proved the role of graphene-based material as counter electrode in dye sensitized solar cell (DSSC) system. One of graphene-based material is reduced graphene oxide (rGO) that have been widely used in optoelectronic devices. Based on it, in this work we prepared the rGO thin films and used as counter electrode (CE) in DSSC with FTO/TiO_2/Ru-dye/CE/FTO structure. The rGO thin films were deposited on FTO (Fluorine-doped Tin Oxide) substrates by drop casting technique. The 0.5 and 1.0 mg/ml of rGO powder were dispersed in two kind of solvents which are dimethylformamide (DMF) and l N-methyl-2-pyrrolidinone (NMP) for each concentration. The rGO thin films that produced from two different solvents have a differ visible wavelengths absorbance, although prepared from the same concentration. Form scanning electron microscopy (SEM) observation, the morphology of rGO thin film that prepared using NMP is more homogenous than DMF solvent. This situation may relate to the ability of each solvent in dispersing rGO powder. Fourier Transform Infra-Red spectroscopy was used to identify the presence of certain functional groups both in powder and thin film. The carboxyl group of C-O and C=O is appeared in thin film. The rGO thin film as counter electrode in DSSC was studied from current-voltage (I-V) characteristics. This work revealed that rGO thin film is able to use as alternative counter electrode in Pt-free DSSC since it produced 2.22% in efficiency (under 36.5 mW/cm^2 light irradiation).

1. Introduction

Dye-sensitized solar cell (DSSC) is one of emerging photovoltaic devices that offers a good alternative in solar cell technology. Dye sensitized based solar cell is a device that converts visible light into electricity (charge motion) based on sensitization of wide bandgap semiconductors [1]. DSSC was invented by Brian O'Regan and Michael Gratzel in 1991. Dye-sensitized solar cells (DSSCs) have been considered as a next-generation renewable power source owing to relatively high solar-to-electricity conversion efficiency (η), simple fabrication process, and low cost. The challenging issues for developing DSSC are to achieve high efficiency and to reduce production cost. Recent work for reducing production cost is focusing on replacement of the expensive catalytic Pt layer on counter electrodes (CEs) with other cheaper materials, such as graphite, fullerene, amorphous carbon, active carbon, carbon black, carbon nanotubes and graphene [2].

Graphene is a carbon 2D allotrope composed of a stable hexa-atomic honeycomb structure with a theoretical thickness of only 0.34 nm and has sp^2 hybridized carbons [3]. The high optical transmittance (~97.7%) and good conductivity of graphene made it as an excellent candidate for transparent conductive electrodes [4]. The most promising path for producing graphene related compounds, from both scale up and cost perspectives, is starting from graphene oxide (GO) obtained by a chemical oxidation method and reduction of oxygen contain by either chemically or thermally process [4]. The reduced graphene oxide (rGO) possesses some oxygen-containing functional groups (−OH, C=O), −COOH and carbonyl groups decorating the periphery of the planes. These functional groups which are exist along lattice surface defects produced during the oxidation of graphene sheets, are believed to be responsible for the electrocatalytic activity [5].

Reduced graphene oxide (rGO) thin film with the thicknesses ~10 nm deposited on Transparent Conducting Oxide (TCO) has a low electrical resistance (~0.84 kΩ/sq), 85% transmittance (λ = 550 nm), conductivity of 1.190 S/cm, high *surface-to-volume* ratio, high carrier mobility, and high catalytic activity [5-7]. In this work, we investigate the drop-coated rGO thin film to study and to prove graphene material as good alternative counter electrode in DSSC.

2. Experiments

FTO substrates were cleaned with detergent and subsequently washed by ethanol and isopropanol in ultrasonic bath. TiO_2 mesoporous T/SP (Solaronix SA) was deposited twice on FTO by screen printing technique and annealed up to 500°C for 30 minutes. Single layer of TiO_2 MC/SP (Solaronix SA) was also deposited on top of TiO_2 mesoporous with the same technique and annealing treatment as deposition of TiO_2 mesoporous. Dye solution was prepared from ruthenium dye N719 (Solaronix SA) and chenodeoxycolicacid was dissolved in ethanol with 1:10 wt/wt. The FTO/TiO_2 was immersed in dye solution for 24 hours. The FTO/TiO_2/Ru was washed by acetonitrile several times for remove unabsorbed dye particles.

We used DMF (dimethylformamide) and NMP (N-methyl-2-pyrrolidone) to dispersed rGO powder (Graphenea) with concentration 0.1 mg/ml for each solvent. Before deposition of rGO layer, cleaned FTO substrates were treatment previously by dropping the solvent on top of FTO during annealed at 65°-100°C (for DMF) and 100°C (for NMP). Afterward, rGO dispersed solution was drop-coated on prepared FTO and annealed at the same heating temperature according to each solution. Finally, the photoanodes (FTO/TiO_2/Ru) and counter electrode (FTO/rGO) were assembled using thermal adhesive (Surlyn, 25 μm thicknesses) as spacer to construct a sandwich structure (FTO / TiO_2 / dye/ electrolyte / RGO / FTO). The electrolyte (mosalyte, Solaronix SA) was injected via cavity that available in rGO/FTO side's. The prototype DSSC have an active area approximately of 0.2 cm^2.

The absorbance spectra of rGO thin film as counter electrode was measured by UV-Visible spectrophotometer. Fourier Transform Infrared (FT-IR) spectroscopy was also carried out to investigate the functional groups of hydroxyl and carbonyl in rGO powder and rGO thin film. The surface morphology and element composition of rGO thin film were analyzed using scanning electron microscope (SEM) and energy dispersive spectrum (EDS). Solar cell performance was characterized by current-voltage (I-V) using DC voltage integrated with digital multimeter under 36.5 mW/cm^2 light irradiation.

3. Results and discussion

The rGO thin films show high transmittance in visible wavelength, therefore they cannot be distinguished with bare FTO. We measured UV-Vis absorbance spectra from each sample to ensure that the rGO thin film was successfully deposited. Figure 1 shows UV-visible absorption spectra for all samples. The higher absorption of rGO/FTO as compared with bare FTO indicates that rGO thin film has been successfully deposited. The films prepared from NMP dispersion medium has higher absorption than that of DMF dispersion. Figure 2 shows SEM micrograph of FTO/rGO prepared from NMP and DMF. The dark images are identified as rGO sheets, whereas the bright round clusters are

IOP Conf. Series: Journal of Physics: Conf. Series **1080** (2018) 012025 doi:10.1088/1742-6596/1080/1/012025

FTO surface. Homogenous rGO layer from regular stacked rGO sheets should have transparent layer. The dark images in figure 2(a) exhibit irregular stacked rGO sheets due to not well dispersed of rGO in DMF. Meanwhile, transparent images of surface in figure 2(b) show that rGO sheets are well dispersed in NMP. This result might be related to higher polarity and solubility of rGO in NMP than that of DMF [8, 9].

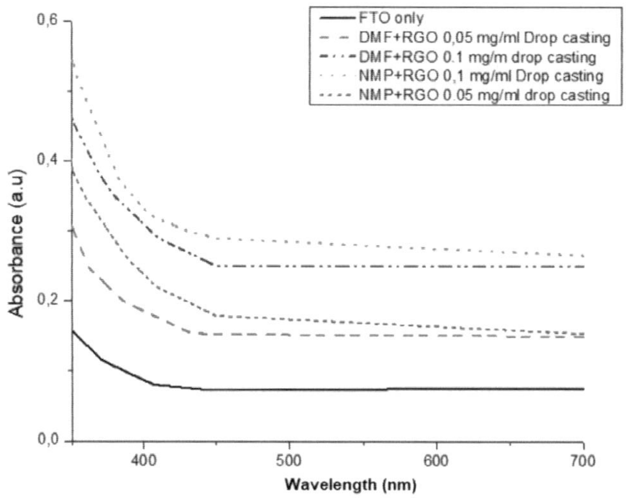

Figure 1. Absorption spectra for RGO thin films deposited by drop casting technique.

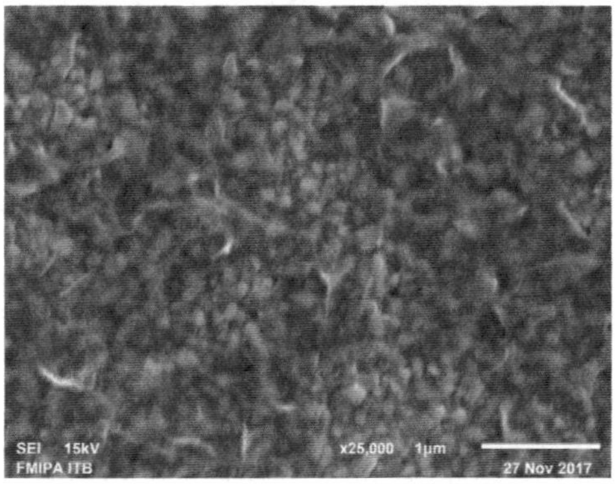

(a)

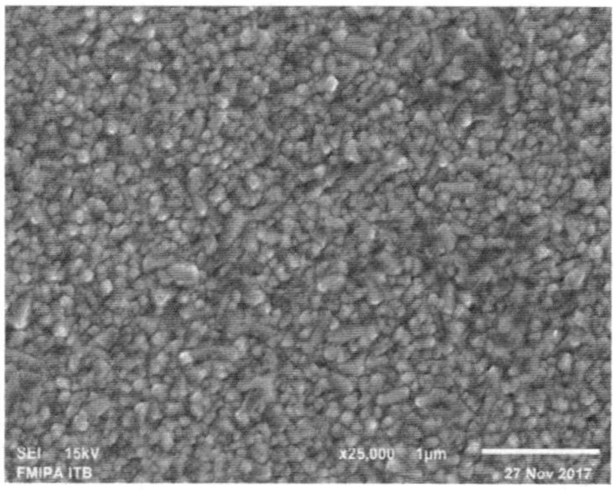

(b)

Figure 2. SEM micrograph of FTO/rGO thin film prepared from different solvent, (a) DMF and (b) NMP.

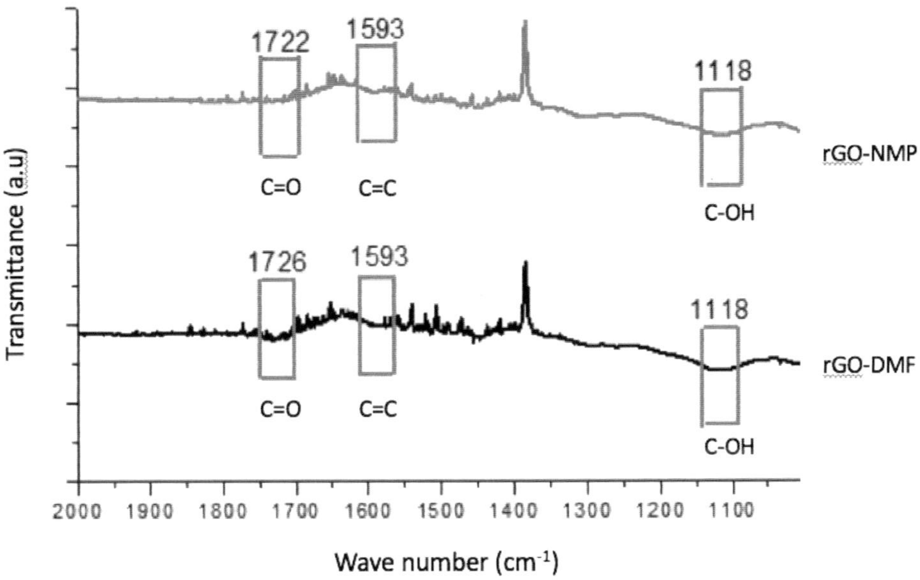

Figure 3. FT-IR spectra of rGO powder and thin film dissolved in NMP and DMF.

Figure 3 shows Fourier Transform Infra-red (FTIR) spectra of rGO thin films prepared from DMF and NMP dispersion. All spectra have main functional groups C-OH at 1118 cm^{-1}, C=C at 1593 cm^{-1} and C=O at 1722 cm^{-1}. These spectra are good agreement with other studies [10, 11], indicates that rGO films have been formed.

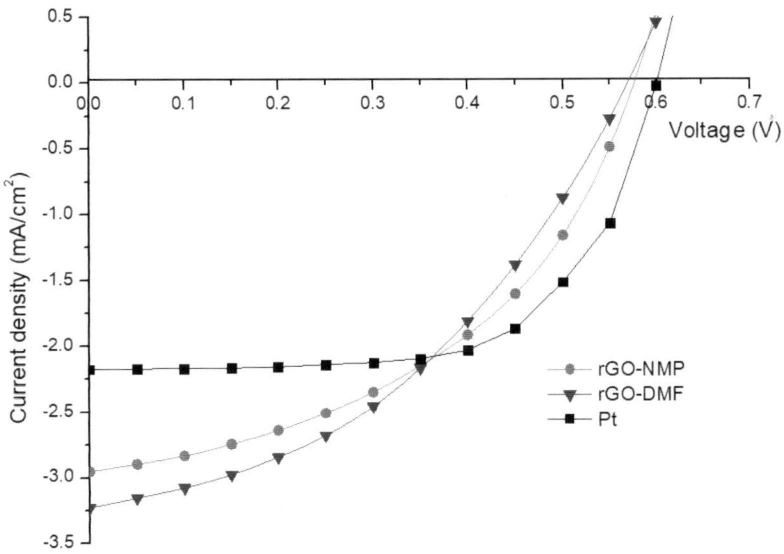

Figure 4. Photocurrent density-voltage (J-V) curve of DSSC with the structure of $FTO/TiO_2/Ru/mosalyte/rGO/FTO$.

Figure 4 shows current density-voltage (J-V) characteristic curves using rGO thin films as counter electrode. J-V curve of DSSC using platinum (Pt) as counter electrode is also plotted for comparison. The solar cell parameters such as open circuit voltage (V_{oc}), short current density (J_{sc}), Fill Factor (FF) and efficiency (η) are tabulated in table 1. J_{sc} values of devices using rGO-NMP and rGO-DMF are higher than that of Pt as counter electrode, which shows that rGO has higher catalytic property than Pt. Meanwhile, all devices have comparable V_{oc} values. However, the FF and η values of rGO based DSSC are lower than that of Pt. This result is related to higher value of series resistance of rGO thin film due to low quality of the rGO layer. Further optimization of rGO thin film is highly required to improve the performance of DSSC. This study shows that rGO thin film can be used as counter electrode of DSSC.

Table 1. Solar cell parameters of DSSC with different counter electrodes.

Counter Electrode	V_{oc} (V)	J_{sc} (mA/cm^2)	P_{max} (mW/cm^2)	FF (%)	Eff (%)
rGO-DMF 0.1	0.60	3.23	0.76	39.22	2.08
RGO-NMP 0.1	0.58	2.96	0.767	44.75	2.10
Pt	0.6	2.17	0.84	64.43	2.3

4. Conclusions

The reduced graphene oxide thin films as counter electrode in DSSC with structure of FTO/TiO_2/Ru/Mosalyte/rGO/FTO was successfully prepared. The rGO thin films were prepared from dispersed rGO powder in NMP and DMF. The rGO thin film prepared in NMP has more homogenous surface morphology than that of in DMF. The functional groups of rGO do not depend on solvent. The efficiency of devices using rGO thin films as counter electrode are comparable to device using Pt. This result shows that rGO thin film can be used as alternative counter electrode of DSSC.

Acknowledgment

This work was funded by Academic Leadership Grant (ALG) Universitas Padjadjaran contract number 855/UN6.3.1/PL/2017

References

[1] Lai W H, Su Y H, Teoh L G and Hon M H 2008 *Journal of Photochemistry Photobiology A: Chemistry* **195** 2–3 307
[2] Yeh M -H Lin L-Y, Chang L-C, Leu Y-A, Lin J-J and Ho K-C 2014 *Chem. Phys. Chem.* **15** 6 1175
[3] Patel M, Shil S and Patni N 2014 *Conference Proceeding: International Conference on Multidiciplinary Research & Practice* **I VII** 143
[4] Yuliasari F, Aprilia A, Syakir N, Safriani L, Saragi T, Risdiana, Hidayat S, Bahtiar A and Siregar R E 2017 *IOP Conference Series Material Science Engineering* **196** 012049
[5] Pei S and Cheng H M 2012 Carbon N Y **50** 3210
[6] Cruz R P, Tanaka D A and Mendes A 2012 *Solar Energy* **86** 716
[7] Ho C Y and Wang H W 2015 *Applied Surface Science* **357** 147
[8] Santosh G V, Mark R E and Alan G M 2012 *April Journal American Society for Mass Spectrometry* **23**
[9] Konios D, Stylianakis M M, Stratakis E and Kymakis E 2014 *Journal of Colloid and Interface Science* **430** 108
[10] Noor Azwen N A, Ahmad A A B, Norhana A, Sarada I, Abdul Rahman M, Aidil A H 2017 *Applied Surface Science* **392** 1134
[11] Young J O, Jung Joon Y, Yong I K, Jae K Y, H Na Y, Jong-Huy K, Seung B P 2014 *Electrochimica Acta* **116** 118

Effect of palm oil temperature on thermal losses of PTC sensor designed for overfill protection using current-voltage characteristic

L Umar*, R N Setiadi and Y Hamzah

Department of Physics, Faculty of Mathematics and Natural Sciences, University of Riau, Pekanbaru 28293

*lazuardi@unri.ac.id

Abstract. Characteristic of positive temperature coefficient (PTC) sensor can be described using current-voltage curve which depends on thermal properties of medium. The PTC sensor for overfilling protection designed for palm oil (CPO) container cannot identify the medium liquid with high temperature and high viscosity due to the overlapping of current voltage curve. For this case, a signal of worst-case alarm is generated. This paper presents the effect of CPO temperature on thermal losses of the PTC sensor represented by thermal resistance factor between the sensor and the medium surrounding it. A type of EPCOS B59050D1100 PTC sensor was characterized in two mediums; CPO and still air, to simulate "full" and "empty" status. The sensor responds was then observed through the current-voltage curve at different temperature from 300 K to 360 K. The thermal resistance were then obtained from the modeling of current voltage curve, resulting the values of 125 K/W for CPO and 343 K/W for still air with 63.5% in difference between both mediums. Using proposed two points method, the effect of medium temperature on thermal resistance can be eliminated and reduced, where the Rw variation is less than 3.15% for palm oil and 4.5% for still air. This result is within the normal operation tolerance below 5% and can be used safely to determine the status whether the sensor is in still air or in palm oil medium.

1. Introduction

The palm plantations in Indonesia are mostly 70% located on the island of Sumatra and result the large number of oil palm factories operating in the region. Fresh processed Crude Palm Oil (CPO) has high temperatures and is stored in giant containers before shipping. Level sensor for CPO detection system in the tank is needed to determine contain of the tank and also to prevent the overfilling of CPO in the process of filling, which causes environmental pollution. Level measurements control the filling process using the pump automatically, where the measured material can be gaseous, liquid or solid.

The method for level measurement of liquids in the tank can be divided into continuous and non-continuous monitoring. Santhosh [1] and Reverter [2] have developed a continuous level measurement system of liquid in the tank based on capacitive sensor. Two electrodes are installed in the tank so that the liquid level can be detected based on the capacitance value of the material between the two electrodes. Techniques like these are very expensive and have low responses [3]. A wide-range sensor levels using fiber-optic Fabry-Perot interferometers that allow for high sensitivity measurements was developed using Bragg fiber-optic lattices with fluid variations [4, 5]. These measurements provide very precise results for continuous fluid measurements but have a high price [6]. The research presented by Zixiao *et al.* [7] has designed sensors that work using a single mode laser ring in determining fluid levels while Vismanath *et al.* [8] used ultrasonic method.

Thermistor sensor with Positive Temperature Coefficients (PTC) is designed as safety industrial instrument for overfilling protection of the CPO in tanks to prevent from overfilling. The detection is evaluated by the altering of power dissipation with respect to the changes of the thermal losses of the medium surrounding it. An encapsulated sensor at the bottom of the probe detect the liquid surface, an actuator is closed which switches off the pump and stops the filling process.

Current PTC sensor for overfilling protection tank measures their resistance as a function of the medium temperature. In self-heated mode, the properties of an electrically loaded PTC sensor are well described by the current-voltage model which depends on thermal losses between the sensor and the medium surrounding it. Previous work in [9] reported that the sensor status "immersed" and "not immersed" in liquid with high temperature cannot be properly recognized status; two incorrect states are encountered: false alarm and worst case.

Crude palm oil has high viscosity and temperature; therefore it can shift significantly the current-voltage characteristic of PTC sensor with conventional method. This problem is solved through modeling of the current-voltage (I-U) characteristic and direct measurement of the thermal losses represented by thermal resistance independent of the medium temperature. In this paper we presented the effect of CPO temperature on the stability of the thermal resistance using modeling of I(U)-curve and new method based on two points measurement to eliminate the effect of temperature.

2. Experimental Methods

2.1. Overfill Protection Sensor for Palm Oil Container

The palm oil processing produce vegetable oil in the form of crude palm oil (CPO), obtained from the results of fruit meat extraction (Mesocarp). Palm oil is semi solid, this is because the oil has a high melting point of 25°C to 60°C [10]. Temperature 25°C is the standard temperature measurement method that also describes the condition of room temperature, while the temperature 60°C is the temperature of the tank filling process and the drainage of CPO in the pipe [11, 12]. Chemical composition, structure, moisture content and binding of the water, temperature and thermal history of the material are the key factors affecting thermo-physical properties of food materials. The application of overflow sensor for palm oil is shown in figure 1.

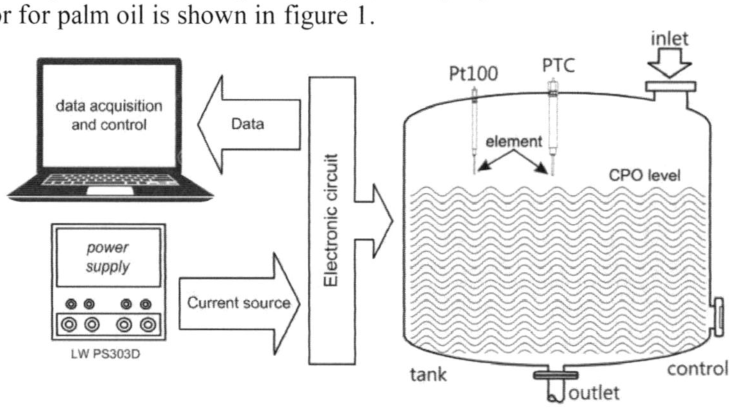

Figure 1. Overfill protection sensor with PTC sensor for CPO container.

Due to self-heating mode, the sensor temperature (T_S) is much higher than CPO temperature (T_M), so that the heat transfer to the environment is depend on thermal losses between sensor element and filling medium, represented by thermal resistance (R_W). In practical applications, the sensor characteristic can be explained by the equation (1) [13, 14].

$$R = R_{T0} e^{B\left(\frac{T_S - T_C}{T_S}\right)} + R_0 \tag{1}$$

where R_{T0} is the sensor resistance at temperature T_0, R_0 is the unloaded resistance, T_C is Curie temperature, T_S is temperature sensor and B is the temperature coefficient of the sensor.

Heat transfer from the PTC sensor to the surrounding medium can be explained by mechanism of heat conduction and convection. For conduction, the relationship between a heat transfer, spatial temperature gradient $\partial\vartheta/\partial n$ in a spatial direction n of time frame Δt is described by the Fourier Law of Heat Transport [15].

$$\vec{j}_q = -\lambda \frac{\partial \vartheta}{\partial \vec{n}} = -\lambda \cdot grad\ \vartheta \tag{2}$$

The constant λ is the material-dependent proportionality constant of Fourier's law whereby the density of the energy-transferring molecules in still air remain low so that the thermal conductivity is particularly low. If the sensor is considered as a cylinder with single layers, then the heat transport of the sensor is expressed as equation (3).

$$\dot{Q} = \frac{2\pi \cdot L \cdot \lambda}{\ln\left(\frac{r_2}{r_1}\right)} \cdot (T_1 - T_2) \tag{3}$$

The relationship between the thermal resistance R_W of the sensor in the measuring substance that depends on the value of the thermal conductivity and the heat transfer coefficient α_w, the surface area A of the sensor can be explained in (4) [16].

$$R_W = \frac{1}{\alpha_W \cdot A} \tag{4}$$

Various filling substances with different heat transfer coefficient have specific Rw, which can be used as the sensor status within medium.

2.2. Sensor Characterization

Characterization sequence of the PTC sensor has been integrated using microcontroller based circuit with a preprogrammed digital to analog converter (DAC) served constant current, as shown in figure 2. An EPCOS B59050D1100 PTC sensor immersed in CPO samples was then tested in self-heating mode by providing a programmable voltage ranging from 0 V to 30 V. Effect of CPO temperature variation was measured by heating the palm oil sample in the Heraus Oven T6060 to keep the sample temperature stable.

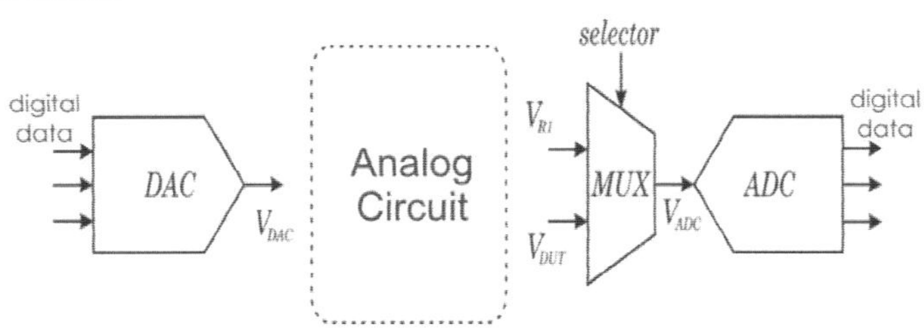

Figure 2. Data acquisition circuit for sensor characterization consist of digital to analog converter, signal processing circuit and analog to digital converter.

When current flows through the PTC-thermistor, the sensor will heat up and transfer the thermal energy to the medium. Heat transfer mechanism depends not only on the load applied, but also on the thermal losses between sensor and medium, represented by R_W factor. The condition of equilibrium state of the electrical power and the delivered thermal output can be written as equation (5) [9].

$$U_T^2 \cdot \left[R_{T0} e^{B\left(\frac{T_S - T_C}{T_S}\right)} + R_0 \right]^{-1} = \left(T - T_M\right) \cdot R_W^{-1} \tag{5}$$

In the experiment, the EPCOS sensor with dimensions of 25mm long and 2.8mm diameter was modified with an encapsulated housing made of brass served as mounting to the tank wall. Figure 3 shows the sensors dimension and the housing.

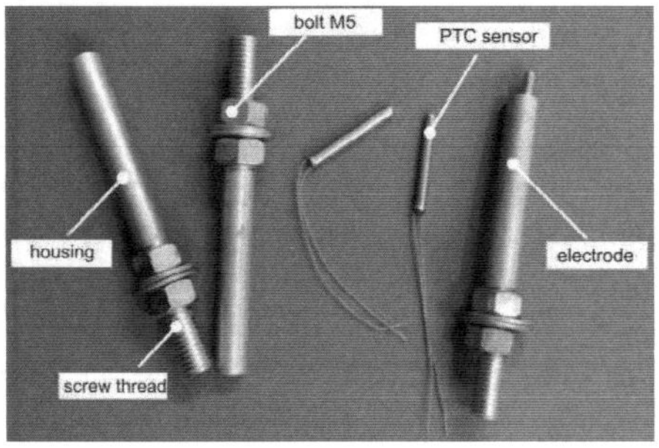

Figure 3. Sensor housing and mounting.

2.3. Modeling of I(U) Curve and Two-Points Method

Sensor parameters R_{T0}, R_0, T_C, B, T_S and thermal resistance R_W in equation (5) are obtained simultaneously by modeling I(U)-curve using Jacobi Matrix method, where the determinant function must not be equal to zero [11].

$$FD(x_i, x_1, \ldots, x_n) = \begin{vmatrix} \frac{\partial g_1}{\partial x_1} & \cdots & \frac{\partial g_1}{\partial x_n} \\ \cdots & \cdots & \cdots \\ \frac{\partial g_n}{\partial x_1} & \cdots & \frac{\partial g_n}{\partial x_n} \end{vmatrix} \neq 0 \tag{6}$$

The estimated values are selected as initial values for the unknown sensor parameters and also thermal resistance, changed iteratively until an optimum of the quality function is completed. The calculation process use least-square-method, where the result is an estimate of the unknown coefficients of the model.

$$S = \sum_{i=1}^{n} r_i^2 = \sum_{i=1}^{n} \left(y_i - \hat{y}_i\right)^2 \tag{7}$$

where n is the number of data points.

Since Rw is obviously used for the sensor status "immersed" or "not immersed" in liquid, it is expected to be constant and not affected by CPO temperature. An algorithm was developed to eliminate the medium temperature using two-points measurements on the I(U)-curve, $(U_1; I_1)$, $(U_2; I_2)$ resulting R_1 and R_2, respectively. Using this data and predetermined sensor parameters, the Rw can be calculated directly derived from equations (5-7).

$$R_W = \left(T_{S1} - T_{S2}\right) \left[\left(\frac{U_1^2}{R_1}\right) + \left(\frac{U_2^2}{R_2}\right) \right]^{-1} \tag{8}$$

where

$$T_{S1} = \frac{T_C}{1 - \frac{ln\left(\frac{R_1 - R_0}{R_{T0}}\right)}{B}} \text{ and } T_{S2} = \frac{T_C}{1 - \frac{ln\left(\frac{R_2 - R_0}{R_{T0}}\right)}{B}} \quad (9)$$

T_{S1} and T_{S2} are refers to the sensor temperatures at the measured points, as shown in figure 4.

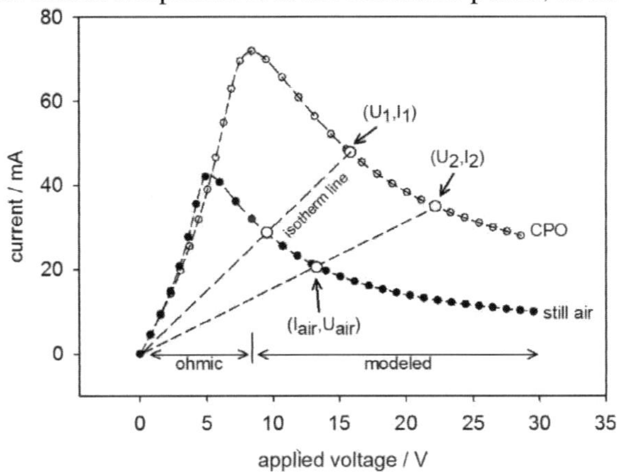

Figure 4. Isotherm curve with 2-points method for direct R_W determination.

3. Results and Discussion

Figure 5(a) shows I(U) characteristic of the EPCOS PTC sensors immersed in CPO with medium temperatures varying from 300 K to 360 K. At the initial applied voltage up to 5 V, the curve step increase linearly in an ohmic region and reach maximum at the Curie point, followed by a decrease exponentially of current consumption. High temperature of CPO will decrease current consumption, since the sensor does not require much power to attain the thermal equilibrium.

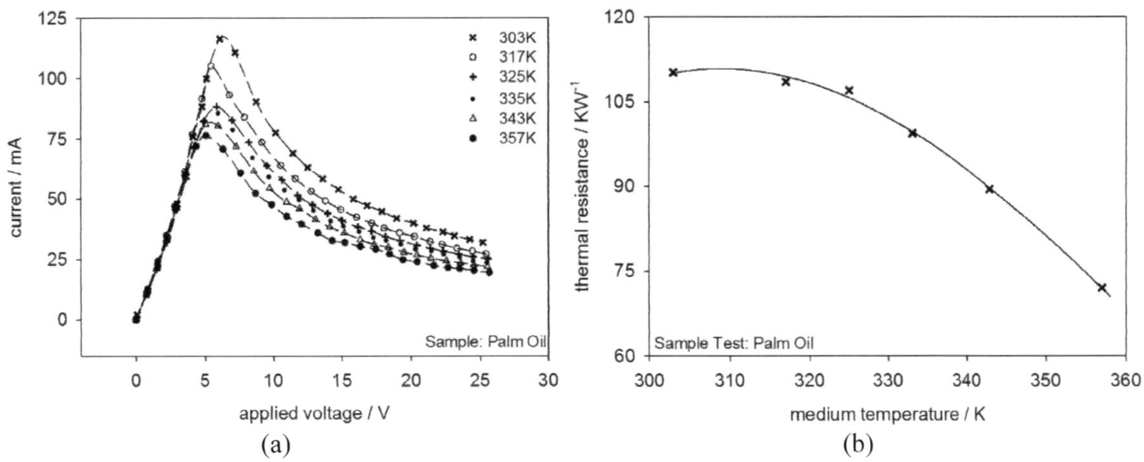

Figure 5. The variation of I(U) curve of the PTC sensor influenced by CPO temperature from 303 K to 357 K in (a) and the R_W value changes about 35% from initial temperature at 300 K (b).

Table 1 shows the results of the I(U) modeling which described the nature of the sensor for a long period of operation. The obtained sensor parameters are then stored as history and used for two-point algorithm to determine the thermal resistance value without having to measure the overall curve I(U) and independent of the medium temperature.

Table 1. Sensor parameters obtained from modeling of the I(U)-curve.

No	Sensor Parameters		
1	Sensor resistance at temperature T_0	R_{T0}	23.0Ω
2	Unloaded resistance	R_0	18.0Ω
3	Temperature coefficient	B	80.0
4	Curie temperature	T_C	380.2K
5	Medium temperature	T_M	300K to 360K

From figure 5(b) and table 1 is shown that the CPO has a good conductance properties. When the oil temperature rise, the heated palm oil will degrade into polar material and the chemical property varies during heating. As a consequence, the heat properties will change in the form of R_W parameter. Heating mechanism will also encourage a reaction of free fatty acids and lead to the formation of a new substance that will change the thermal properties of the CPO [17]. Calculated R_W as shown in figure 5(b) was unstable due to CPO temperatures and results decreasing of R_W from 110.2 K/W to 72.1 K/W or about 34.65%. This result cannot be accepted to determine the sensor status, since the thermal resistance range is out of normal tolerance within 5%.

To overcome this problem, the application of the two-points method is proposed to minimize the changing of R_W value due to medium temperature. Figure 6 shows the results using two-point measurement method.

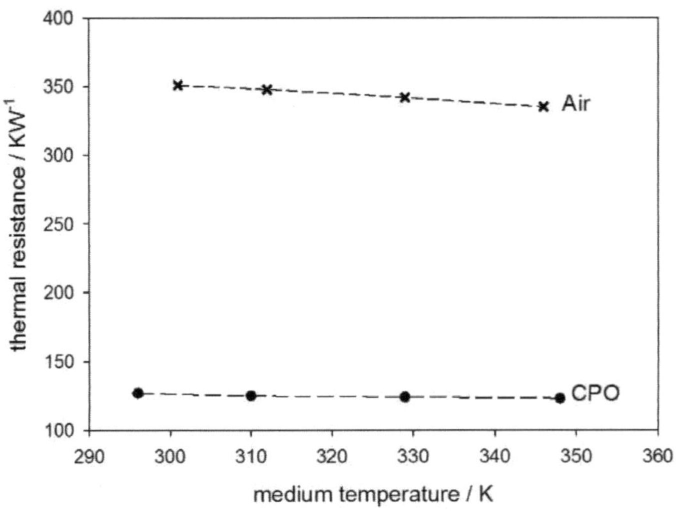

Figure 6. Two points method for thermal resistance measurement results a tolerance below 5% for each filling mediums and 63.5% difference between CPO and still air.

Detection algoritm can be used to reject the temperature influence on the measurement. Results showed only a small deviation of the R_W observed for each medium; for CPO, R_W varies from 127 K/W to 123 K/W with 3.15% and for still air, from 351 K/W to 335 K/W or 4.5% deviation. Since the calculated thermal resistance between the mediums was relative high about 63.5% in the difference, this range is able to avoid the overlapping status "full" or "empty".

From this results, the sensors can decide filling medium within normal tolerance below 5%, and about 63.5% between both CPO and still air. This can be programmed in the microcontroller for the application and to determine the status whether the sensor is in still air or in palm oil.

4. Conclusion

Thermal resistance of PTC sensor used for level sensor application shows the unstable parameter due to high medium temperature. This R_W and sensor parameters are obtained from modeling of I(U) curve simultaneously and shows 34.65% difference in accordance with CPO temperature variation. Using two-points method of I(U) curve measurement, a relatively stable R_W value with low instabilities about 3.15% for CPO and 4.5% for still air was produced, whereas the difference between CPO and air can be achieved up to 63.5%. This stable parameter can be safely used as a determinant of status sensors for the filling process of the CPO containers.

Acknowledgement
The research was funded by the University's Research Funded UR 2017.

References

[1] Santhosh K V and Binoy K R 2015 *Measurement and Control* **48** 217
[2] Reverter F 2007 *Sensor and Actuator A: Physical* **138** 1
[3] Xinpu Z, Wei Peng Z, Liu Z G 2015 *Principle and Optical Interferometry* **6** 6801108
[4] Wenhua W and Fang L 2014 *Optics and Lasers in Engineering* **52** 201
[5] Bobo G, Wenliang Q, Yanyan Z, Zhifang W, Perry P S and Feng L 2014 *Optics Express* **22** 11834
[6] Yun B, yang N and Cui Y 2007 *IEEE Photonics Technology Letters* **19**
[7] Zixiao W, Tan Z, Xing R, Liang L, Qi Y and Jian S 2016 *Optics & Laser Technology* **84** 59
[8] Viswanath S, Belcastro M, Barton J, O'Flynn B, Holmes N and Dixon P 2015 *International Journal on Smart Sensing And Intelligent Systems* **8** 26
[9] Lazuardi U 2010 *Proc. Int. Conf. on Autonomous and Intelligent System (AIS)* Povoa de Varzim Portugal p 21
[10] Wulandari N, Muchtadi T R, Budijanto S and Sugiyono 2011 *J. Teknol. dan Industri Pangan* **22** 177
[11] Horn M and Lazuardi U 2002 *Proc. of IEEE IMTC Conference*, Anchorage Alaska USA p 1
[12] Lazuardi U 2004 *Untersuchungen zur Selbstüberwachung von Füllstandssensoren* (VDI Reihe, Duesseldorf, Germany)
[13] Perovich S M and Calasan M P 2016 *Proc. 5th Mediterranean Conference on Embedded Computing (MEeo 2016)* Montenegro p 247
[14] Schruefer E 2014 *Elektrische Messtechnik.* (Fachbuchverlag Leipzig im Hanser Verlag).
[15] Kolahi K 2001 *Technisches Messen* ***68*** 220
[16] VDI3519 2010 *Füllstandmessung von Flüssigkeiten und Feststoffen (Schüttgütern)* VDI/VDE Verlag
[17] Fasina O and Colley Z 2008 *International Journal of Food Properties* **11** 738

IOP Conf. Series: Journal of Physics: Conf. Series **1080** (2018) 012027 doi:10.1088/1742-6596/1080/1/012027

Generation of nanoporous model using sequential annealing and largest cluster selection method

Y Rosandi* and B Wijatmoko

Department of Geophysics, Faculty of Mathematics and Natural Sciences, Universitas Padjadjaran, Jatinangor, Sumedang, Indonesia 45363

*rosandi@geophys.unpad.ac.id

Abstract. Using molecular dynamics method, we develop a robust algorithm to generate nanoporous material with considerably natural porous structure. The method consists of sequential heating, relaxation, and quenching of Lennard-Jones meta-material. We vary the binding strength, the equilibrium distance, and the cut-off radius of the interaction potential of each sequence, in order to obtain reasonable melt structure, and agglomeration of the particles into nanowires of the nanofoam. The resulted foamy material has good void and wire size distribution, hence can be applied as target material model on computational study of the physical behaviour of porous materials. The method enables fast creation of the porous media with tuneable values of porosity.

1. Introduction

The importance of porous material spreads from technological application to the phenomena and processes which occurs in nature. Using porous materials one can create light weight substances with tuned physical properties strength, weight, as well as the optical phenomenon such as surface plasmon excitation [1, 2]. In chemical application, the porous material is also used very frequently, for instance in catalytic activities, chemical filters and ion traps [3]. These importance originated from the nature that the medium has large surface area. Usually all the application mentioned above is made possible by laboratory-made porous medium. The procedures involve ion bombardment and laser ablation of materials [4].

In nature, such medium is created by mainly thermal process which consists of heating and cooling of material. Porous rocks is formed by sudden cooling of lava that may be formed when the earth has opening region such as volcanoes and ridges. The porosity of rock is one of decisive parameters, for instance in the surveillance of oil or water reservoir. In this geological and geophysical fields, methods are available to identify porosity, however, the detailed characteristics and structures are still not clearly known.

The theoretical understanding of physical and chemical properties of porous medium is still immature. Studies concerning the properties of such medium is one of the hot topics, especially in the nano-meter scale [5, 6]. In this work we propose a method to produce nano-material model, which can be used for further studies of the properties of porous medium.

2. Method

We used molecular dynamics method to create porous material. The Lennard-Jones meta-material of size 64x64x64 lattice constant was used as the initial crystal with FCC structure. The crystal has the density of 0.89 (in reduced unit) which is the equilibrium density, where the material is in the minimum pressure. We applied periodic boundary condition to all the side of the crystal. Initially the system consist of 1048576 atoms, and then half number of the atoms is removed in order to induce porosity, remaining 523935 atoms in the system. The position of the removed atom was chosen randomly. In this work we consider a porosity of 50%.

We let the atoms interacting with a high binding energy $\varepsilon = 20$ and $\sigma = 1$, also used in [5], with cut radius at 2.5 σ. This is to make the particles sticking to each other, and prevent too many gas phase formation. The trajectories of particles was followed up to 5000 steps. In this study, we assume, the actual time of the simulation system is not relevant with the real physical condition. After some defined time step we took the crystal to be processed as follows. Firstly, we used the cluster detector algorithm to select the largest connected group of atoms. The smaller clusters are removed from the target system, in order to get one percolated foamy structure. Secondly, the crystallite is relaxed, and let the surface of the filaments in the structure smoothen by the work of surface tension. The relaxation is done through the quenching process to a very low temperature using NPT ensemble. This procedure is applied to snapshots of simulations taken sequentially every 1000 time step. We performed the sequential annealing to various annealing temperature T* from 0.1 to 1.0, around the melting temperature of the meta-material, to observe the effect of temperature to the produced porous structure. The method to find the number of clusters inside the simulation system is created following the prescription from [7]. This method has been used in previous works, such as [8, 9, 10]. Let i and j define the indices of particles inside a system having randomly distributed clusters. C and A represent the enumerated clusters, where the members of a cluster is determined as the connecting particles with maximum inter-particle distance of r_{cut}^{cl}. Hence, the cluster criteria is defined as,

(i) If $i \in C$ and $r_{i,j} < r_{cut}^{cl}$ then $j \in C$

(ii) If A is any satisfying criteria (1), and if i is in both A and C, then the intersection $A \cap C = C$

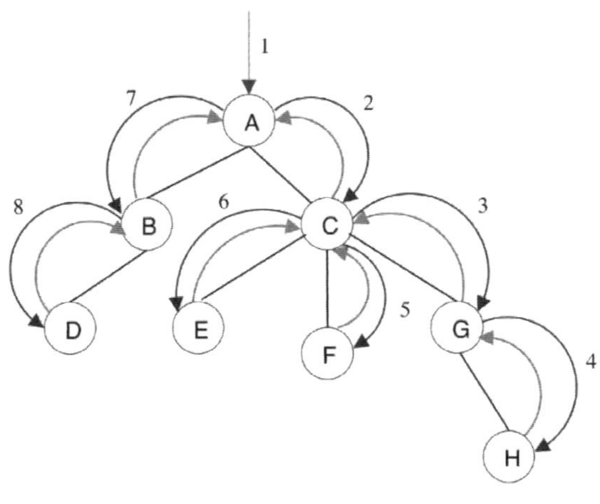

Figure 1. The diagram of the binary tree search algorithm to define the cluster of particles in the simulation system. Blue arrow denotes the visit, and the red arrows denotes the return of a visit. The numbers on the blue arrows determine the number of visit, which is the number of particles in one cluster.

The cluster detector was written using the binary tree algorithm as can be seen in figure 1. In this algorithm we take a random particle in the system, and calculate the distances between the particle with the neighbouring particles. When the distance is less than or equal the predefined radius, r_{cut}^{cl}, the pointer of atoms is moved to the neighbour particle. If the visit sequence ends at the first particle and no other neighbouring particles which has $r < r_{cut}^{cl}$, all the members of one cluster is counted. The list of particles is then removed from the searching list, and the procedure is restarted from a new randomly picked particle. To make the calculation of distances faster, we utilizes the neighbour-list, which is standard algorithm in molecular dynamics simulation [11].

The resulted porous model can be used as a platform to create crystalline porous target of any material, by mapping the position to the new bulk of crystal. We used LAMMPS code, a scalable molecular dynamics program to run the simulation [12] and *Ovito* software [13] to do the processing, such as the filling factor and the surface area of the porous media.

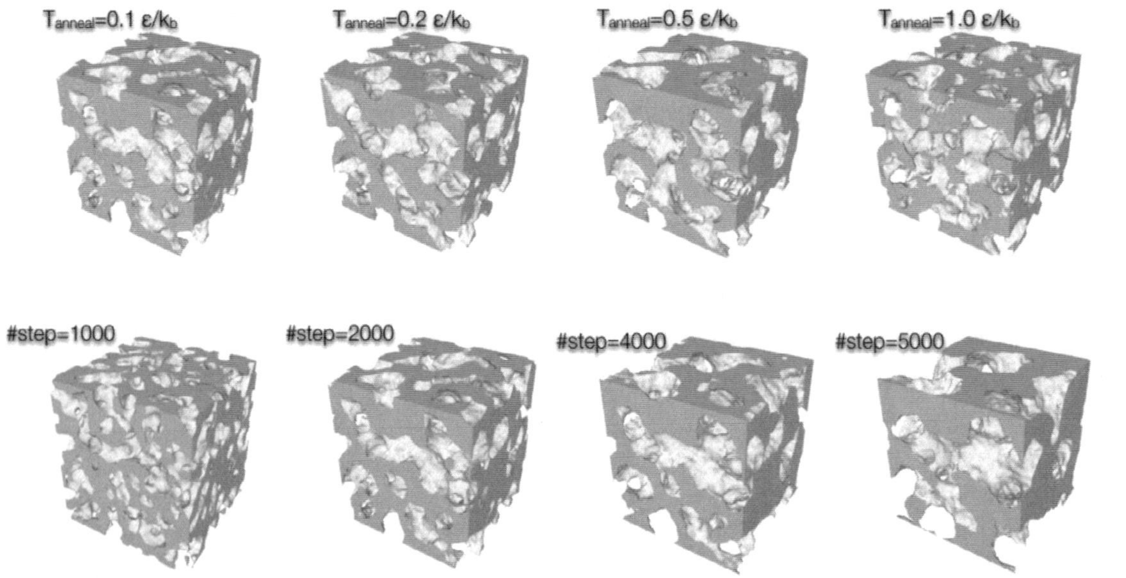

Figure 2. Animation of snapshots of the generated porous medium model. Upper row: varied annealing temperature (T*) at #step=1000, lower row: varied annealing time (#step), at T* = 0.1.

3. Result

Figure 2 shows the snapshots of the resulted porous medium model, with varied annealing temperature and sequential time step. On the upper row, we observe the effect of the annealing temperature to the model. Here we find that the porous structure of the model looks similar. Contrary, on the lower row, we see different structure with the same value of porosity. The animation of these snapshots was taken from the annealing simulation with T* = 0.1. The snapshots show that the porous structure depends only weakly to the annealing temperature, but having strong dependence on the annealing time.

Figure 3 shows the surface area and the filling factor of the model. The surface area is defined as the area of the solid surface in the porous model (the yellow surface), and the filling factor is the ratio between the agglomerates (wires) enveloped by the yellow surface, and the void. It is shown in figure 3(a) that the decrease of the surface area follows the Arrhenius law, which decays with the step. The overlap of the plots indicates that the annealing temperature does not affect very much the surface area. Nevertheless, we observe a small rainbow at the higher temperature, which might be due to the thermal behaviour of the meta-material. Figure 3(b) shows the similar trend of the increase of filling factor with the step. This facts demonstrates the small effect of the annealing temperature in the process. The deviation of the filling factor seems to be caused by the thermal expansion.

We now evaluate the effect of the annealing temperature to the filling factor at the longer simulation time. We observe the value at #step=5000, as shown in figure 4. We find a systematic increase of the filling factor, even though, the difference is small. Between T* = 0.1 and T* = 1.25 we have 0.008 filing factor difference, which is insignificant. We assume that this difference is merely due to the larger volume of material at a higher temperature.

IOP Conf. Series: Journal of Physics: Conf. Series **1080** (2018) 012027 doi:10.1088/1742-6596/1080/1/012027

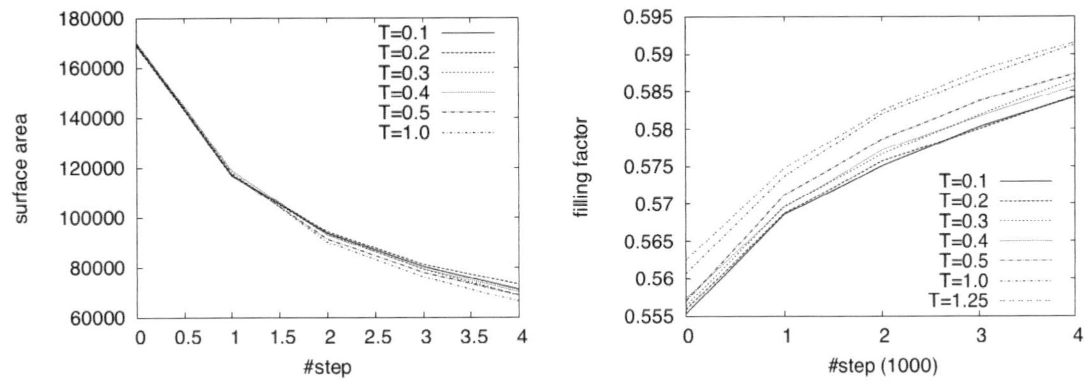

(a) (b)

Figure 3. The surface area of the porous medium model as a function of (a) annealing time, (b) the filling factor, plotted in several annealing temperature T* as stated in the legend. The surface area is measured in reduced unit σ^2.

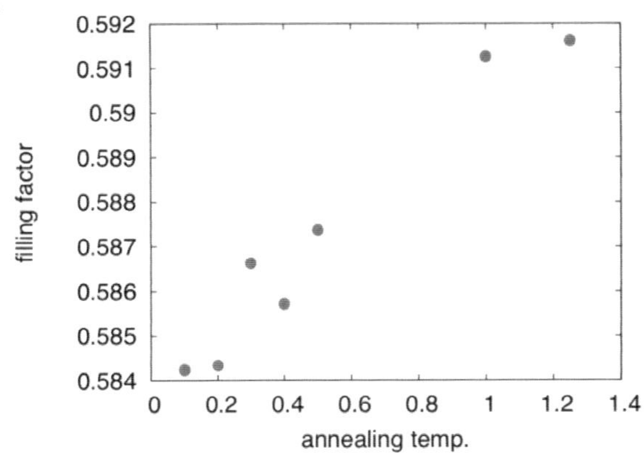

Figure 4. The filling factor plotted as a function of annealing temperature, at the same time, #step=5000. The annealing temperature is measured in reduced unit.

In order to get more insight into the dependence of porous model to the annealing temperature and time, we measured the distribution of the solid length. This measurement gives a rough view to the size of wires in the porous structure. We simply create three dimensional grids overlaying the porous model, and count the number of cells which is occupied by particles. The counting stops whenever an empty cell is encountered, and started when a filled cell is found. Assuming the isotropic structure of the model, we only calculate in one axial direction (z-direction). The counting result is shown in figure 5.

The peaks in figure 5 indicate the most probable wire size in the model. For #step=2000 we found the sizes distribute at around length $l = 10\ \sigma$, which indicate the population of small voids and wires. This can be confirmed by the snapshot on the lower row of figure 2. The distributions of different annealing temperature T* seems overlapping. Figure 5(b) shows the situation at #step=5000, where the porous model has coarsened. Even though the lengths experience fluctuation, the peaks are clearly visible. We can see that the wires are dominated by lengths about $l = 12\ \sigma$, $l = 30\ \sigma$, and $l = 40\ \sigma$. This finding shows the coarsening process of the porous model, from the small and homogeneous voids, to the larger in-homogeneous porosity, with wire diameter around 30 σ and 40 σ. This indicates clearly that the thermal process induces coarsening of the porous medium models.

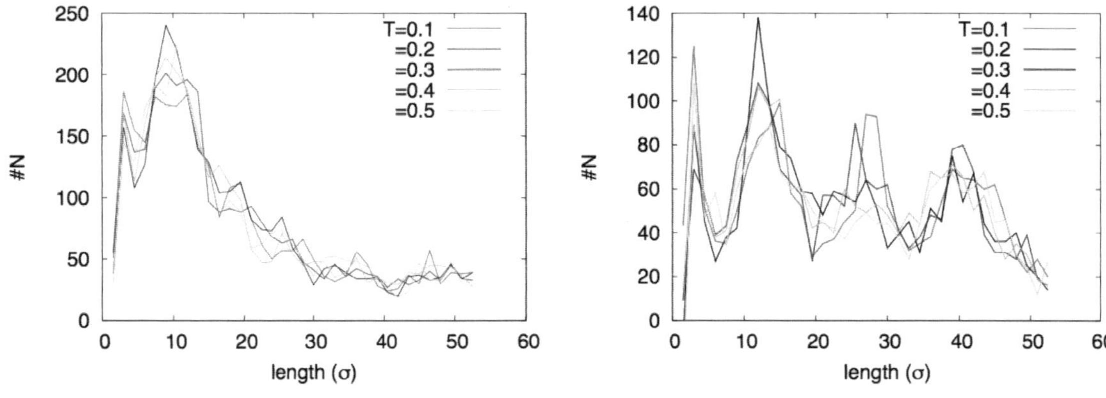

Figure 5. Size distribution take at (a) #step=2000 and (b) #step=5000, for varied annealing temperature T*.

4. Conclusion

We found that the porous structure of the model created by annealing process depends strongly on the annealing time, instead of the annealing temperature. The annealing process induces coarsening of porous media, which leads to larger void and filament (wire) size. This shed light on the coarsening mechanism of the porous media, induced by annealing process, which can be potentially applicable on the fabrication porous material with wide range of void size distribution.

Acknowledgments

YR is grateful for funding from the Indonesian Directorate General of Higher Education (RISTEKDIKTI), contract No. 718/UN6.3.1/PL/2017. BW is grateful for funding from the Internal Research Grant, Universitas Padjadjaran, contract No. 855/UN6.3.1/PL/2017. The simulation was performed using the Guriang-Cluster facility, Universitas Padjadjaran (http://guriang.unpad.ac.id).

References

[1] Halas N J, Lal S, Chang W s, Link S and Nordlander P 2011 *Chem. Rev.* **111** 3913
[2] Maaroof A I, Gentle A, Smith G B and Cortie M B 2007 *Journal of Physics D: Applied Physics* **40** 5675
[3] Jani A M M, Losic D and Voelcker N H 2013 *Progress in Materials Science* **58** 636
[4] Castelo A, Afonso C N, Pesce E and Piscopiello E 2012 *Nanotechnology* **23** 105603
[5] Gunkelmann N, Rosandi Y, Ruestes C J, Bringa E M and Urbassek H M 2016 *Comp. Mat. Sci.* **119** 27
[6] Rovere M, Ricci M a, Vellati D and Bruni F 1998 9859 11 (Preprint 9803248) URL http://arxiv.org/abs/cond-mat/9803248
[7] Stoddard S D 1978 *Journal of Computational Physics* **27** 291
[8] Nietiadi M L, Rosandi Y, Lorincik J and Urbassek H M 2012 *Nuclear Instruments and Methods in Physics Research Section B: Beam Interactions with Materials and Atoms* **303** 205
[9] Nietiadi M L, Rosandi Y, Kopnarski M and Urbassek H M 2012 *Nuclear Instruments and Methods in Physics Research, Section B: Beam Interactions with Materials and Atoms* **289** 97
[10] Rosandi Y and Urbassek H M 2010 *Applied Physics A: Materials Science and Processing* **101** 509
[11] Grest G S, Dunweg B and Kremer K 1989 *Computer Physics Communications* **55** 269
[12] Plimpton S 1995 *J. Comp. Phys.* **117** 1
[13] Stukowski A 2009 *Modelling and Simulation in Materials Science and Engineering* **18** 015012

IOP Conf. Series: Journal of Physics: Conf. Series **1080** (2018) 012028 doi:10.1088/1742-6596/1080/1/012028

Preparation of graphene oxide thin films using voltage cyclic electrochemical deposition method and its characterization

V Marcelina[1], H Addini[1], N Syakir [1], A Bahtiar[1], S Wyantuti[2], Y W Hartati[2] and Fitrilawati[1*]

[1]Department of Physics, Faculty of Mathematics and Natural Sciences, Universitas Padjadjaran, Jl. Raya Jatinangor km 21 Sumedang, West Java, Indonesia
[2]Department of Chemistry, Faculty of Mathematics and Natural Sciences, Universitas Padjadjaran, Jl. Raya Jatinangor km 21 Sumedang, West Java, Indonesia

*fitrilawati@phys.unpad.ac.id

Abstract. Graphene oxide (GO) is frequently used as a precursor to obtain 'graphene like' materials. We study preparation of GO thin film using voltage cyclic electrochemical deposition. GO thin films were obtained from 0.5 mg/ml GO-water dispersion on ITO glass substrate in the range of -1.6 to 0 V at scan rate of 75 mV/s. The electrochemical deposited GO thin films were characterized using UV-Vis spectroscopy, the GO morphology and oxygen containing groups were obtained by using SEM, EDX and FTIR, respectively. Cyclic voltammetry (CV) was employed to measure its capacitive characteristic. We obtain that thickness of GO thin films can be controlled by number of voltage cycles applied during deposition. From cyclic voltammetry measurements, we obtained hysteresis curve that imply its potential application as supercapacitor electrode.

1. Introduction

Graphene is one atom-thick layered crystalline with a honeycomb structure of sp^2-bonded carbon. Because of its excellent properties, graphene has been widely studied for many applications [1]. Graphene oxide (GO) has received a lot of attention due to its role as a precursor for cost-effective and mass production of graphene-based materials. Hydrophilic property of GO makes this material easily dispersed in water and deposited on substrates [2]. The most attractive property of GO is that the material can be (partly) reduced to graphene-like sheets by removing its oxygen-containing groups and recovery its conjugated structure [3].

Various coating methods have been used for GO depositions, such as spin coating, solution casting, dip coating and electro-deposition. Previously we prepared thin films of GO from GO dispersion in water using spin coating [4, 5] and drop casting [6] methods. It was reported one-step electrodeposition of GO in glassy carbon which the film deposited and reduced electrochemically [7, 8]. In this paper, we report preparation of GO film from GO that dispersed in water using voltage cyclic electrochemical deposition on indium tin oxide (ITO) substrate and its characterization.

2. Experimental

2.1 Preparation of Samples

For GO film preparation, we employed a commercial GO suspension (Graphenea SA ES A75022608) with concentration of 0.5 mg/ml that dispersed in water. Prior to use, GO dispersion was sonicated for 1 h at room temperature. Indium Tin Oxide (ITO) glass with size of 2.5 cm x 1.25 cm that used as conductive substrate was cleaned with teepol and then rinsed respectively with aquadest, ethanol and isopropanol using ultrasonication for 15 minute at 50°C and dried before used.

The GO samples were electrochemically deposited in three-electrodes electrochemical cell contained GO/milliQ water dispersion with a clean ITO glass substrate as a working electrode, a platinum plate as a counter electrode, an Ag/AgCl as a reference electrode, and was carried out by different number of voltage cycles deposition. The GO films were deposited on the ITO glass substrates in the voltage range of -1.6 to 0 V at scan rate of 75 mV/s. The voltage range was setting in respect to the reference electrode.

2.2 Sample Characterizations

Electrochemically prepared GO films were characterized using UV-Vis spectroscopy (T70+ UV-Vis Spectrometer) in the wavelength range of 200 nm to 700 nm and FTIR in order to inspect deposited GO in the substrates. Scanning electron microscope (SEM) and EDX of GO samples were measured using HITACHI SUE3500. Cyclic voltammetry (CV) measurements of samples were carried out using a MetrOhm Autolab potentiostat in 1 M KCl.

3. Results and Discussions

3.1. Electrochemical Deposited GO Films

Visually we can observe a color change of ITO substrate after one cycle of voltage scanning with a scan rate of 75 mV/s in range of -1.6 V–0 V. It indicates a deposition of GO on ITO substrate has been taken place. In order to follow a deposition process, we run an in-situ current-voltage (CV) measurement in the GO-milliQ water dispersion. Figure 1 shows the current change during voltage scanning of GO films deposition in ITO glass substrate, from one cycle until six cycles. We found that a shape of CVs changed with increasing number of cycle. The change of peak currents with increasing number of voltage scans confirm that the deposition of GO on ITO substrate has taken place during the cycle.

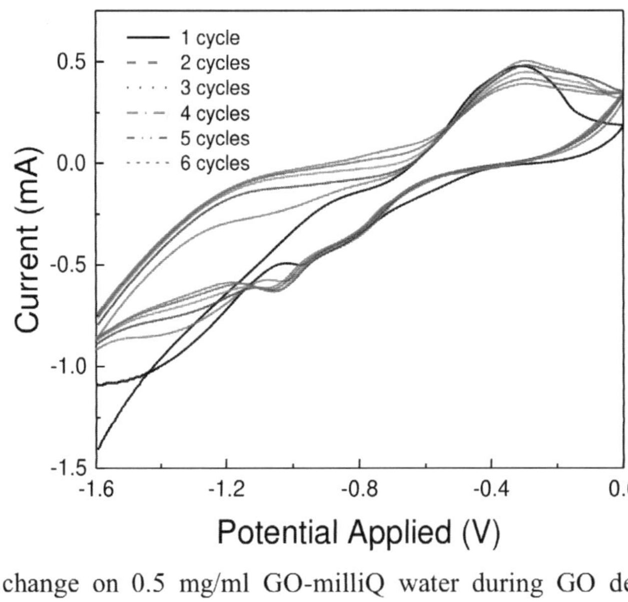

Figure 1. Current-Voltage change on 0.5 mg/ml GO-milliQ water during GO deposited on ITO substrate at varied number of cycles.

Quantitatively an increasing of GO films thickness with increasing number of deposited cycles were characterized by using UV-Vis spectroscopy at range wave length 200 – 700 nm. We observed an increasing of absorbance of GO films with increasing number of voltage cycles applied during the deposition. Figure 2 shows the change of absorbance of samples that deposited at certain number of voltage cycles evaluated at several wavelengths. Figure 2 also shows that the absorbance change at higher wavelength is more pronounced than at shorter wavelength. It is obvious that the absorbance increased with increasing number of deposition cycles. The change of absorbance confirmed that number of voltage cycles applied during a deposition process influenced the film thickness. We found that until six cycles, the GO films deposited using higher number of deposition cycle will produce thicker GO films.

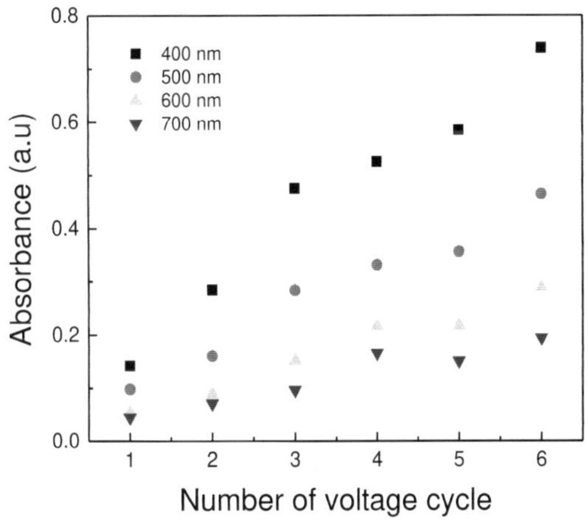

Figure 2. Absorbance of GO films that deposited at varied number of voltage cycles.

From EDX analysis GO samples on ITO substrate have 49.34 C atomic percentage and 50.66 O atomic percentages. The oxygen contain in the EDX result are contributed by GO and ITO substrate. The morphology of GO as shown by SEM images in figure 3 has wrinkled paper-like sheets. These geometric wrinkling is caused by nanoscale interlocking of GO sheets, providing enhanced mechanical properties, reduced surface energy, increased surface roughness and area [1, 2].

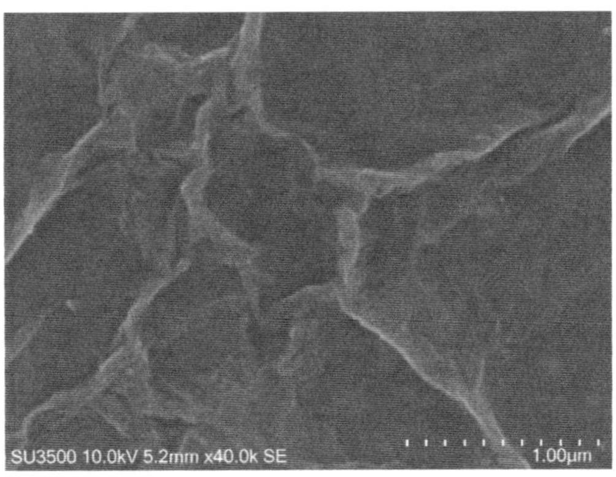

Figure 3. SEM Image of electrochemical GO film on ITO glass substrate.

The FTIR spectrum of the GO film that prepared using six voltage-cycles deposition is shown in figure 4. The GO spectrum has peaks at 3400 cm^{-1} that corresponding to O-H stretching, at 1630 cm^{-1} related to C-C vibration, 1575 cm^{-1} related to C=C, 1382 cm^{-1} related to C-O carboxyl, and 1090 cm^{-1} related to C-O stretching. This spectrum is slightly different compare to the spectrum of GO film that prepared using drop casting [6] and spin coating [9]. The IR spectrum of the electrochemical deposited

GO film has no peak at 1735 cm^{-1} that related to carbonyl and carboxyl C=O group. Also the spectra of electrochemical sample have a weak peak at 1575 cm^{-1} that related to C=C. It is slightly similar to IR spectra of rGO film prepared using casting technique that has peak at 1575 cm^{-1} but no peak at 1735 cm^{-1} [6, 9]. This result shows that in electrochemical deposition technique, beside deposition of GO, the reduction process also slightly taken place.

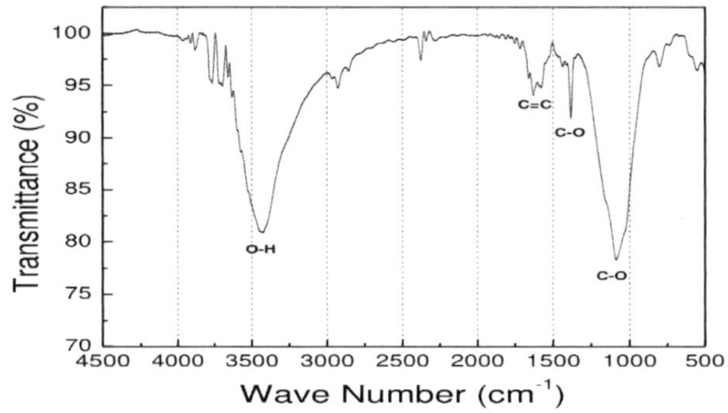

Figure 4. FTIR Spectra of electrochemical deposited GO film.

3.2. Electrochemically Prepared GO Films

The electrochemical measurement of GO film were carried out using cyclic voltammetry measurement with three electrode configuration, that RGO film on ITO as working electrode, a platinum as counter and Ag/AgCl as a reference electrode and 1 M KCl as electrolyte medium. Measurement were performed in voltage range of 0 V to 0.9 V at varied scan rate from 25 mV/s to 125 mV/s.

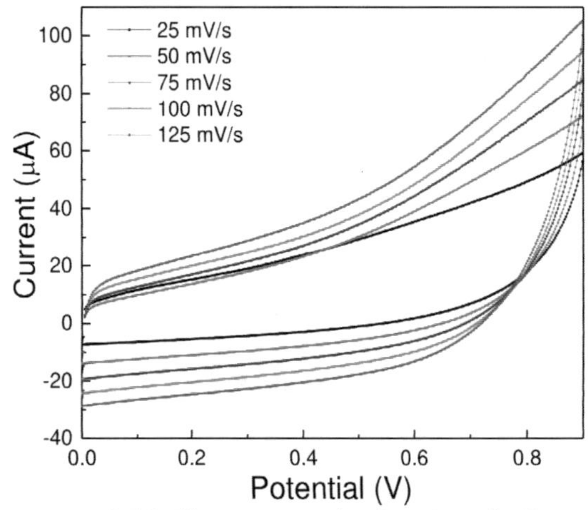

Figure 5. Cyclic voltammograms of GO films prepared using 6 cycle Deposition at varied scan rate in 1 M KCl.

As shown in figure 5, cyclic voltammograms of GO film that prepared using six voltage-cycles has a rectangular-like shape. It is indicated an electrochemical double-layer capacitance characteristics of the sample [10]. It is also obvious that the voltammograms in figure 5 do not have any peaks indicated the supercapacitive behavior of the electrochemical GO film. It is clearly that electrochemical GO film free from redox reactions or purely based on the electrostatic mechanism [11]. CV hysteresis area of these samples increases at the higher scan rates. CV hysteresis area are calculated according to equation (1) [12], where P is CV hysteresis area or power (μWatt), i is current respond and dV is step potential (which is 0.00244 V). The specific capacitance of RGO electrode obtained by CV measurements is calculated according to equation (2) [12], where, C is specific capacitance of RGO

(F/g), m is mass of electroactive materials (g), ΔV is potential window (V), s is scan rate (mV/s) and i is current response in a given potential V.

$$P = \left(\int_{-0.2}^{0.8} i\,dV + \int_{0.8}^{-0.2} i\,dV\right) \quad (1)$$

$$C = \frac{1}{2m \cdot \Delta V \cdot s}\left(\int_{-0.2}^{0.8} i\,dV + \int_{0.8}^{-0.2} i\,dV\right) \quad (2)$$

We can see that the CV hysteresis area is increasing as the scan rate increased (figure 6(a)). It relates to current response of GO electrode, which at the higher scan rate the GO electrode can give the higher current response [13]. The calculation result of specific capacitance of GO electrodes at varied scan rates is shown in figure 6(b). The specific capacitance of GO is decreased at higher scan rate. It is well known that at very low scan rates, capacitance are higher because the ions have much longer time to penetrate and reside in the electrode pores and form electric double layers, which are needed to generate higher capacitance [11].

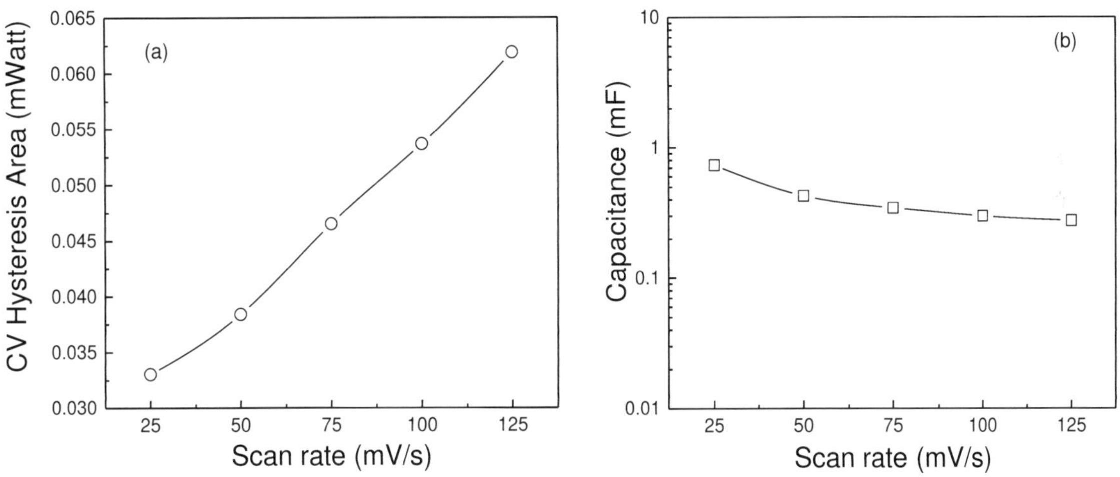

Figure 6. Hysteresis area (a) and capacitance of electrochemical deposited GO film at varied scan rate.

4. Conclusion

Graphene oxide (GO) can be deposited on ITO glass substrate using voltage-cyclic electrochemical deposition using voltage in range of -1.6 V to 0 V using scan rate of 75 mV/s. Thickness of GO film can be controlled by number of voltage cycle applied during deposition. The prepared GO film has lower oxygen contain compared to film prepared using casting. The voltammogram shows that the prepared GO has reduced electrochemical double-layer capacitance characteristic.

Acknowledgements

This work was funded by Penelitian Unggulan Perguruan Tinggi (PUPT) contract no. 718/UN6.3.1/PL/2017 April 17th 2017 and partially support by Academic Leadership Grant (ALG) Universitas Padjadjaran with contract no : 855/UN6.3.1/PL/2017.

References

[1] Joon Y *et al.* 2014 *Electrochimica Acta* **116** 118
[2] Park S and Ruoff R S 2009 *Nature Nanotechnology* **4** 217
[3] Pei S and Cheng H 2011 *Carbon* **11** 1
[4] Fitrilawati, Syakir N, Aprilia A, Liu Z, Feng X and Bubeck C 2015 *Material Science Forum* **827** 317
[5] Marcelina V, Syakir N, Wyantuti S, Hartati Y, Hidayat R and Fitrilawati 2017 *IOP Conference Series Material Science and Engineering* **196** 012034

[6] Fitrilawati *et al.* 2017 *IOP Conf. Series: Materials Science and Engineering* **196** 012027
[7] Chen L, Tang Y, Wang K, Liu C and Luo S 2011 *Electrochemistry Communication* **13** 133
[8] Hilder M, Winther-jensen B, Li D and Macfarlane D R 2011 *Physical Chemistry Chemical Physics* **13** 9187
[9] Yuliasari F, Aprilia A, Syakir N, Safriani L, Saragi T, Risdiana, Hidayat S, Bahtiar A, Siregar R and Fitrilawati 2017 *IOP Conf. Series: Materials Science and Engineering* **196** 012049
[10] Yang J and Gunasekaran S 2012 *Carbon* **51** 36
[11] Chen Z, Ren W, Gao L, Liu B, Pei S and Cheng H M 2011 *Nature Materials* **10** 424
[12] Ramadoss A, Saravanakumar B and S. J. Kim 2015 *Nano Energy* **15** 587
[13] Farma R, Deraman M, Talib I A, Omar R, Manjunatha J G and Ishak M M 2013 *International Journal of Electrochemical Science* **8** 257

Development of new method of drug abuse detection based brain computer interface

A Turnip[1*], D E Kusumandari[1], T Hidayat[2] and T Hidayat[1]

[1]Technical Implementation Unit for Instrumentation Development, Indonesian Institute of Sciences, Bandung, Indonesia
[2]Faculty of Medicine, Universitas Padjadjaran, Bandung, Indonesia

[*]jujhin@gmail.com

Abstract. A methods of drugs abuse detection is urgently needed. Currently a conventional method such as urine test, hair test, Functional Magnetic Resonance Imaging, Positron Emission Tomography, and others are used. Those methods, instead of expensive and complicated, it is require a medical experts. Furthermore, this tool cannot be used to detect new variations of drug. In this study, a new method of drugs abuse detection based brain computer interface (BCI) which recorded with an electroencephalogram is developed. In principle, as long as the subject is still be able to remember, then the tool will be able to detect whether ever use a drug or not. In the experiment 15 subjects are tested. In the signal processing, the EEG signal is filtered with baseline correction and band-pass filter and extracted using wavelet symlet method. The brain maps of damaged brains are in sharp contrast to those of brains which have not been subjected to drug abuse. This studies could be used to identify drug abusers and alternative tool for rehabilitation.

1. Introduction

Electroencephalography (EEG) equipment is now becoming more accessible in the general market, allowing a variety of research related to brain activity easier and wider. The brain computer interface (BCI) is a user method to communicate their brain signal activity with the external device without any help of muscle contraction [1]. The brain signal activity used in BCI can be measured using invasive or noninvasive techniques. The EEG equipment is one of the noninvasive techniques.When many neurons which produce a small electrical voltage become active in the brain, their combination can be detected on the scalp surface with the help of electrode sensors or EEG. These very small electrical signals are amplified and recorded for the latter known as brain waves (represent brain activity that occurs in different areas of the brain). The BCI-based technology developers strive to make BCI more user-friendly, real-time with high accuracy, suitable for voluntary general users such as clinical patients and those with disabilities. Also it must be noted that the current use of BCI is not only used by patients but also by healthy users with various applications [2-12]. Thus, the researcher will do the best to provide an improvement until the tool can be easily found in the market and is expected to operate well with just one channel. In the end anyone who developed the BCI should be really useful for those who need it. In this paper, a medical instrument based BCI for early detection of drug abuse is developed.

Consumption of drugs causes losses in various things including economic and social for human. Furthermore, it is harmful to health both physically and physically. Drug effects on life in particular on brain activity have already been reported [13-19]. Weakness and health abnormalities due to drug abuse will persist for a long time and tend to increase the consumption dose. In addition, it can cause accidents while driving a vehicle and even cause bad behavior in the community. Interesting evidence is found that drug abuse can cause disfunction of cortical orbitals, altering the frontal and temporal areas, and tend to make users more addicted. Generally, it is associated with a cognitive deficits which directly effecting the user in the emotional and motivation regulations, concentration, working and learning ability, and a respons to make decision. Therefore, needed a tool that is able to detect early drug users so the family can do early treatment before it gets worse. So far, very rare research on the development of early detection of drugs is done.

In this paper, an application of signal processing before the brain mapping is proposed. The Brain mapping highly important for the neurophysiological evaluation such as caused by drug abuser. It is used to discriminate the drug effects on the brain and also to evaluate the relationship between the brain and behavior. All involved subjects in the experiment were provided with a complete brain map analysis. The analysis results indicated that there are an enhancement of brain disturbances in the drug abusers compare with the control subjects. The dependents can be measured and detected using the proposed developed EEG tools. This kind of disturbances are not obtain in as great a number of normal subjects with similar psychiatric problems. It is found that the drug abuser had the greatest number of total brain abnormalities. Moreover, the higher abnormalities were detected in the more severe drug abuser than the normal subjects.

2. Experiment Method

An EEG is a test that measures and records the electrical activity of the brain subject. An electrode sensors are attached to subject head and connected by wires to a computer. While the computer records, the brain's electrical activity is diplayed on the screen. In the experiment, 10 male rehabilitaion subjects were hired from Hasan Sadikin General Hospital (HSGH) of Bandung-Indonesia[20-24]. For control 5 male subject were hired from Indonesian Institute os sciences. Before the experiment, all subject was interviewed by the medical doctor according to their personal data and time period information of the drug abuse. Then the subject completed the questionnaires. Also, urine tests were performed on all subjects. During the experimental period it has been determined that no subjects are taking prescription. The experiment was supported by the Ethical clearance from RSHS Committee to make sure that the conducted experiment was based on the clinical ethic. Moreover, all subject was provided by written informed consent and rewarded with IDR 150.000 for transport compensation.

Participants (i.e., has been informed about the entire experiment procedure) were requested to contribute in an experiment of the EEG signals record of the brain activities. The experiemnt consist of three session which are before, ten minutes and one hours after methadone consumption, respectively. The EEG technologist will attach several flat metal discs (electrodes) to different places on the head, using a sticky paste to keep contact between the electrodes and the scalp. During the experiment, the electrodes impedance was retained under 5 KΩ such that a high-quality of the EEG data is obtained. After prerparation, each subject were sitting relaxed on a chair while concentrate to the flashed stimuli on the monitor. They were asked to count how many time the drug picture was appear. The sequence appearance of the flashed picture (shown in Fig. 1) were randomly displayed at the separate but sequential experiment. In this paper, we are only considering the channels of Fp1, Fp2, F7, F3, Fz, F4, F8, P3, and P4. The recorded EEG signals were amplified and preprocessed with embended softwares on WinEEG system with sampling rate about 500 Hz. The experiment process is shown in Fig. 2 (from the topto the right: interview by the docter, explanation by experimenter, the EEG setup, and experiment start up). The stimuli was run on a different computer with the EEG machine was attached to record and save the data.

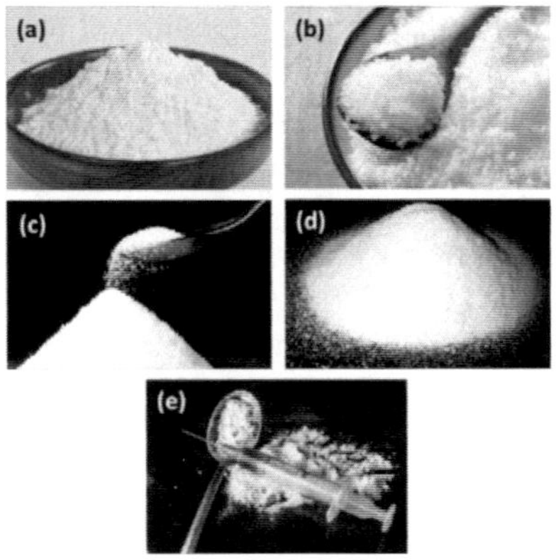

Figure 1. Flashed similar stimuli with (e) as a target [11]

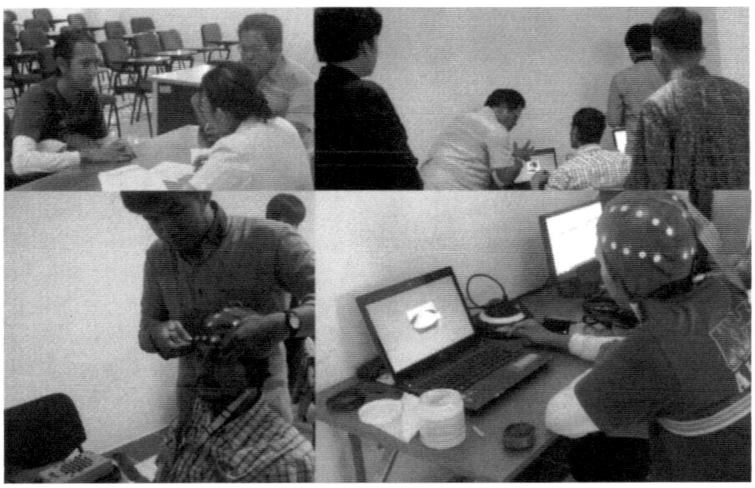

Figure 2. Experiment setup

3. Results and Discussions

The raw data, filtered, and extracted of the EEG signals of the rehabilitation subjects from 10 channels (Fp1, Fp2, F7, F3, Fz, F4, F8, P3, Pz, P4) is described in figure 3. Before feature extraction and brain maps is started, a 6th-order band pass filter (BPF) with cutoff frequencies of 3 Hz and 30 Hz was applied. The filtered EEG signals then decomposed by the decomposition of the signal with noise on wavelet bases where the wavelet coefficient hold an information. Denoising (separating the wavelet coefficients with the download threshold) is obtained by applying the thresholding wavelet coefficients technique. The wavelet group is used to pass the signal through wavelet transform blocks for the decomposition process into wavelet coefficients (i.e., in the thresholding and denoising). The thresholding technique is applied to reduce the noise and keep the information in the maximum signal. After rearrange each trial, the peak amplitude and the latency of P300 component for each considered channels is obtained. The average maximum amplitudes and latencies before and after one hours of the methadone consumption is provided in figure 4. Figure 4 shows the extraction results of the P300 signal: (a) at the craving condition (shortly before taking methadone) and (b) after one hour of taking methadone, respectively.

The maximum amplitude of about 4 micro volts when craving condition tends to be higher when compared with the amplitude after one hour of drinking methadone which is about 1.25 micro volts. The high amplitude of the target is also followed by high amplitude of the nontarget, which means that the brain activity is much higher overall and tend to not focus. The differences can be seen in the post-

IOP Conf. Series: Journal of Physics: Conf. Series **1080** (2018) 012029 doi:10.1088/1742-6596/1080/1/012029

drinking methadone (figure 4(b)) where the target amplitude is significantly higher than nontarget. The high amplitude of the target are triggered by the flashed image of the drug, while the high nontarget is suspected to be triggered by the craving condition. Latency also looks significantly different before and after taking methadone ie each about 390 ms and 240 ms. This difference can be interpreted as a change in the condition of the subject body including brain as a whole that is subject conditions become more comfortable after taking the drug. Moreover, subjects tend to be more focused and especially more responsive to stimuli inputs. The difference between these two features seems to be sufficient to describe the effect of drugs on brain activity.

Figure 5 shows brain activity maps which obtained from the P300 signal from two of the 10 methadone subjects when before taking the drug or methadone (craving condition) and after one hour of taking the drug. Red color indicates more dominant the brain activity than other areas. The first that can be seen from figure 5 is that in the craving condition the brain activity tends to spread rather than after taking the drug (more active in the frontal lobe area). The spread of brain activity occurs in subject 1 which is the second line of the condition after taking the drug. These area is an occipital lobe and the increase in activity is thought to be due to visual effects of the stimuli. For both subjects, changes in brain activity in the frontal lobe appear to change from the outer layer to the deeper layers. this pattern of change is most clearly seen in subject 1 i.e., the third line craving condition and the second line after taking the drug.

Similar to the P300 signal occurs in the brain map of the beta wave before and after taking the drug, the concentration of brain activity occurs in the second row for each subject as shown in figure 6. The brain activity at 10 minutes after taking the drug is not significantly change compared with craving conditions. It can be concluded that the drug is very decisive to the dynamics changes of the brain activity. The reasons why the results may not be helpful caused by movement of the subject during experiment, consuming some medicines, coffee, soda, tea, or other foods that contain caffeine before the test which effected the recorded brain activity. The limitation of the used algorithm for processing also support the final results of the extaction. The EEG records changes in the brain waves that may not be in just one area of the brain. A case affecting the entire brain such as drug intoxication or metabolic disorders that change the chemical balance in the body and the brain may cause these kinds of changes.

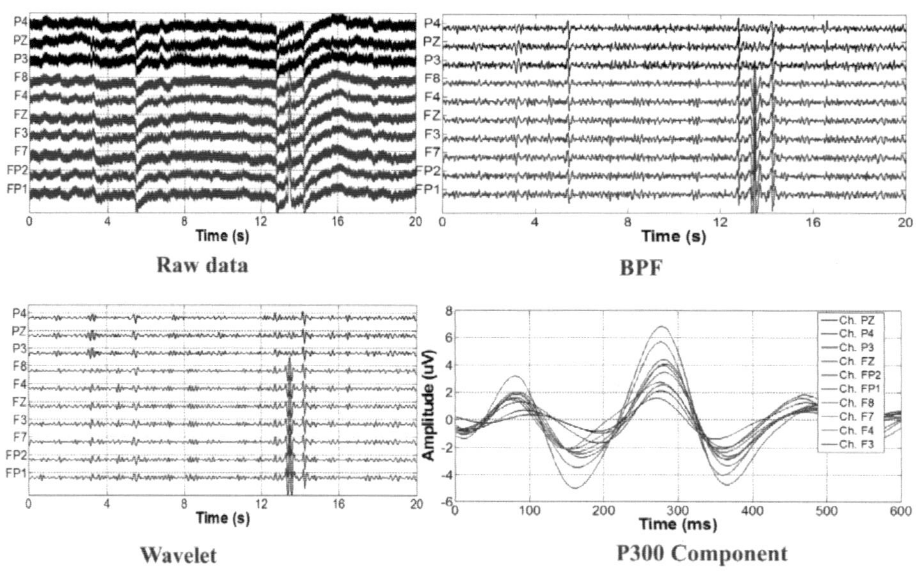

Figure 3. Raw data, filtered, and extracted of P300 Component.

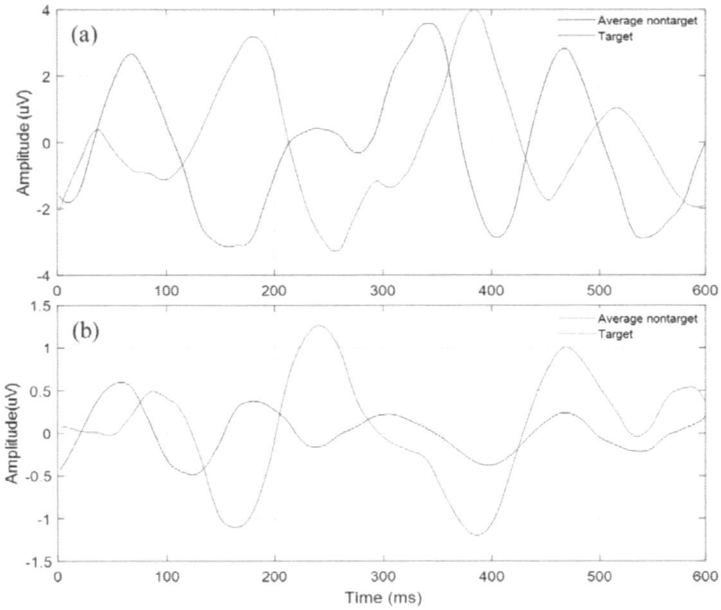

Figure 4. The average maximum amplitudes and latencies before (a) and after one hours (b) methadone consumption.

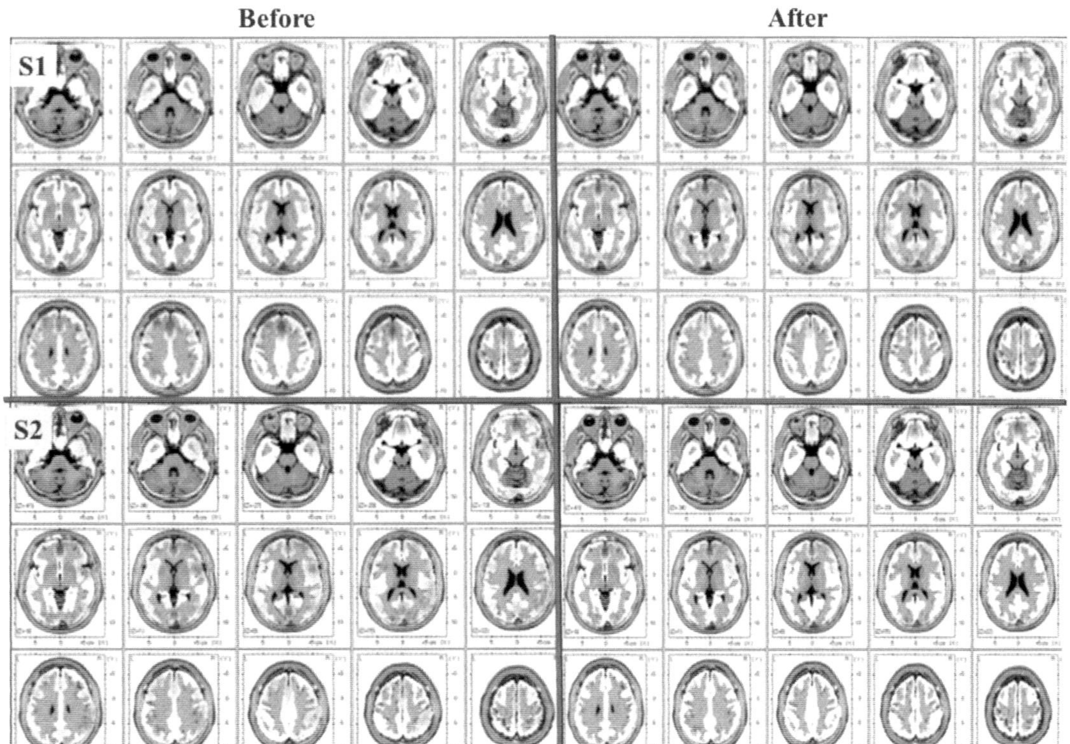

Figure 5. The P300 brain activity maps of the two subjects before (a) and after one hours (b) of taking methadone.

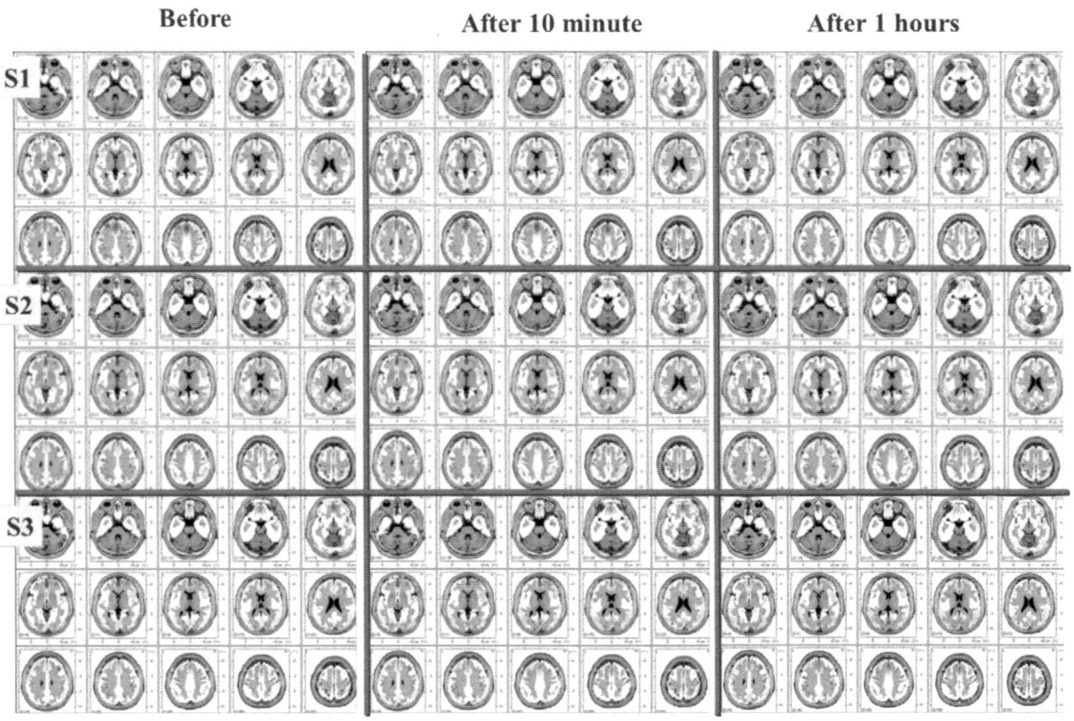

Figure 6. The beta brain activity maps of the three subjects before (a), and after ten miute (b), and after one hours (b) of taking methadone.

4. Conclusions

The comparison effect of the drug abuse to the brain activities before (craving condition) and after taking the drug or methadone is provided. It is obtained that the maximum average amplitudes is highly higher in craving condition than after taking the drug. The abnormality of the brain activity in the craving condition is highly related with the drug abuse. The drug abusing subjects had the greatest number of total brain abnormalities. Moreover, increasing abnormalities were observed in the more severe drug abuser or dependents compared to the normal controls.

Acknowledgment

This research was supported by the Technical Implementation Unit for Instrumentation Development, Indonesian Institute of Sciences, Faculty of Medicine (Padjajaran University) and, Insentif Riset Sistem Inovasi Nasional (INSINAS), Kementerian Riset, Teknologi, dan Pendidikan Tinggi (Ristekdikti), Indonesia.

References

[1] Wolpaw J R *et al.* 2000 *IEEE Trans. Rehabil. Eng.* **8** 164
[2] Turnip A, Kusumandari D E, Amri M F, Simbolon A I, Suhendra M A, IsKandar I and Wirakusumah F F 2017 *IOP Conference Series: Materials Science and Engineering* **220** 012042
[3] Ray S, Haney M, Hanson C, Biswal B and Hanson S J 2015 *Neuropsychopharmacology* **40**
[4] Liang X, He X, Salmeron B J, Gu H, Stein E A and Yang Y 2015 *J Neurosci* **35** 8081
[5] Turnip A, Hong K S and Jeong M Y 2011 *BioMedical Engineering OnLine* **10** 83
[6] Turnip A and Hong K S 2012 *Int. J. Innov. Comp. Inf. Control* **8** 5839
[7] Turnip A, Hutagalung S S, Pardede J and Soetraprawata D 2013 *International Journal of Brain and Cognitive Sciences* **2** 63
[8] Turnip A and Siahaan M 2014 *Advanced Science Letters* **20** 2034
[9] Turnip A and Kusumandari D E 2014 *Journal of Computer* **9** 688

IOP Conf. Series: Journal of Physics: Conf. Series **1080** (2018) 012029 doi:10.1088/1742-6596/1080/1/012029

[10] Turnip A, Haryadi, Soetraprawata D and Kusumandari D E 2014 *Advanced Science Letters* **20** 80
[11] Turnip A, Simbolon A I, Amri M F, Setiadi R H and Mulyana E 2017 *Internetworking Indonesian Journal* **9** 53
[12] Turnip A, Amri M F, Suhendra M A and Kusumandari D E 2017 *Electronic and Computer Engineering* **9** 107
[13] Hu Y, Salmeron B J, Gu H, Stein A E and Yang Y 2015 *JAMA psychiatry* **72** 584
[14] Wisner K M, Patzelt E H. Lim K O and MacDonald A W 2013 *Am J Drug Alcohol Abuse* **39** 403
[15] Lebel C *et al.* 2013 *Psychiatry Res.* **213** 161
[16] Konova A B, Moeller S J, Tomasi D and Goldstein R Z 2015 *Brain Res.* **1628** 147
[17] Mitchell M R *et al.* 2013 *Am J Drug Alcohol Abuse* **39** 392
[18] Spronk D B, van Wel J H, Ramaekers J G and Verkes R J 2013 *Neurosci Biobehav Rev.* **37** 1838
[19] Potvin S, Stavro K, Rizkallah E and Pelletier J 2014 *Journal of addiction medicine* **8** 368
[20] Turnip A, Kusumandari D E and Hidayat T Signals *Journal of Physics: Conference Series* **1007** 012060
[21] Turnip A, Kusumandari D E, Hidayat T and Hidayat T 2018 *Journal of Physics: Conference Series* **1007** 012059
[22] Turnip A, Kusumandari D E Simbolon A I, Hidayat T and Wirakusumah F F 2017 *Information Systems and Electrical Engineering* (ICITISEE) 244
[23] Kusumandari D E, Hidayat T, Turnip A, Hidayat T and Istiqomah A N 2017 *The 2nd International Conference on Automation, Cognitive Science, Optics, Micro Electro-Mechanical System, and Information Technology (ICACOMIT)* 12
[24] Turnip A, Suhendra M A, Kusumandari D E and Sihombing P 2017 *5th International Conference on Instrumentation, Control, and Automation (ICA)* 121

IOP Conf. Series: Journal of Physics: Conf. Series **1080** (2018) 012030 doi:10.1088/1742-6596/1080/1/012030

Set-up of prompt gamma neutron activation analysis system at Kartini reactor

Syarip[1*] and Z Abidin[2]

[1]Center for Accelerator Science & Technology, National Nuclear Energy Agency (BATAN), Jl. Babarsari, Yogyakarta 55281
[2]Polytechnic Institute of Nuclear Technology, Jl. Babarsari, Yogyakarta 55281

[*]syarip@batan.go.id

Abstract. Setup and testing of prompt gamma neutron activation analysis (PGNAA) system by using neutron source from the radial beam-port of Kartini reactor have been done. The method used is an experimental characterization of neutron flux in the radial beam-port then setting up of PGNAA system and determination of elements in a bulk solution sample of salt (NaCl). The PGNAA system configuration consists of the collimator with L/D of 28.9 made from lead placed inside the beam-port, bulk sample with 20cm x 20cm x 43cm in dimension located at 30 cm from outer surface of the beam-port, and NaITl detector. The distance of sample to NaITl detector is 20 cm. The experimental test results show that the elements contained in the sample solution of salt can be detected well. The neutron flux in the outer surface of beam-port was measured using activation foil detector, the result was in the order of 10^6 n cm^{-2} s^{-1} with a cadmium ratio of 14.

1. Introduction

The previous study in development of the use of Kartini reactor's beam port for Prompt Gamma Neutron Activation Analysis (PGNAA) experimental facility have been reported [1-4]. The beam characterization of the three kinds of beam ports available to select the most prospective one and beam modeling using the selected beam port to get neutron beam, meeting the specified criteria. The study was conducted by means of simulations using MCNPX code. The results conclude that tangential and radial beam ports was considered to be feasible for the purpose of PGNAA experimental facility [3-4]. The study was performed by means of simulations using MCNPX code.

PGNAA is another variant of NAA techniques, working based on the gamma rays emitted promptly following the decay of the compound nucleus formed in the excited state. The intensity of the gamma-rays emitted is directly proportional to isotopic (elemental) concentrations. This technique has enabled the in situ and on-line application such as those used in coal mining; cement industries, etc. [5-8].

PGNAA technique, basically can work with the whole range of neutron energy spectrum, depending on the neutronic properties of the sample to be analyzed. However for most research application it is preferred to use neutrons of lower energy range, since activation reactions occur mostly with neutrons of low energies. An exception is for nuclides having very low thermal neutron capture reaction such as carbon, it is preferred to use fast neutron, throughout the neutron inelastic scattering reaction [6].

The main components of PGNAA system at Kartini reactor consisted of radial beam-port with neutron collimator, radiation shielding and beam shutter, and gamma spectrometry system. A good PGNAA system need a suitable neutron source with a good collimation. Therefore, the prompt gamma spectrometry with low noise is expected. The preliminary performance test of installed PGNAA is conducted using NaITl detector and bulk sample of salt solution.

2. Material and Method

Schematic diagram of the PGNAA system using neutron source from radial beam-port of Kartini reactor is illustrated in figure 1. Based on figure 1 the working principle of PGNAA can be described in three zones as follow: core zone, beam-port zone and instrumentation zone. In the core zone the neutron beam has multiple energies. In the beam-port zone, neutron beam undergoes moderation process into thermal neutron, then passes collimation 1 to obtain parallel thermal neutron beam. Shutter is used as a gate to close or open the thermal neutron beam from the beam-port of Kartini reactor, for PGNAA purpose. In the instrumentation zone or PGNAA zone the neutron beam is passed through 2nd collimation to direct the neutron parallel to the PGNAA counting chamber. The neutron particle that leak from the chamber will be captured by the beam stopper so as not to cause noise in the counting system.

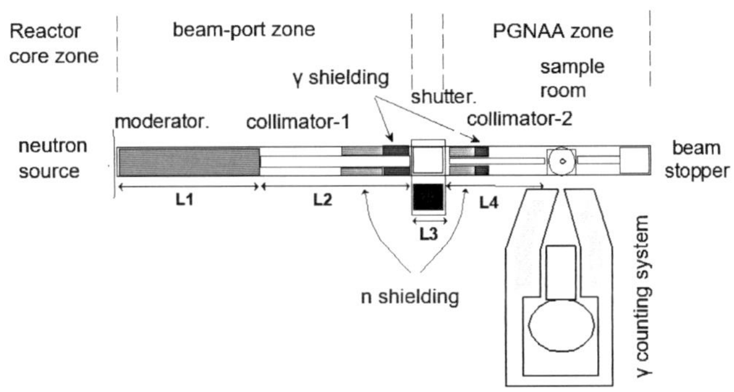

Figure 1. Schematic diagram of PGNAA system at the beam-port of Kartini reactor.

The next step is the installation of neutron collimator in the radial beam-port. The basic geometrical model of collimator as represented in figure 2. The collimator consists of three main sections i.e. inner, middle and outer sections each with the length of 20 cm, 80 cm, and 50 cm respectively. Lead (Pb), with density of 11.34 g/cm^3 is good for suppressing gamma rays, and it has also low neutron capture cross section which is relatively transparent against neutron passage.

Another material having similar properties with lead is bismuth (Bi) with density of 9.780 g/cm^3. It has low neutron capture cross section and high neutron elastic cross section that can improve the neutron transmission. Polyethylene of high density (HDPE) with density of 0.95 g/cm^3 has low neutron capture cross section and high elastic scattering cross section which could improve the moderation of fast neutrons.

Further step is a measurement of neutron flux at the outer surface of beam-port with collimator installed, and neutron flux at the sample position. The activation foil detectors of gold (Au) and indium (In) are used. The neutron flux measurement is done at nominal power of 100 kW.

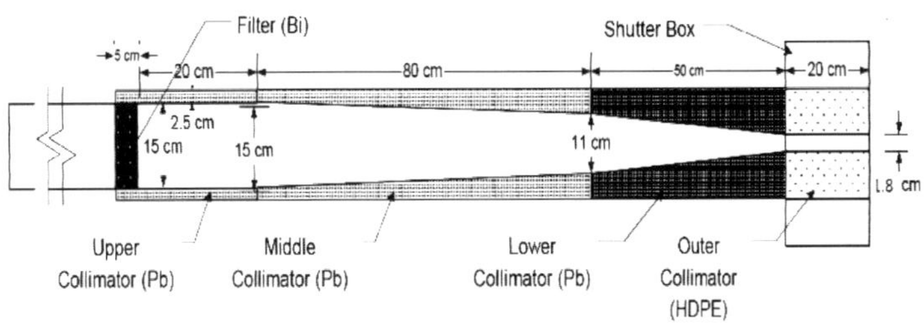

Figure 2. Neutron collimator at the Kartini reactor beam-port.

3. Results and Discussion

3.1. PGNAA installation result

The conical neutron collimator was installed at the radial beamport of Kartini reactor. The required L/D ratio of collimator, where D is the diameter of the collimator and L is the length has been determined previously i.e. L/D = 28.9 [1]. L/D is a characteristic parameter of a collimator that defines the degree of divergence of the neutron beam, the greater the L/D the more coherent the neutron beam. The neutron flux measurement results at the outer surface of the collimator installed in the beamport is described in table 1, it is shown that the average thermal neutron flux is 3.54×10^6 n/cm^{2}s, the epithermal (fast) neutron flux is 1.66×10^5 n/cm^{2}s and the average cadmium ratio is 14. The value of cadmium ratio is greater than 10 is very suitable for PGNAA system.

The neutron source of the radial beamport of Kartini reactor can produce thermal neutron intensity at the sample surface PGNAA system of 3×10^6 n/cm^{2}s in average which is appropriate for PGNAA system. Based on IAEA data [10-11] neutron intensity (flux) from some PGNAA facilities in some countries such as BARC reactor in India, Korea (KAERI), Dalat research reactor in Vietnam, and Budapest research reactor in Hungary, indicating that the appropriate thermal neutron flux for PGNAA is in the order of 10^6 to 10^8 n/cm^{2}s. There are even those who can reach a neutron flux in the order of 10^9 n/cm^{2}s like similar facilities in Argentina's RA-3 reactor [12].

The installation of collimator to the radial beamport of Kartini reactor and radiation shielding for PGNAA is shown in figure 3.

Table 2. The neutron flux measurement results at the outer surface of the collimator installed in the beamport.

Total (n/cm^2 s)	**Fast (n/cm^2 s)**	**Thermal (n/cm^2 s)**	**Cadmium ratio**
3,82 x 10^6	2,68 x 10^5	3,55 x 10^6	14,25
3,81 x 10^6	2,64 x 10^5	3,54 x 10^6	14,22

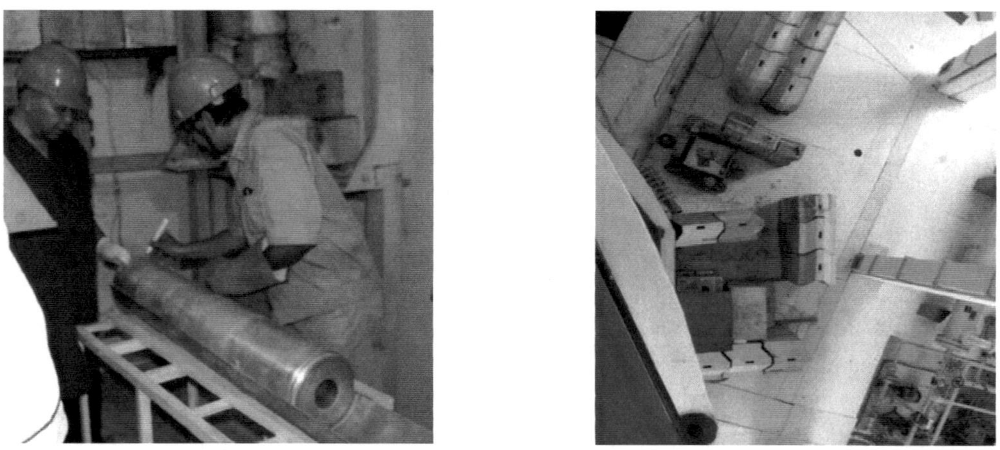

(a) (b)

Figure 3. The collimator is being (a) installed to the beam port and (b) installed radiation shielding.

3.2. PGNAA analysis result

The PGNAA system for this this experiment set-up was used NaITl gamma detector and a bulk sample of NaCl solution (20% concentration) contained in the plastic box with the dimension of 20 cm x 20 cm x 43 cm. The Eu^{152} gamma radiation source was used for γ energy calibration detection system. The γ spectrometry of Eu^{152} radiation source is shown in figure 4 and the efficiency calibration result is shown in figure 5, while the γ energy and channel calibration is shown in figure 6. The performance test of the PGNAA system was done at the condition of Kartini reactor in operation at 100 kW power level for 1 hour. The sample is located at 30 cm from the outer surface of collimator and the distance of sample to NaITl detector is 20 cm. The result of γ spectrometry of bulk NaCl sample is shown in figure 7, where the PGNAA counting was done 3 times, and the detected elements is shown table 3. It is shown in table 3 that Au^{194}, Cd^{117}, Na^{24}, Cl^{38} can be detected, its means that the all elements of NaCl i.e. Na and Cl can be well enough detected.

The average radiation exposure measured around the PGNAA system is 0.6 mR/hour or 52.6 mSv/year or 1.67 x 10^{-9} $Gy.s^{-1}$. The safe radiation dose limit based on the Regulatory Body (BAPETEN) is 20 mSv/year or 2.78 x 10^{-9} $Gy.s^{-1}$ [13], therefore the shielding of PGNAA system was good enough and the system is safe for continuous operation. The results also prove that in order to improve the signal to noise ratio different collimators and a filter have to be placed between the neutron source and the sample object [14].

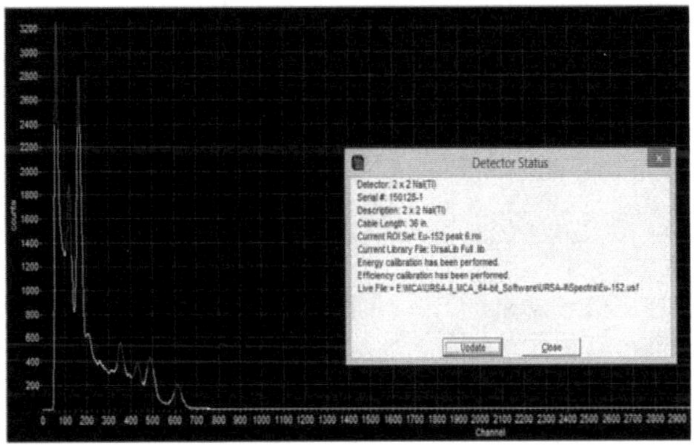

Figure 4. The γ spectrometry of Eu^{152} radiation source.

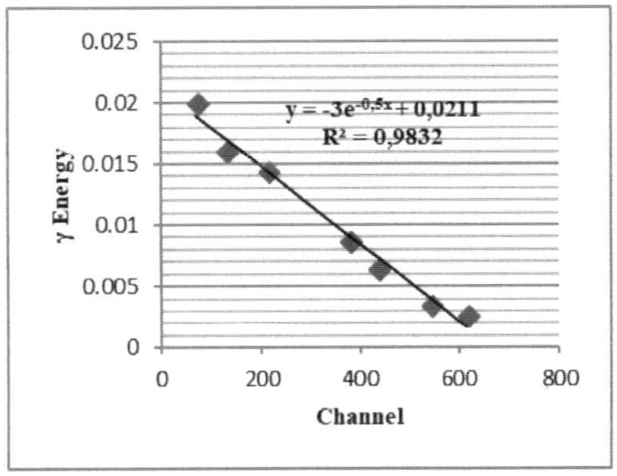

Figure 5. The γ energy efficiency of detection system.

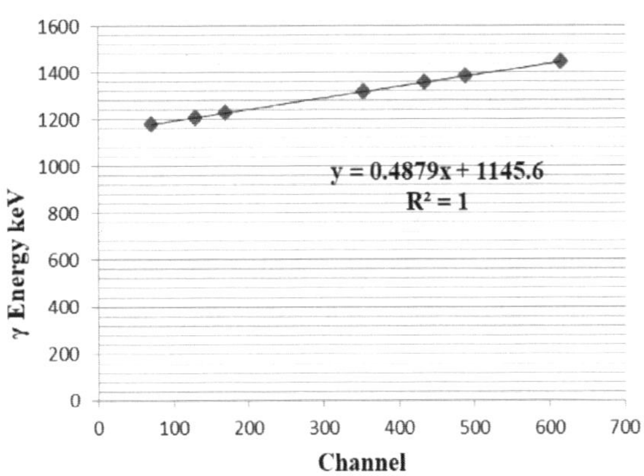

Figure 6. The γ energy – channel calibration.

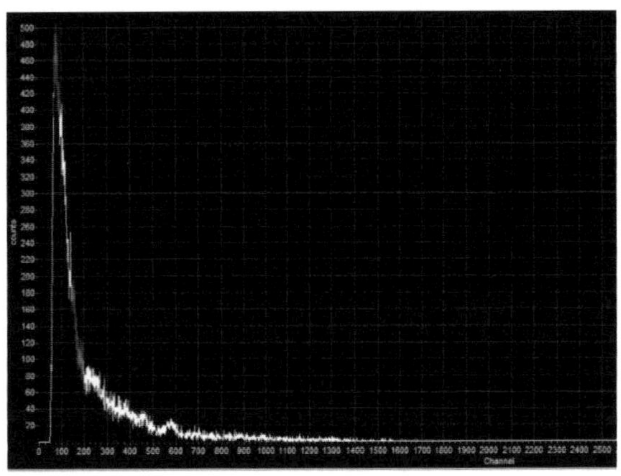

Figure 7. The PGNAA γ spectrometry of bulk NaCl sample.

In order to improve the gamma spectrum resolution, it is planned to perform the PGNAA system by using HPGe gamma detector.

Table 3. The PGNAA gamma spectrum of NaCl bulk sample.

Sample identity	**γ energy (keV)**	**Elements**	**Half-life (s)**	**Mass of elements (g)**
Reaktor-1	1180	Au^{194}	142200	$4.92x10^{-13}$
	1593			
	1196	Cd^{117}	8964	$4.20x10^{-12}$
	1271	Na^{24}	54000	$5.55x10^{-12}$
	1363			
Reaktor-2	1641	Cl^{38}	2232.6	$1.13x10^{-11}$
	1183	Au^{194}	142200	$8.47x10^{-11}$
	1325			
	1192	Cd^{117}	8964	$2.24x10^{-12}$
	1270	Na^{24}	54000	$3,77x10^{-14}$
Reaktor-3	1361			
	1639	Cl^{38}	2232.6	$6.05x10^{-13}$
	1185	Au^{194}	142200	$3.61x10^{-12}$
	1323			
	1190	Cd^{117}	8964	$2.71x10^{-12}$
	1269	Na^{24}	54000	$2.06x10^{-12}$
	1360			
	1637	Cl^{38}	2232.6	$9.24x10^{-11}$

4. Conclusion

The PGNAA system has been successfully installed at the Kartini reactor by using neutron source from the radial beamport. The average thermal neutron flux at the sample location is 3.54 $x10^6$ n/cm^{2}s, and the average cadmium ratio is 14, this neutron flux is already meet the requirements. The performance test results shows that the PGNAA system can detect the elements of the bulk sample of NaCl solution with sufficient efficiency and resolution according to the ability of the NaITl detector type.

Acknowledgments

The authors would like to thank the Director of Center for Accelerator Science and Technology Yogyakarta, the Head of Reactor Division and all staffs for their supports.

References

[1] Triyono E B S and Syarip S 2017 *Indonesian Journal of Nuclear Science & Technology (JSTNI)* **18** 1
[2] Syarip S 2002 *Ganendra Majalah Iptek Nuklir* **5** 23
[3] Sutondo T 2015 *Ganendra Majalah Iptek Nuklir* **18** 107
[4] Sutondo T and Syarip S 2014 *Ganendra Majalah Iptek Nuklir* **17** 83
[5] Dariush R and Mohammad M 2012 *JTPC* **1** 11
[6] Tarek Q Q, Al-Etaiwi T and Al-Khasheb M 2011 *Proc. of the 2011 International Conference on Industrial Engineering and Operations Management* (Kuala Lumpur, Malaysia p 1219)
[7] Darrell L 2014 *PGNAA Improves Process and Quality Control in Cement Production* (Accelerating Science, Thermo Scientific)
[8] Akhtar A N *et al.* 2012 *Journal of Nuclear Science and Technology* **46** 737
[9] Alexandre S L, Daniel C and Zsoltrevay 2011 *Proc. IAEA-CN-188-A19* (Rabat Morocco: International conference on research reactor safe management and effective utilization p 1)
[10] International atomic energy agency 2007 *Database of Prompt Gamma Rays From Slow Neutron Capture For Elemental Analysis* (Vienna: STI/PUB/1263)
[11] Manuel L *et al.* 2011 *Applied Radiation and Isotopes* **69** 1928

IOP Conf. Series: Journal of Physics: Conf. Series **1080** (2018) 012030 doi:10.1088/1742-6596/1080/1/012030

[12] Abdi M R A, Rezaee K H, Shahayan P and Farazneh A 2012 *Middle-East Journal of Scientific Research* **11** 648
[13] BAPETEN *Peraturan Kepala Badan Pengawas Tenaga Nuklir Nomor 4 Tahun 2013* Jakarta
[14] Fantidis J G, Nicolaou G E, Potolias C, Vordos N and Bandekas D V 2011 *J. Radioanal Nucl. Chem.* **290** 289

IOP Conf. Series: Journal of Physics: Conf. Series **1080** (2018) 012031 doi:10.1088/1742-6596/1080/1/012031

Magnetized deformation of neutron stars

R Rizaldy* **and A Sulaksono**

Department of Physics, Faculty of Mathematics and Natural Sciences, University of Indonesia, Kampus UI Depok, Indonesia 16424

*ryan.rizaldy@sci.ui.ac.id

Abstract. The high magnetic field in neutron stars leads to interesting consequences. One of them is the deformation of neutron stars. The correction equations due to deformation effect that complemented the spherical Tolmann-Oppenheimer-Volkoff equations can be calculated from Einstein field equations by including the magnetic contribution on the energy-momentum tensor of neutron star matter. Here, the multipole expansion is carried out on energy-momentum tensors containing magnetic fields as well as the corresponding metric around Schwarzschild one. The corresponding multipole expansions are performed only up to second order. We have shown that the effect of deformation appears more significantly in neutron stars with small masses. If we decrease the mass starting from maximum mass into the smaller one, the shape of neutron stars' changes from spherical into oblate.

1. Introduction

The observations indicate that normal pulsars usually have a magnetic field at surface with intensity around $B_s \sim 10^8 - 10^{13}\ G$ [1]. There are some particular pulsars detected by Soft Gamma Repeaters (SGRs) [2], Anomalous X-Ray Pulsars (AXPs) and the most recent report is from Repeating Fast Radio Burst [3, 4]. They have magnetic field value at the surface $B_s \sim 10^{15}\ G$. This magnetic field is significantly higher than those of normal pulsars called Magnetar. The value of magnetic field in the neutron stars core is not observed yet. However, it can be predicted theoretically around $B_c \sim 10^{18} - 10^{20}\ G$ [5].

The origin of high magnetic field inside neutron stars is still not clear until now. Magnetohydrodynamic (MHD) is the most accepted theory for describing the origin of magnetic fields on neutron star surface. The theory was developed by Duncan and Thompson based on an additional magnetic field is generated from a combination of rapid rotation and convective processes in the plasma during the proto-neutron star phase of the corresponding neutron stars [6].

The study of magnetic fields in neutron stars has been done by many researchers. In general, the presence of a magnetic field can be theoretically studied from microscopic and macroscopic aspects of the stars. Microscopically, the presence of a large magnetic field that significantly affects energy levels of charged particles is caused by Landau quantization. This affects the equation of state (EoS) for matter inside neutron stars. The effect of magnetic fields on the EoS model to describe hyperon stars [7, 8], quark stars [9] and hybrid stars [10] has been investigated. Macroscopically, the present of magnetic field can affect the structure of the star geometry. By solving the Einstein's field equations using the corresponding metric, it shows that neutron star can be deformed due to magnetic field. Solutions for the later problem have been proposed by Konno *et al.* using an analytical method for magnetized relativistic stars [11]. In otherwise, Mallick and Schramm using semi-numerical methods for hyperon stars [12].

This study is based on the method of Mallick and Schramm for energy-momentum tensor perturbation [12]. We obtain a rather different result on the corresponding solution of Einstein field equations. It was occurred because in our calculation, the pressure in energy-momentum tensor is expanded up to the quadrupole term while the energy density is expanded only up to the monopole term.

2. Formalism

The basic foundation of this study is based on the method Mallick and Schramm [12], by expanding the energy-momentum tensor in a form of multipolar expansion and using the Schwarzschild metric modification tensor of Hartle where the expansion is up to second order [13], we also using magnetic field profile ansatz proposed by Chakrabarty [7].

2.1. *Multipole expansion in Tensor Energy-Momentum*

We assumed the energy-momentum tensor in ideal fluid form. The additional magnetic field will be aligned along the z-axis, it is written in the energy-momentum tensor form as

$$T_{\mu\nu} = \begin{pmatrix} -\varepsilon & & & 0 \\ & P_\perp & & \\ & & P_\parallel & \\ 0 & & & P_\parallel \end{pmatrix} \tag{1}$$

with ε is the total energy density, $P_\perp$ and $P_\parallel$ is a perpendicular and parallel component of total pressure in respect to the direction of the magnetic field. The explicit expression of each term in equation (1) is $\varepsilon = \varepsilon_m + \mathrm{B}^2/8\pi$, $P_\perp = P_m - MB + \mathrm{B}^2/8\pi$ and $P_\parallel = P_m - \mathrm{B}^2/8\pi$ [12], with ε_m and P_m are the energy density and pressure of the matter, MB is the magnetization and $B^2/8\pi$ is a magnetic stress. Mallick and Schramm argue that the magnetic effects in matter EoS and magnetization term not significant [12]. Therefore, their contributions can be negligible. In this way each tensor element reduced into $\varepsilon = \varepsilon_m + \mathrm{B}^2/8\pi$, $P_\perp = P_m + \mathrm{B}^2/8\pi$ and $P_\parallel = P_m - \mathrm{B}^2/8\pi$ [12]. For the pressures parts, they can written based on the direction of the magnetic field angle as equation (2) [12].

$$P = P_m + \frac{B^2}{8\pi}(1 - 2\cos^2\theta) \tag{2}$$

Equation (2) can be also written as equation (3-4) [12].

$$P = P_m + \frac{B^2}{8\pi}\left[\frac{1}{3} - \frac{4}{3}P_2(\cos\theta)\right] \tag{3}$$

$$P = P_m + [p_0 + p_2 P_2(\cos\theta)] \tag{4}$$

where $P_2(\cos\theta)$ is the second order Legendre polynomial, $p_0 = B^2/24\pi$ is as monopole contributions and $p_2 = -4B^2/24\pi$ is as quadrupole contributions. So, the new energy-momentum tensor is written as

$$T_{\mu\nu} = \begin{pmatrix} -\varepsilon & & & 0 \\ & P & & \\ & & P & \\ 0 & & & P \end{pmatrix} \tag{5}$$

where

$$\varepsilon = \varepsilon_m + 3p_0 \tag{6}$$

$$P = P_m + p_0 + p_2 P_2(\cos\theta) \tag{7}$$

IOP Conf. Series: Journal of Physics: Conf. Series **1080** (2018) 012031 doi:10.1088/1742-6596/1080/1/012031

We can solve Einstein field equations using modified energy-momentum tensor that previously explained a multipolar expansion of metric tensor is also required. Similar to Ref. [12], we use the Hartle metric tensor expansion [13]. From these approximations we can calculate the deformation of magnetized neutron stars.

2.2. *Hartle Approximation for Tensor Metric*

Previously, we assume neutron stars as spherically symmetric. The interior solution of a static spherically symmetric object can be written in the Schwarzschild coordinate form t, r, θ, ϕ.

$$ds^2 = -e^{\nu}dt^2 + e^{\lambda}dr^2 + r^2(d\theta^2 + \sin^2\theta\, d\phi^2) \tag{8}$$

where the metric function for ν and λ is a function for r only, with solution $e^{\lambda(r)} = e^{-\nu(r)} = [1 - 2Gm(r)/r]^{-1}$, where G is the gravitational constant and $m(r)$ is the mass within the radius r spherical objects. The author of reference [14] formulates the metric configuration for the slow rotational equilibrium of stationary and symmetric systems axially written until the second order (by excluding rotation term) as follows.

$$ds^2 = -e^{\nu}(1 + 2h)dt^2 + e^{\lambda}\left[1 + \frac{e^{\lambda}}{r}m\right]dr^2 + r^2(1 + 2k)(d\theta^2 + \sin^2\theta\, d\phi^2) \tag{9}$$

for h, m and k are functions of r and θ defined as follows [15].

$$h(r,\theta) = h_0(r) + h_2(r)P_2(\cos\theta) + \cdots \tag{10}$$

$$m(r,\theta) = m_0(r) + m_2(r)P_2(\cos\theta) + \cdots \tag{11}$$

$$k(r,\theta) = k_0(r) + k_2(r)P_2(\cos\theta) + \cdots \tag{12}$$

where h_0, m_0, h_2, m_2, k_0 and k_2 are correction function up to second order, we set the value of $k_0(r) = 0$ according to coordinate transformation [14, 15]. To synchronize equation (9) with equation (6) and (7), the new metric tensor can be written up to quadrupole term as follows.

$$\begin{aligned} ds^2 = &-e^{\nu}\{1 + 2[h_0(r) + h_2(r)P_2(\cos\theta)]\}dt^2 \\ &+e^{\lambda}\left\{1 + \frac{e^{\lambda}}{r}[m_0(r) + m_2(r)P_2(\cos\theta)]\right\}dr^2 \\ &+r^2[1 + 2k_2P_2(\cos\theta)](d\theta^2 + \sin^2\theta\, d\phi^2) \end{aligned} \tag{13}$$

Using these definitions of metric and energy-momentum tensor, we can obtain the solution for Einstein field equations.

2.3. *Solution of Einstein Equations*

Using equation (5) and equation (13), we can obtain from the corresponding Einstein equations the following first order differential equations.

$$P_m' = -\frac{G}{r^2}(\varepsilon_m + P_m)(m + 4\pi r^3 P_m)\left(1 - \frac{2Gm}{r}\right)^{-1} \tag{14}$$

$$m' = 4\pi r^2 \varepsilon_m \tag{15}$$

Equation (14) and (15) are known as TOV equations with the correction for monopole term:

$$m_0' = 12\pi r^2 p_0 \tag{16}$$

$$p_0' = 4\pi G r e^{\lambda} p_0 + \frac{1}{r}G\nu' e^{\lambda} m_0 + \frac{1}{r^2}G e^{\lambda} m_0 \tag{17}$$

while for quadrupole term:

$$h_2' + k_2' = h_2\left(\frac{1}{r} - \frac{\nu'}{2}\right) + \frac{e^{\lambda}}{r}\left(\frac{1}{r} + \frac{\nu'}{2}\right) \tag{18}$$

IOP Conf. Series: Journal of Physics: Conf. Series **1080** (2018) 012031 doi:10.1088/1742-6596/1080/1/012031

$$h_2 + \frac{e^\lambda}{r} G m_2 = 0 \tag{19}$$

$$h_2' + k_2'\left(1 + \frac{r}{2}\nu'\right) = 4\pi G r e^\lambda p_2 + \frac{1}{r^2} G e^\lambda m_2 + \frac{1}{r} G \nu'^{m_2} + \frac{3}{r} e^\lambda h_2 + \frac{2}{r} e^\lambda k_2 \tag{20}$$

From the conservation of total energy-momentum, we obtained:

$$\nu' = -\frac{2}{\varepsilon_m + P_m} P_m' \tag{21}$$

for monopole term is

$$p_0' = -2\nu' p_0 - (\varepsilon_m + P_m) h_0' \tag{22}$$

while for quadrupole term is

$$p_2 = -(\varepsilon_m + P_m) h_2 \tag{23}$$

$$p_2' = -\frac{1}{2}\nu' p_2 - (\varepsilon_m + P_m) h_2' \tag{24}$$

We can solve these differential equations simultaneously by using Runge-Kutta method. The corresponding boundary conditions are $P_m(0) = P_c$, $P_m(R) = 0$, $m(0) = 0$, $m(R) = M$ and for the multipole factor according to reference [12,14], we use $h_0(0) = 0$, $m_0(0) = 0$, $h_2(0) = 0$ and $k_2(0) = 0$.

2.4. Magnetic Field Profile

The magnetic field profile of the neutron star is assumed depending on baryon density as follows [7]

$$B(\rho) = B_s + B_0\left\{1 - \exp\left[-\beta\left(\frac{\rho}{\rho_0}\right)^\gamma\right]\right\} \tag{25}$$

The ansatz is satisfying the general physical condition where the magnetic field in a star is not uniform. The model is constructed with condition the magnetic field in the center of the star that must be larger than the surface. ρ is a baryon density and ρ_o saturation density with value $\rho_o = 0.15\, fm^{-3}$. Parameter β and γ control how fast the center of the magnetic field (B_c) fall to the asymptotic value at the surface (B_s). Here we set at $\beta = 0.01$ and $\gamma = 2$. According to the observations data, a magnetic field at the surface is ~ 10^{15} G. The value of magnetic field in the center is estimated at 10^{18} - 10^{19} G assuming the dynamo effect inside the star [6], for asymptotic value $B_o = 4 \times 10^{18}$ G, where the density in the center is not greater than eight times the saturation density of symmetric nuclear matter, thus making the magnetic field value at the center is less than $B_c = 1.75 \times 10^{18}$ G [12], which is the value of B_c obtained from equation (25) when we set ρ at the $R = 0$.

3. Results

From the expansion parameters, we obtained the total mass and radius [12,15]:

$$M_{total} = M_0 + \delta M, \tag{26}$$

$$R = R_0 + \delta R, \tag{27}$$

where M_o is mass which exclude the magnetic field, $\delta M = m_o$ as an additional correction mass due to the magnetic field, R_o is radius without magnetic field and δR is an additional radius due to magnetic field effect. The later explicitly is defined as

$$\delta R \equiv \xi_0(r) + (\xi_2(r) + r k_2)\, P_2(\cos\theta), \tag{28}$$

Therefore, for the specified angle $R_e \equiv \xi_0(R) - (\xi_2(R) + Rk_2)/2$ and $R_p \equiv \xi_0(R) + (\xi_2(R) + Rk_2)$, R_e and R_p are equatorial and polar radius respectively, R is the radius of the spherical stars, ξ_0 and ξ_2 are defined as

$$\xi_0(r) \equiv -\frac{r[r-2Gm(r)]}{G[4\pi r^3 P_m + m(r)]} h_0, \tag{29}$$

$$\xi_2(r) \equiv -\frac{r[r-2Gm(r)]}{G[4\pi r^3 P_m + m(r)]} h_2. \tag{30}$$

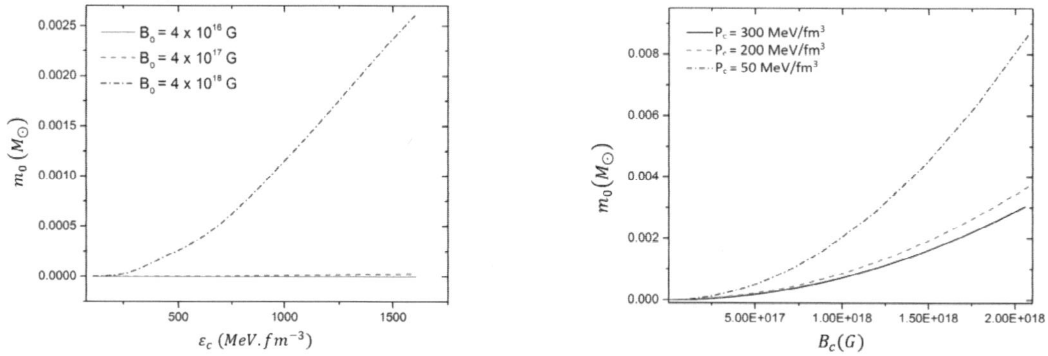

Figure 1. (a) Relation between m_0 and energy density with different set value of B_0 (b) relation between m_0 and Magnetic field in the center of star for different set value P_c.

The polar and equatorial radius of the star has the contribution of three parameters, which is ξ_0, ξ_2 and k_2. The contribution ξ_i derived from the strong magnetic field at the surface of the magnetars, and k_2 is contribution of the integrated magnetic pressure inside the star [12]. The degree of deformation on a star is given by the deformation parameter called eccentricity or ellipticity defined as

$$e = \sqrt{1 - \left(\frac{R_p}{R_e}\right)^2} \tag{31}$$

In this study we used EoS model from reference [16] where the hyperons are taken into account. In figure 1(a) we present a plot of energy density at the center of the star as a function of the mass correction term. For each fixed B_0 value, the mass correction increases with increasing energy density at the center of the star. The effect becomes significant if $B_0 > 4 \times 10^{18}$ G. This indicates, a mass increasing if the stars are deformed, and according to equation (26), the value of maximum mass will increase along the increasing energy density in the center of the star. However, if we see the mass correction is lesser two orders magnitude than M_0, then the maximum mass is not significantly increased. Figure 1(b) describes the relation between m_0 and Magnetic field in the center of star for three different set value of fix pressure in center of the stars (P_c). As we see, m_0 increases with increasing magnetic field in the center of star, following the density energy in the center of the star trend. The Value of m_0 increase for $P_c = 50\ MeV/fm^3$ than the others rapidly, indicating significant deformation that occurs at a low value of P_c.

Figure 2(a) describes the relation between the radius and the density energy in the center of the star with variant of magnetic field intensity B_0. The solid line represents a spherical symmetric radius for a neutron star without magnetic field. With the presence of a magnetic field term the polar and equatorial radius will be different where the difference can be observed also from equation (28). Along with the increase of density energy in the star center, the value of R_e always higher than the spherical radius, and the value R_p always below the value of the spherical radius so that the implication is an equatorial radius higher than the polar radius which geometrically makes the star tend to be oblate due to the existence of magnetic field. When we look at the changes in each set of values B_0, there is no significant change in R_e for every value of B_0, but R_p increases in respect to increasing energy density in the center of the star for $B_0 = 4 \times 10^{18}$ G. Figure 2(b) shows that an increasing value of ellipticity leads to decreasing the total mass. If we refer to equation (31) where the value e is near to 0 means there is no deformation and if it near to 1, deformation is maximum to oblate shape, then so it occurs at small total mass values significantly. The effect becomes more pronounced for $B_0 = 4 \times 10^{18}$ G.

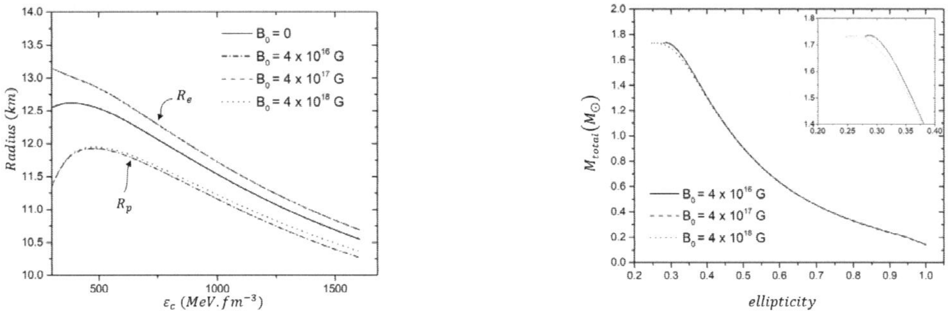

(a) (b)

Figure 2. (a) relation between radius and energy density (b) relation between total mass and ellipticity with different set value of B_0.

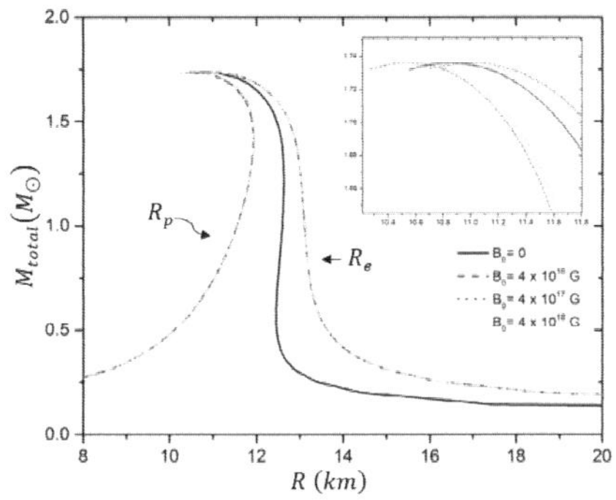

Figure 3. Relation between total mass and radius for different set value of B_0.

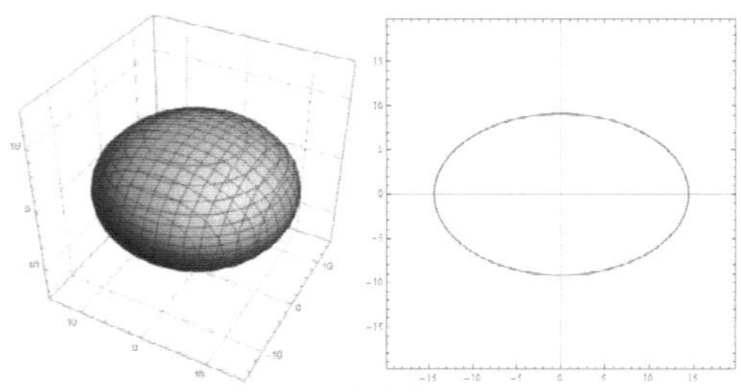

(a) $M = 0.36\ M_\odot$

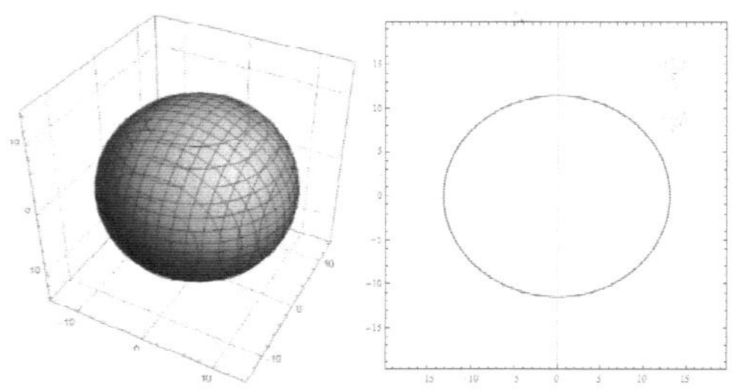

(b) $M = 0.96\ M_\odot$

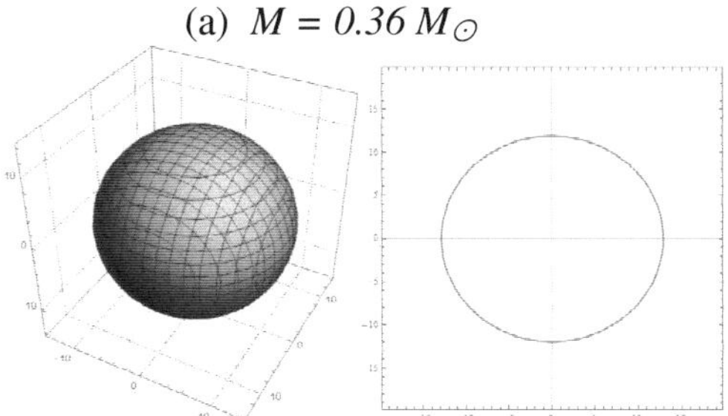

(c) $M = 1.42\ M_\odot$

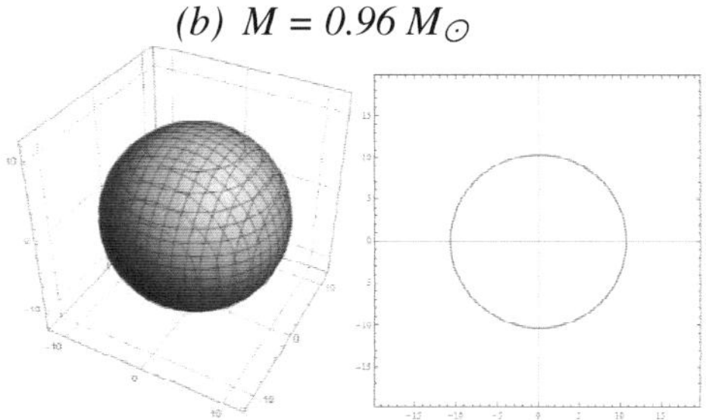

(d) $M = 1.78\ M_\odot$

Figure 4. Illustration of 3-dimensional deformation of neutron stars in each particular mass at $B_0 = 4 \times 10^{18}$ G.

In figure 3 solid lines represent radius without taken into account magnetic field, and we also showed R_e and R_p with different set value of B_0. At the maximum mass of stars, it can be seen R_e and R_p overlap each other, indicating almost no deformation of the mass. Along with the maximum mass reduction of stars, there are differences among the three lines which is represent different value of magnetic field, the value of R_p decreased to a low radius rapidly, otherwise R_e increased to the high value. We also showed the 3-dimensional illustration of star profiles on the figure 4, the horizontal line represents R_e and the vertical line represents R_p.

In figure 4 we make 3D illustration of neutron stars in each particular mass at $B_0 = 4 \times 10^{18}$. Figure 4(a) we show the deformation profile for small mass i.e., $M = 0.36\ M_{\odot}$, where $R_e = 14.42\ km$ and $R_p = 9.15\ km$. This result shows a very significant deformation. The star has an oblate shaped. For a higher mass of (b) $R_e = 13.12\ km$, $R_p = 11.5\ km$ and (c) $R_e = 12.88\ km$, $R_p = 11.95\ km$, the equatorial star radius decreases, but the polar radius increases so the star still oblate, but the effect becomes less dramatic. Similarly, it happens for figure 4(c). Finally, at the maximum mass as shown figure 4(d) $R_e = 10.36\ km$, $R_p = 10.36\ km$, shows that the star becomes spherical.

4. Summary

We have studied numerically the deformation of neutron stars due to magnetic field using proposed method by reference [15]. The correction factors in mass and radius due to magnetic fields indicate the deformation of neutron stars occurs significantly for small mass. Furthermore, the effect of magnetic field is more pronounced for $B_0 > 4 \times 10^{18}$ G. Physically, for the small mass of the star causing the balanced between the gravity and magnetic field is significantly different for different directions. Therefore, the star becomes oblate-shaped due to the gravity pull of magnetic field on the z-axis more significantly than for other axes.

Acknowledgments

This study was supported by UI's PITTA grant No. 622/UN2.R3.1/HKP.05.00/2017 2017 from Directorate of Research and Community Engagement, University of Indonesia.

References

[1] Kaspi V M 2010 *Proc. Nat. Acad. Sci. USA* **107** 7147
[2] Thompson C and Duncan R C 1995 *Monthly Notices of the Royal Astronomical Society* **275** 255
[3] Usof V V 1992 *Nature* **357** 472
[4] Spitler L G *et al.* 2016 *Nature* **531** 202
[5] Ferrer E J, de la Incera V, Keith J P, Portillo I and Springsteen P L 2010 *Phys. Rev.* **C82** 065802
[6] Thompson C and Duncan R C 1993 *Astrophysical Journal* **408** 194
[7] Chakrabarty S, Bandyopadhyay D and Pal S 1997 *Phys. Rev. Lett.* **78** 2898
[8] Sinha M, Mukhopadhyay B and Sedrakian A 2013 *Nucl. Phys.* **A898** 43
[9] Isayev A A 2015 *J. Phys. Conf. Ser.* **607** 012013
[10] Dexheimer V, Negreiros R, Schramm S and Hempel M 2013 *AIP Conf. Proc.* **1520** 264
[11] Konno K, Obata T and Kojima Y 1999 *Astronomy and Astrophysics* **352** 211
[12] Mallick R and Schramm S 2014 *Phys. Rev.* **C89** 045805
[13] Hartle J B 1967 *The Astrophyical Journal* **150** 1005
[14] Regge T and Wheeler J A 1957 *Phys. Rev.* **108** 1063
[15] Hartle J B and Thorne K S 1968 *The Astrophyical Journal* **153** 807
[16] Sulaksono A and Agrawal B K 2012 *Nuclear Physics A* **895** 4458

IOP Conf. Series: Journal of Physics: Conf. Series **1080** (2018) 012032 doi:10.1088/1742-6596/1080/1/012032

Electrical and electrochemical properties of reduced graphene oxide film for secondary battery electrode

H Aprida, S Hidayat, N Syakir, R Siregar and Fitrilawati*

Department of Physics, Faculty of Mathematics and Natural Science, Universitas Padjadjaran, Jl. Raya Bandung Sumedang km 21 Jatinangor, Sumedang, Indonesia 45363

*fitrilawati@phys.unpad.ac.id

Abstract. We prepared thin layer of reduced graphene oxide (rGO) from reduced graphene oxide powder that dispersed in deionized water and dimethylformamide (DMF) using drop-casting technique. We used carboxymethyl cellulose (CMC) as a binder in order to obtain better surface of rGO film with varied concentration of rGO. We obtained that rGO film prepared using rGO dispersed in deionized water has better surface compare to the ones dispersed in DMF. An addition of CMC into rGO has improved its film surface and increasing amount of rGO resulted higher conductivity of the sample. The cyclic voltammogram (CV) of the rGO film has shown a hysteresis curve that correspond to charging and discharging process. From CV measurement we obtain that rGO-CMC film that prepared using 3 mg/ml rGO dispersed in DI water and blend with CMC has higher specific capacitance of 400 mF/g.

1. Introduction

A battery is an electrochemical cells that consists of two electrodes i.e. anode and cathode that separated by an electrolyte [1]. It will provide a current at a voltage for a time Δt. Lithium-ion rechargeable battery has enabled a revolution of global communication.

The battery performance such capacity and cycle-ability depend on the anode material. An ideal anode material can be absorbed a large amount of lithium per-unit cell. An anode must has a good electronic and ionic conductivity as well as has a stable and reversible cycle [2]. If the anode material has a good electrical conductivity it can rapidly activate electrons and ion transport, so the charging process of battery becomes faster [3].

Graphite is a material that often used as an anode. In order to improve a battery performance, characteristics of anode materials should be upgraded. Graphene is a carbon monolayer of graphite that has 2D structure. Graphene received a lot of attention due to its excellent mechanical, electrical, thermal and optical properties. Graphene has better property compare to graphite with a thermal conductivity of 4840 W/mK - 5300 W/mK and an electrical conductivity of 2000 S/cm, which is classified as a conducting material [4].

For application, it is necessary to provide a large quantities of graphene. In relation to that, graphene oxide (GO) is mostly used as a precursor material to obtain graphene. Graphene Oxide (GO) does not have unique physical and electronic characteristics such as pure graphene properties. By eliminating or reducing the oxygen functional group will produce a structure such as pure graphene or called reduced Graphene Oxide (rGO). The oxygen group from the GO sheet can be removed by chemical reduction, electrochemical or thermal reduction. More of oxygen functional group are reduced that showing properties increasingly similar to pure graphene, in figure 1 [5].

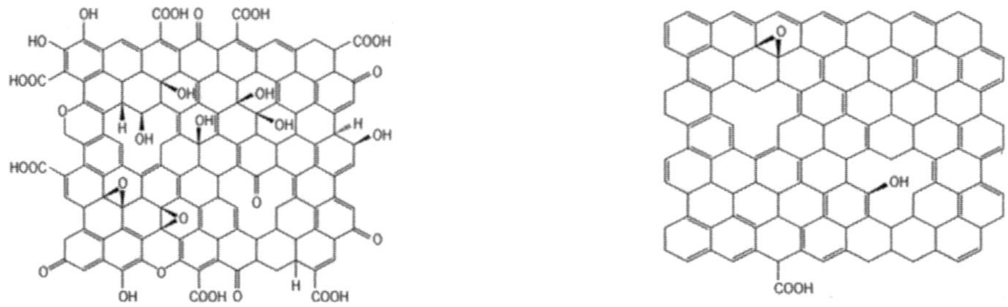

(a) (b)

Figure 1. Representation of sheet structure (a) Graphene Oxide (GO) and (b) reduced Graphene Oxide (rGO) [6].

In this research, we used reduced Graphene Oxide (rGO) powder with electrical conductivity ~ 667 S/m [7], thermal properties 110W/mK - 1100 W/mK at room temperature and also has an excellent mechanical resistance properties [8].

The binder material is an important part of the electrode batteries. Binder material changes the structure of rGO to binds each other. The role of binder material is related to the energy density of the active material. A good binder for ion batteries has a criteria such as having stable chemical and electrochemical properties in electrode and electrolyte systems, and also environmentally friendly [9]. One of the best binder for binding anode material is Carboxymethyl-Cellulose (CMC) [10]. The advantages of CMC are soluble in H_2O, environmentally friendly and cheaper than Polyvinylidene fluoride (PVDF) binder with using toxic N-methyl-2-phyrrolidone solvent [10]. The best properties of electrode are achieved by mixing a binder as low as possible in the active materials. An optimum concentration of CMC in active material is 2% [11]. This research intends to determine the optimum concentration of rGO in suitable solvent to analyze electrical properties to improve secondary battery performance.

2. Experimental

2.1. Preparation of samples

We prepared thin films of rGO on Indium Tin Oxide (ITO) substrates using commercial rGO powder (Graphenea) dispersed in deionized water (DI Water) and dimethylformamide (DMF). Prior to use film preparation, the rGO was dispersed for 5 hours at room temperature using ultrasonic bath. We prepared the rGO samples mixed with carboxymethyl cellulose (CMC) as a binder with various concentration of rGO. Before mixing with rGO dispersion, CMC was dissolved with solvent and stirred for 1 hour at 60°C. The rGO-CMC mixture was dispersed for 30 minutes using ultrasonic bath. Thin films were obtained using drop-cast method and dried in the hotplate at 40°C. We prepared five type of samples as shown in table 1.

We used glass substrate (2 cm x 2 cm) for conductivity measurement and ITO or Cu foil substrate (1.25 cm x 2 cm) for cyclic voltammogram (CV) characterization. For electrolyte, we used 1M $LiClO_4$ solution with propylene carbonate as solvent.

Table 1. The samples variation.

Sample	rGO (mg)	DMF (ml)	DI Water (ml)	CMC Binder (mg)
1	1	-	1	-
2	1	1	-	-
3	1	-	1	0.1
4	2	-	1	0.1
5	3	-	1	0.1

2.2. Characterizations
We used four-line probe method for electrical characterization and Scanning Electron Microscope (SEM) for morphology observation. The samples for SEM and conductivity measurements were deposited on glass substrate. The probe for conductivity measurement consist of four parallel line of platinum wire that connected with multimeter and DC voltage/current source. The electric current was flowed to the sample by through the probe, and the voltage was measured between two parallel probes.

Cyclic voltammetry measurement (MetrOhm Autolab) was performed using two-electrode system with platinum as counter electrode (CE), rGO sample as working electrode (WE), and 1M $LiClO_4$ solution as an electrolyte. Scan rate of CV measurement was varied from 25 mV/s - 150 mV/s. To study stability of the rGO-CMC electrode, we performed CV measurement on rGO-CMC sample with rGO content of 3% for many cycles at scan rate of 50 mV/s.

3. Results and discussion
Since number of oxygen functional group has reduced in rGO, it takes more time to disperse rGO powder in DI water and DMF. After dispersion rGO in DI water for a certain time with concentration up to 3 mg/ml we have a homogeneous rGO solution without aggregate. In contrast, rGO dispersed in DMF at the same concentration shows some aggregates. It was reported that rGO can be dispersed homogeneousely in DMF (organic solvent) at low concentrations of <0.1 mg/ml [7]. Based on this result we only used DI water as dispersion medium of rGO.

Compare to rGO film that prepared from rGO dispersion without CMC, we found that sample prepared from rGO-CMC mixture has a better surface. An addition of CMC into rGO has improved film quality since it become more flat and homogeneous. We varied rGO content of the sample with concentration of 1 to 3 mg/ml. Surface morphology of rGO-CMC mixture with varied rGO concentration are shown in figure 2. The morphology of rGO-CMC sample is appeared similar to surface morphology of rGO sample. Amount of 0.1 mg CMC that mixed with 0.1 mg rGO seem does not change the surface morphology of the rGO film significantly. The CMC might only slight cover rGO and does not close surface of rGO layer.

Figure 2(a) is a surface morphology of rGO-CMC film prepared using rGO dispersion with concentration of 1 mg/ml and 0.1 mg CMC. The image shows that rGO-CMC film consist of a stacking of many layers of rGO with a width about 1 μm for 1 layer. Figure 2(a.2) is the same image of the sample with a magnification of 40,000 times. At the higher rGO concentration it seems the surface area get larger with a width around 3 μm, as shown in figure 2(b, c). The second image of each igure are a magnification image of 40,000x. The role of CMC as a binder is filling the voids between layers thus extending the width of the layer. The lithium ions can easily change over the electrode. With the wider surface of rGO film, lithium ions can fill the pores of the rGO layer then the electrons will be easier to move between layers.

We measured conductivity of all rGO samples including rGO sample that prepared from rGO dispersed in DI water (rGO-DI water), rGO sample dispersed in DMF (rGO-DMF) and sample rGO with CMC binder. We applied constant current from 1mA to 10 mA and measured the voltage between two probes. All measured samples show a linear I-V curve. From the I-V data, probe distance and sample dimension, the conductivity (σ) were calculated using equation (1). I in the equation (1) is the current input from DC voltage/Current Source, ΔV is the voltage between two probes, d is the probe distance, w is the sample wide, and t is the sample thickness. The values of conductivities of all prepared samples are shown in table 2.

$$\sigma = \left(\frac{I}{\Delta V}\right)\left(\frac{d}{w.t}\right) \quad (1)$$

Table 2 shows that the rGO sample prepared using DI water as dispersion medium has higher conductivity compare to the one using DMF. It is related to the quality of the resulted rGO film. Sample prepared using rGO dispersion in DMF was not flat since rGO dispersion in DMF with a concentration of 1 mg/ml has an aggregate.

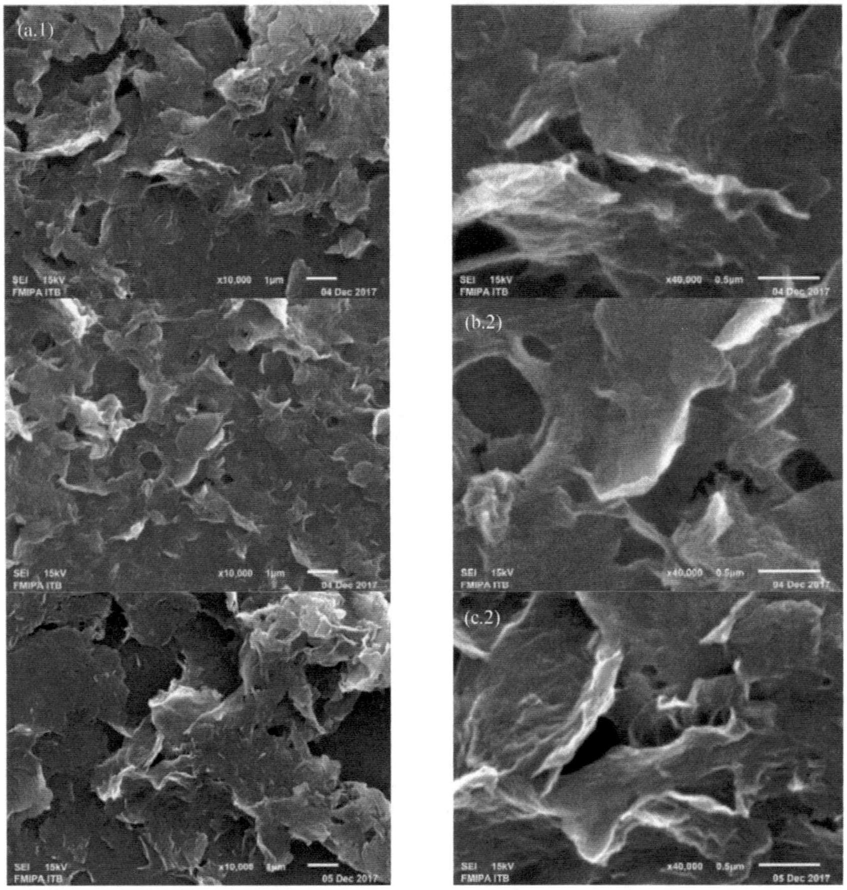

Figure 2. SEM images of rGO films prepared using CMC binder with rGO concentration of (a) 1 mg/ml with magnification of 10,000x (a.1) and 40,000x (a.2), (b) 2 mg/ml with magnification of 10,000x (b.1) and 40,000x (b.2), (c) 3 mg/ml with magnification of 10,000x (c.1) and 40,000x (c.2).

An addition of CMC binder has an effect on decreasing of a conductivity of rGO since CMC is an isolator material. When the concentration of rGO was increased double or triple times, the conductivities of the rGO samples were increased significantly. The result shows that, compare to the pure rGO film, the conductivity of rGO-CMC sample does not only depend on rGO concentration, but also depend on the surface morphology of the film. The highest conductivity value of sample also related with the condition of film that has a good surface morphology with widely area. Data on table 2 shows that sample with higher rGO concentration has a higher conductivity.

Table 2. Conductivity of rGO samples with and without CMC binder.

Samples	Conductivity (S/m)
rGO – DI Water 1mg/ml	2036.67
rGO – DMF 1mg/ml	20.40
rGO – DI Water 1mg/ml - CMC	671.66
rGO – DI Water 2mg/ml - CMC	2396.38
rGO – DI Water 3mg/ml - CMC	4188.91

The CV measurement were performed for all of rGO samples in $LiClO_4$/PC electrolyte at the voltage range of -0.5 V to +1.5 V. The CV curves that measured at scan rate of 50 mV/s were shown

IOP Conf. Series: Journal of Physics: Conf. Series **1080** (2018) 012032 doi:10.1088/1742-6596/1080/1/012032

in figure 3. The figure shows that when the voltage applied from -0.5 V increased to +1.5 V there is an accumulation of charge between the surface of the rGO electrode and $LiClO_4$ electrolyte and result a current. This process is known also as charging. Then when the applied voltage was decreased from +1.5 V to -0.5 V, the resulting current response is not equal to the charging current and formed a hysteresis. This hysteresis area is described as a charge storage. The decreasing applied voltage is called as discharging process. The figure shows a comparison of CV curves of five samples. Compare to other sample, the rGO film prepared using DI water has narrowest hysteresis. While rGO film prepared using DMF has wider hysteresis compare to the one using DI water. Hysteresis area of rGO samples prepared using CMC binder are also wide enough. Its area is depended on the rGO concentration, the higher rGO concentration the wider its hysteresis area.

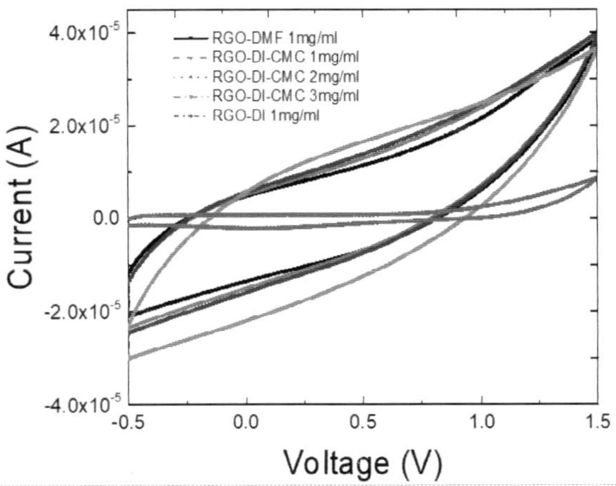

Figure 3. Cyclic voltammetry curve of various rGO samples measured at scan rate of 50mV/s.

From CV data that measured at varied scan rate, we calculate power of the system by using equation (2) [12]. P in the equation is a power (Watt), i is the current response in a given potential V, and the dV is step potential which was 0.00244 V. The specific capacitance C is calculated using equation (3) [12]. C in the equation is the capacitance of rGO (Farad), m is the mass of electroactive materials, ΔV is the potential window, s is the scan rate and i is the current response in a given potential V.

$$P = \left(\int_{-0.5}^{1.5} i\, dV + \int_{1.5}^{-0.5} i\, dV\right) \tag{2}$$

$$C = \frac{1}{2m\,.\Delta V\,.s}\left(\int_{-0.5}^{1.5} i\, dV + \int_{1.5}^{-0.5} i\, dV\right) \tag{3}$$

At the low scan rate, the lithium ions will have a longer time to get into all the pores on the electrode. Then the capacitance of rGO will be increased, but at a low of scan rate the reaction is still not stable. The value of the capacitance is likely to be declined and then stabilized along with higher scan rate value, as shown in Figure 4. Electrode with higher concentration of rGO can store more charge. The highest specific capacitance of the rGO electrode is 399.89 mF/g that achieved at scan rate of 25 mV/s using electrode contain highest concentrate of 3 mg/ml rGO prepared with DI water and CMC binder. At higher concentrations of rGO there is more charge that can be stored by electrodes.

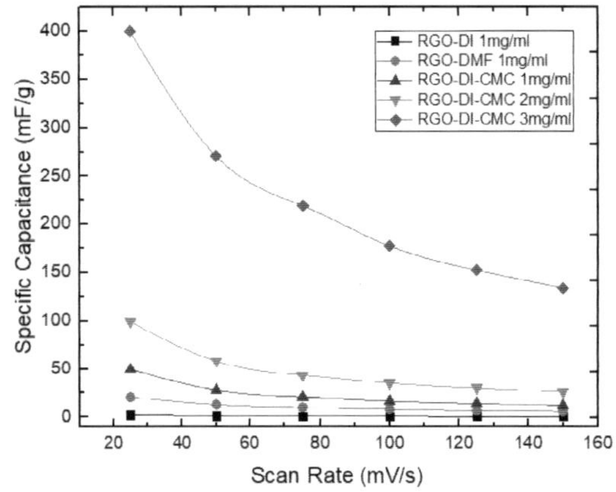

Figure 4. Specific capacitance of rGO electrode measured at varied scan rate.

In relation to its stability, we performed CV measurement of rGO – DI Water 3mg/ml – CMC sample for 12 cycles at scan rate of 50 mV/s. Figure 5 shows hysteresis area of the sample has identical shape as a number of cycle increased. Using the measured curve and interpolation approached we obtain that the rGO-CMC electrode tends to be stable until drop capacity at 98th cycle. It provides information about a stability of the rGO – DI Water 3mg/ml – CMC electrode until 98th cycles.

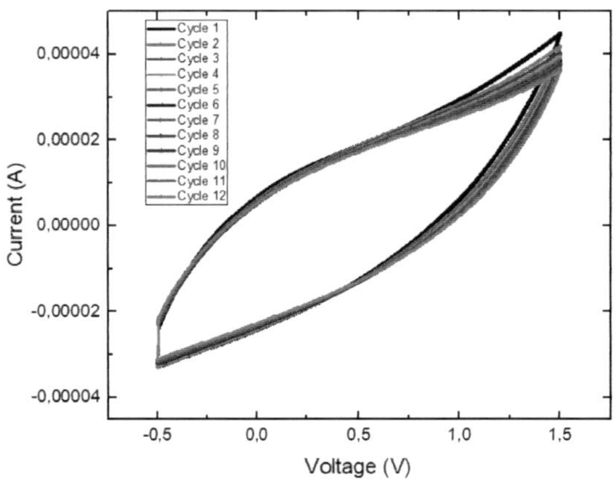

Figure 5. CV curves of rGO – DI Water 3mg/ml – CMC sample after 12th cycles measured at 50 mV/s.

4. Conclusions

We obtained that rGO films prepared using deionized water and CMC as a binder has a better surface quality compare to the other ones prepared with rGO only. The rGO-CMC film that was prepared using 3 mg/ml rGO-DI water dispersion and CMC binder has a highest conductivity of 4.2 x 10^3 S/m. The cyclic voltammogram of the rGO film show a hysteresis curve that correspond to the charging and discharging characteristic of the rGO. Increasing concentration of rGO has increased number of ions can be stored in rGO film. The rGO-CMC film prepared using 3 mg/ml rGO-DI water dispersion has the specific capacitance about 400 mF/g and still has CV identical shape after 12th cycles.

Acknowledgements

The authors acknowledge funding by Academic Leadership Grant (ALG) Universitas Padjadjaran with contract no : 855/UN6.3.1/PL/2017

References

[1] Goodenough J B and Park K 2013 *J. Am. Chem. Soc* **135** 1167
[2] Reddy M V, Rao G V S and Chowdari B V R 2011 *Chem. Rev* **113** 5364
[3] Li N, Chen Z, Ren W, Li F and Cheng H-M 2012 *Proc. Natl. Acad. Sci* **109** 17360
[4] Raccichini R, Varzi A, Passerini S and Scrosati B 2014 *Nat. Mater* **14** 271
[5] Data Sheet: Ossila 2017 *Reduced Graphene Oxide Powders*
[6] Griggs C S and Medina V F 2016 *McGraw Hill Encyclopedia of Science & Technology, 11th ed.*
[7] Graphenea 2016 *Product Datasheet Graphenea Reduced Graphene Oxide (rGO)* 22
[8] Ahn H N et al 2015 *Exp. Therm. Fluid Sci* **60** 361
[9] Qiu L *et al* 2014 *RSC Adv* **4** 24859
[10] Drofenik J, Gaberscek M, Dominko R, Poulsen F W, Mogensen M, Pejovnik S and Jamnik J 2002 *Sci. Direct Electrochimica Acta* **48** 883
[11] Hidayat S, Cahyono T, Rahayu I and Alamsyah W 2016 *Prosiding Seminar Nasional Fisika* p 61
[12] Chen Y, Zhang X, Zhang D, Yu P and Ma Y 2011 *Carbon N. Y.* **49** 573

IOP Conf. Series: Journal of Physics: Conf. Series **1080** (2018) 012033 doi:10.1088/1742-6596/1080/1/012033

Retention time of ozone at various water condition

R Andoyo*, I A P Prawitasari, E Mardawati, Y Cahyana, E Sukarminah, T Rialita, M Djali, Zaida, I Hanidah and I S Setiasih

Faculty of Agro-Industrial Technology, Universitas Padjadjaran, Jl. Raya Bandung-Sumedang km 21 Jatinangor, Sumedang, Indonesia 45363

*r.andoyo@unpad.ac.id

Abstract. Ozone is a molecule composed of three atoms of O_2 (oxygen) formed by the merger of oxygen free radicals with molecular oxygen, ozone is unstable and reactive. Ozone serves as a disinfectant, decolorization, as well as to remove organic materials. Ozone can be formed by electrolysis, chemical radiation and corona discharge. The formation of ozone by corona discharge most widely used today. Ozone that is unstable and reactive has a retention time depending on the environment such as temperature, pH and mineral content. Ozone can be applied to foodstuffs solid or liquid food ingredients, including water, milk, vegetables and fruits that have variations in time, temperature, pH value and concentration of minerals. This study aimed to evaluate the influence of temperature, pH, type and concentration of mineral to the retention time of the ozone. In this research, we varied temperatures (5°C, 25°C and 40°C), type of water (distilled, aquabidest, and tap water), concentration of mineral Ca (0 ppm, 125 ppm and 250 ppm) and pH (5.5, 7 and 9.2). The result was analyzed descriptively using different t-test analysis to compare each treatment. The process of ozonation at low temperatures (5°C) has a longer retention time compared to high temperature, samples that contain low minerals have a longer retention time than samples containing high mineral, the addition of mineral (Ca) did not affect the retention time of ozone. Ozonation at acidic pH has a longer retention time compared to alkaline pH. The duration of ozonation 5 and 10 minutes have no different in retention time.

1. Introduction

Ozone is a molecule consisting of three oxygen atoms and has the chemical symbol of O_3 [1]. Ozone (O_3) is formed by the incorporation of an oxygen-free radical with molecular oxygen [2]. Ozone is an unstable gas and reactive, therefore the amount of ozone in the atmosphere tends to have low concentrations. Ozone has many uses, such as for disinfection of microorganisms because ozone is a strong oxidizing agent. Ozone is commonly used in the drinking water industry to kill microorganisms contained in drinking water. Currently ozone is used to kill algae and oxidize organic matter, thus eliminating unwanted taste, odor, and color caused by the reaction of organic matter.

Ozone has different stability depending on environmental conditions, such as temperature [3]. At low temperatures, ozone has a longer retention time compared to high temperatures. The retention time of ozone define as the time when ozone can dissolve in water. The stability of ozone is influenced by the retention time in water under certain conditions. Ozone has a strong oxidizing properties that can reduce microorganisms, therefore suitable for sterilization purposes [4]. Ozone utilization has been widely applied in areas such as in America and Japan ozone used for sterilizing raw foodstuffs as well as preservation of foodstuffs [5]. Foods that can be treated by ozone include solid foods such as vegetables and fruits, wheat, as well as hydrocolloid foods such as milk and fruit juices [2]. Foodstuffs that can be treated by using ozone are complex foodstuffs with different pH, mineral and temperature.

The stability of ozone is affected by its environment, such as pH, temperature, type water and mineral content. Therefore, it affects the retention time of ozone in water. At high temperatures, resistance and the presence of ozone in the water will decrease, whereas at low temperatures the presence of ozone in the water will be longer [6]. Another parameter that may affect the presence of ozone in water is pH. Water with acidic pH can inhibit ozone decomposition, because low pH indicates high alkalinity. Alkalinity is the presence of carbonate ions and bicarbonate in the water [7]. Water has a different amount of mineral content depending on the source and water treatment. The level of mineral contained in water can affect the stability of ozone. Research on how the mineral in the water affect the rate of ozone decomposition has not been much explored, therefore the current study will discuss the stability of ozone in water at different temperature, pH, water type and several variations of mineral content. Water with different mineral concentrations was achieved by using different types of water e.g. aquabidest, aquadest and tap water. Aquabidest is processed by double stage distillation, so it does not contain minerals. Meanwhile, aquadest are only once passed through the distillation process, so it contains less mineral. In contrast, tap water contains higher minerals compared to the other. In addition, to evaluate the effect of ozone stability as affected by mineral, variation of minerals content was added to aquadest [8]. The present study was evaluated the ozone stability in water at different pH, temperature, water type and variations in the amount of mineral content. The results of this study can be used as parameter for ozonation of other liquid-based food.

2. Materials and Methods

The raw materials used in this research were aquadest, aquabidest, tap water. Other materials used are pure oxygen (90-95%), chlorine as the reagent of ozone, $CaCl_2$, NaOH and HCl. Ozonizer TIP-01 was used to produce ozone, ozone test kit, digital camera, CM 400 chromameter.

The research method used was descriptive analysis conducted with two replicates for each sample namely: the dependent variable (X) is the concentration of ozone obtained, while the independent variable (Y) is temperature, pH, water type and mineral concentration.

Table 1. Sample Name of the independent variable (Y).

Different temperature	Different type of water	Different mineral concentrations	Different pH
Y_1 = 5°C	Y_4 = Aquadest	Y_7 = 0 ppm $CaCl_2$	Y_{10} = pH 5,5
Y_2 = 25°C	Y_5 = Aquabidest	Y_8 = 125 ppm $CaCl_2$	Y_{11} = pH 7,2
Y_3 = 40°C	Y_6 = Tap water	Y_9 = 250 ppm $CaCl_2$	Y_{12} = pH 9,2

The parameter observed in this study (X) were the concentration of ozone in water after 5 and 10 minutes of ozonization at various condition as explained in table 1.

3. Result and Discussion

3.1. Ozonation at different temperature

Temperature is one factor that can affect the retention time and ozone solubility in the ozonation process [9]. The results of ozone retention time at various temperature is presented in figure 1. From figure 1, it can be seen at the first 20 min after ozonation process (zone I), ozone concentration was fluctuating and unstable at all-time of ozonation and types of water tested. Ozonation for 5 and 10 minutes resulting almost the same ozone concentration, this is probably due to a slow diffusivity of the ozone gas phase to the liquid phase in the water [10]. The diffusivity of ozone was also influenced by the size of the bubbles produced. The diffusion rate between the ozone and the water depend on the surface area of the ozone bubbles [11]. High number of ozone diffused in water cause the ozone to easily decomposed so that the ozone concentration was easily decreased in a short time, whereas at low ozone concentrations, the decrease in concentration was slower because ozone has room to bind with other.

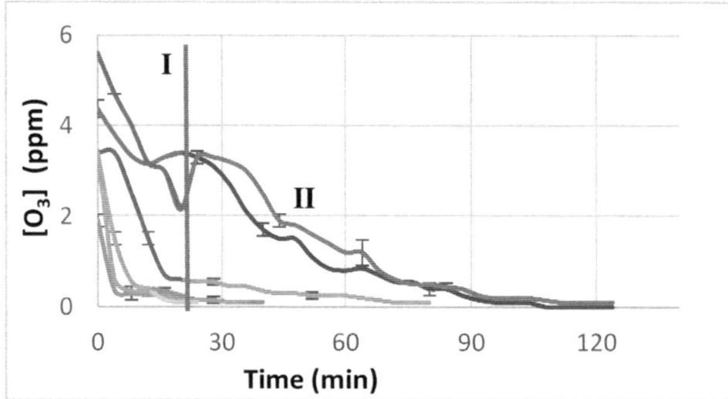

Figure 1. Ozone retention time at various temperatures: --- Aquadest 5°C (5 min), — Aquadest 5°C (10 min), --- Aquadest 25°C (5 min), — Aquadest 25°C (10 min), --- Aquadest 40°C (5 min), — Aquadest 40°C (10 min).

Ozone is a reactive compound that readily binds to other compounds when ozone concentrations are high, nevertheless, at high concentrations ozone tend to be unstable. The ozonation process at 5°C has a longer retention time because ozone is hardly decomposed at low temperatures [12]. At high temperatures, the ozone tends to expand so that the solubility decreases. Solubility of ozone is influenced by temperature, the lower the temperature the higher ozone solubility will be, while at high temperature, ozone insoluble in water [9]. Furthermore, at high temperatures, ozone diffusivity was low because ozone in the gas phase has a lower density than water as compared in a liquid phase, so that ozone is easily released, whereas at low temperatures, ozone has a high diffusivity due to in the gas phase ozone tends not to expand so that ozone was easily to bind with water. Based on t-test ($p < 0.05$) ozonation at 5°C was significantly different from ozonation at temperature of 25°C and at temperature 40°C, while temperature ozonation of 25°C was not significantly different at temperature of 40°C according to t-test ($p < 0.05$).

3.2. Ozonation at different types of water

Mineral content in water is one of the factors that can affect the retention time and solubility of ozone in water. The ozonization process based on type of water uses, by mean of different mineral content, was performed by using aquadest, aquabidest and tap water at ozonation temperature of 5°C. Based on figure 2, in the first 20 min after ozonation for the aquadest samples, ozone concentration was fluctuatively decreased (zone I). The retention time of ozone in aquadest and aquabidest samples was 140 min, while for tap water samples was 20 min. Tap water ozonation process has a shortest retention time as compared to the other treatments, due to groundwater contains many minerals that can accelerate ozone decomposition during the ozonation process [8]. Aquadest and aquabidest do not contain minerals as they have undergone demineralization through distillation. Based on the results of Ca analysis performed by using AAS (Atomic Absorption Spectroscopy), the highest Ca content was presented in tap water (14.46 ppm). Thus, every 1 ppm of minerals contained reduces the ozone retention time for ± 10 minutes. Mineral content affects the diffusion of the ozone bubbles into water, the higher the mineral content, the easier the ozone to be decomposed and thus causes the retention time of ozone to be shorter [13]. According to Von Gunten [14]), ozone firstly decomposed inorganic compound then organic compounds will be decomposed afterwards. In the condition of water that contains many minerals such as tap water, ozone more easily decompose into O_2 (oxygen) and O* (free oxygen ions), while in condition of water with less minerals such as distilled water, ozone was decomposed longer.

According to the statement of Gottschalk *et al.* [13], mineral content affects ozone diffusivity into water. The retention time of ozone in tap water was the shortest due to the minerals containing in tap water so that ozone was easily decomposed by reacting with mineral. On the other hand, in aquadest and aquabidest which contained less minerals and organic compounds, ozone was decomposed at longer time thus it has a longer retention time. Furthermore, ozone did not react with inorganic compounds such as minerals. Figure 2 shows that retention time of ozone at the temperature of 5°C in aquadest and aquabidest sample has a similar property, but it was significantly different from ozone in tap water.

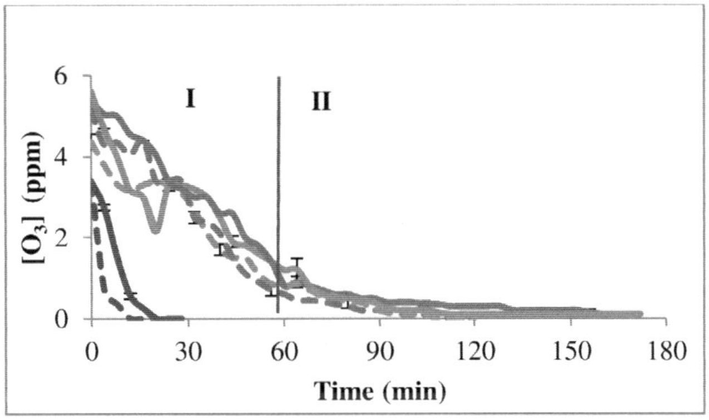

Figure 2. Ozon retention time at various types of water: --- Ozonation *Tap water* (5 min), — *Tap water* (10 min), --- Aquabidest (5 min), — Aquabidest (10 min), --- Aquadest (5 min), — Aquadest (10 min).

3.3. Ozonation at different mineral concentrations

The concentration of minerals or inorganic compounds in water may affect the ozone solubility. Inorganic or salt such as sodium, magnesium, bicarbonate, sulfate, chloride and silica contained at 1-1000 ppm in water [8] may affect diffusion from the ozone bubbles to the water. Oxidation by ozone was also be affected by the presence of mineral [13]. In this study, we added $CaCl_2$ mineral with Ca concentration of 0 ppm, 125 ppm and 250 ppm to aquadest. The results of ozone retention time at different Ca concentration at the temperature of 5°C is presented in figure 3.

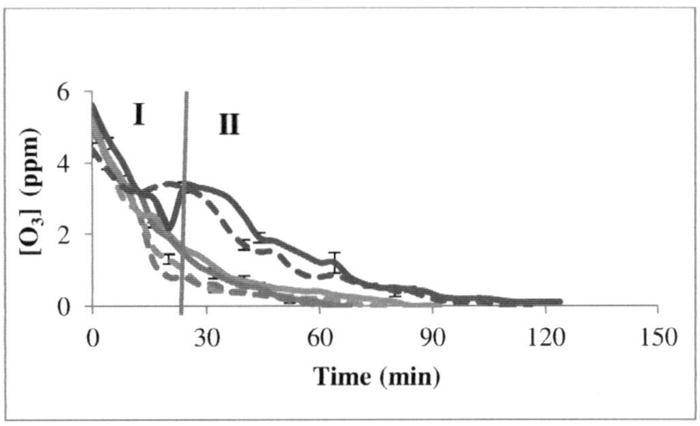

Figure 3. Ozone retention time at various Ca concentrations; --- 0 ppm (5 min), — 0 ppm (10 min), ---125 ppm (5 min), — 125 ppm (10 min), ---250 ppm (5 min), — 250 ppm (10 min).

From figure 3, at the first 20 min after ozonation process (zone I), ozone concentrations from all samples were high and fluctuates. The amount of Ca added affect the retention time of ozone, particularly it can be seen at concentrations of 125 ppm and 250 ppm (zone II). The addition of Ca into water confirmed that minerals can accelerate ozone decomposition. Based on the result above, mineral concentration influences the ozonation process. The increasing of mineral concentration decreased the ozone retention time. As the first compounds to be oxidized by ozone was inorganic compound then followed by organic compounds, the presence of inorganic minerals was mainly affected the ozonation process [14].

IOP Conf. Series: Journal of Physics: Conf. Series **1080** (2018) 012033 doi:10.1088/1742-6596/1080/1/012033

3.4. Ozonation at different pH

The pH value indicates the hydrogen ion activity and it affects the ozone retention time. The presence of hydrogen ions in water reduce the decomposition of ozone, thus results the longer ozone retention time. In this study, the ozonation process was carried out at different condition i.e. alkaline (pH 9.2), neutral (pH 7) and acidic (pH 5.5). The results of ozone retention time at different pH value and at the temperature of 5°C is presented in figure 4.

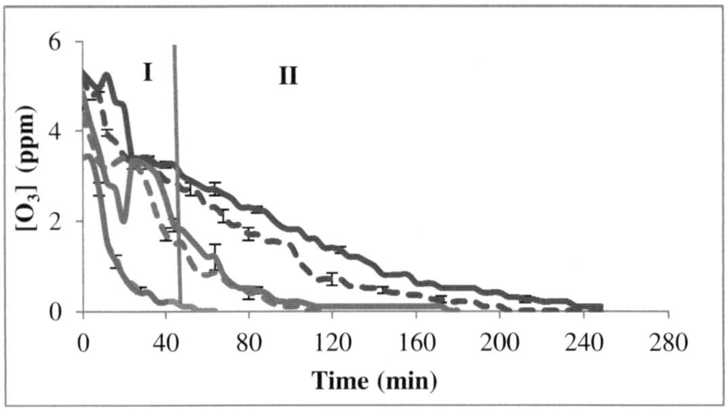

Figure 4. Ozone retention time at various pH: --- pH 5.5 (5 min), — pH 5.5 (10 min), --- pH 7 (5 min), — pH 7 (10 min), --- pH 9.2 (5 min), — pH 9.2 (10 min).

From figure 4, at the first 20 min of ozonation process (zone I) the concentration of ozone was fluctuated especially at pH 5.5 and pH 7. This was occurred due to the unstable ozone properties so that the time of sampling (every 4 minutes after ozonation) was not proportional to the decrease in ozone concentration. The retention time of ozone after 5 min ozonation at acid pH was more than 200 minutes, neutral pH was more than 100 minutes while the alkaline pH has the shortest retention time of 60 minutes. The same results were also observed for 10 min of ozonation. This was in accordance with research conducted by Mandavgane *et al.* [15] the retention time of ozone in acid pH was longer than the alkaline pH. This was occurred due to water with acidic pH can inhibit ozone decomposition and low pH indicates high alkalinity. Alkalinity in water exhibits carbonate (CO_3^{2}) and bicarbonate (HCO_3^-) which cause the ozone half-life was increased and the reduced the reaction of a OH radical [7].

Reaction of carbonate and bicarbonate with OH radical is presented in equation (1, 2). The carbonate and bicarbonate ions will react with the OH radical compound so that the ozone will not quickly decompose, thus the ozone retention time will be longer, whereas at high pH, the OH radical was formed easily and it inhabit ozone decomposition, thus the ozone concentration was declined rapidly [15].

$$OH\bullet + CO_3^{2} \rightarrow CO_3^{2-} + OH^{\bullet} \tag{1}$$

$$OH\bullet + HCO_3- \rightarrow CO_3^{2-} + H_2O \tag{2}$$

According to Tizaoui *et al.* [16], ozone decomposition at a high pH was occurred because the hydroxide ions contributing to ozone decomposition and may cause the reaction as presented in equation (3, 4).

$$O_3 + OH^- \rightarrow HO_2^- + O_2 \tag{3}$$

$$O_3 + HO_2^- \rightarrow OH + O_2^- + O_2 \tag{4}$$

The ozonation at pH 5.5 was significantly different from the ozonation at pH 7 as well as for the pH of 9.2. Based on chemical reaction at equation (3, 4), it can be seen that ozone decomposition converts the OH-group to water (HO_2^-) and more quickly breaks down because the ozone decomposed faster to O_2, thus causing the ozone retention time at higher pH was shorter.

4. Conclusion

The ozonation process was influenced by various factors such as temperature, water type, mineral concentration and pH value. Ozonation at low temperatures has a longer retention time compared to high temperatures. Water that contain high minerals has a shorter retention time compared to low minerals samples, in addition ozonation at low pH has a longer retention time compared to the sample at high pH. To maximize the ozonation process especially in liquid-based food, ozonation shoud be performed at low temperature (5°C), samples should not contain minerals and has acidic pH because these condition could reduce the time of ozone decomposition to increase retention time.

References

[1] Smith W 2011 *Principles of Ozone Generation* (Watertec Engineering Pty Ltd, Australia)
[2] O'Donnell C P, Tiwari B K, Cullen P J and Rice R G 2012 *Ozone in food processing* (Chichester, West Sussex; Ames, Iowa: Wiley-Blackwell)
[3] Khadre M A, Yousef A E and Kim J G 2001 *Journal of Food Science* **66** 1242–1252
[4] Katz J 1980 *Ozone and Chlorine Dioxide Technology for Disinfection of Drinking Water* (Park Ridge, N.J: Noyes Pubns)
[5] Syafarudin A and Novia N 2013 *Jurnal Teknik Kimia* **19**
[6] Rusdi U D and Suliasih N 2002 *Bionatura* **4**
[7] Sari N N, Sururi M R and Pharmawati K Ci 2013 *Jurnal Institut Teknologi Nasional* **1**
[8] Effendi H 2003 Telaah Kualitas Air, Bagi Pengelolaan Sumber Daya dan Lingkungan Perairan (IPB: Kanisius)
[9] Hill A and Rice R G 1982 *Historical background, properties and applications, in Rice, R.G. and Netzer* (Ozone Technology and Applications **1**, Ann Arbor, MI: Ann Arbor Science Publishers; 1–37)
[10] Sauze N L, Laplanche A, Martin N and Martin G 1993 *Water Research* **27** 1071
[11] Mitani M M, Keller A A, Sandall O C and Rinker R G 2005 *Ozone: Science & Engineering* **27** 45
[12] Lenntech 2009 *Water treatment solution, Ozone decomposition* (Technical University of Delft, The Netherlands)
[13] Gottschalk C, Libra J A and Saupe A 2010 *Ozonation of Water and Waste Water* (Wiley- VCH Verlag GmbH & Co. KGaA)
[14] Von Gunten U 2003 *Water Research* **37** 1443
[15] Mandavgane S A and Yenkie M K N 2011 *Rasayan J. Chem* **4** 544
[16] Tizaoui, C. & Zhang, Y 2010 *Chemical Engineering Journal* **162** 557

IOP Conf. Series: Journal of Physics: Conf. Series **1080** (2018) 012034 doi:10.1088/1742-6596/1080/1/012034

Fourier transform of high frequency ultrasonic waves propagated with a transmission mode

D Hidayat*, N S Syafei, B M Wibawa and B Y Tumbelaka

Department of Electrical Engineering, Faculty of Mathematics and Natural Sciences, Universitas Padjadjaran, West Java, Indonesia 45363

*darmawan.hidayat@unpad.ac.id

Abstract. When ultrasonic propagates as a wave, the frequencies contained in the propagated waves may carry information regarding to the medium. The transformation of time-domain propagated signals into those of frequency domain is a common digital signal processing (DSP) technic to extract such important information. This requires quite a high resolution of the detection device which relates to the high frequency ultrasonic waves. The advancement of microprocessor technologies enables the complex and fast digital signal processing in the frequency domain. This allows measurement and evaluation using a broadband frequency domain analysis which is effective in the extraction of information contained in the ultrasonic signals. This work presents the implementation of Fourier transform of propagated ultrasonic signals through air medium. The generation of ultrasonic waves employs two pairs of transmitter-receiver ultrasonic transducers which have the center frequencies of 200 kHz and 1 MHz, respectively. The transmitter transducer was excited by a 107.1-ns negative spike to generate ultrasonic waves propagating through the air and converted into an electrical signal by a receiver transducer. This was amplified and recorded as a digital signal for subsequent off-line digital signal processing.

1. Introduction

Ultrasonic based measurements and evaluations have become one of the versatile methods because of its features of relatively inexpensive, no electromagnetic radiation and high precision. When an ultrasonic wave propagates into a medium, it experiences energy absorption resulting in the reduction of a portion of energy indicated by a reduction of the amplitude of the received signal. More reduction occurs when it propagates at higher frequencies [1]. For the purposes of non-destructive evaluations, frequency content of propagated signals become the importance of interest. They indicate the information of the propagating medium [2, 3]. In addition, ultrasonic waves are useful for the generation of aerosol droplets for material syntheses, such as zinc oxide and barium titanate, using aerosol processes [3, 4].

The extraction of frequency content of time domain signals commonly applies a DSP technic which is so called a discrete Fourier Transform (FT) [5, 6]. This technic requires fast digital processors to perform FT calculation of a recorded digital signal. This analysis frequently applies specific window functions to obtain the desired frequency response related to the applications. The application of window functions improves amplitude and frequency response resolution, such as reduction of noises due to signal discontinuity and sampling process [7].

Many works report the application of frequency domain analysis of ultrasonic signals for non-destructive evaluations and measurements. They include bone density measurements, irregular surface shape thickness measurements [8, 9]. Xu *et al.* (2007) reports the implementation of the spectrum of ultrasonic signals with a frequency center of 0.5 MHz for the measurement of bone density for clinical assessment. Ultrasonic-based thickness measurement of irregular surface reports that the thickness can be calculated accurately by resonant algorithm based on spectral analysis with accuracy up to ±15 μm [10, 11].

This work briefly presents the implementation of FT technic to represent a received digital ultrasonic signal in a transmission-mode in a frequency domain for further signal analysis and interpretation of ultrasonic-based measurements and evaluations. A negative spike high voltage excited a 200-KHz transmitter transducer to generate ultrasonic waves which propagated through air received by the receiver transmitter. The received signals subsequently amplified, displayed and recorded as a digital signal.

2. Method

2.1 Experimental setup

Figure 1 describes the experimental setup of a typical ultrasonic-based non-destructive evaluation system. The excitation negative spike pulse activates a transmitter transducer which converts the electrical signal into ultrasonic waves. The generated ultrasonic waves propagate through the medium and applied for ultrasonic-based application purposes. The ultrasonic propagates as longitudinal wave with a speed determined by the density of propagation medium. It propagates faster in a denser medium. The applications of ultrasonic-based evaluations frequently involve under-test object with the density is higher than that of air. This causes a large propagation speed discrepancy between object and medium, which subsequently causes significant wave impedance mismatch. When it occurs, most portion of ultrasonic energy is reflected at the boundary between media and the under-test object. Underwater ultrasonic evaluation is commonly applied in order to attain the higher impedance matching because of the larger density of water compared to air as medium. Thus, two types of ultrasonic transducers were evaluated, a pair of air-coupled transducer and a pair of water-coupled transducer, respectively.

In the pulse-echo mode, only one transducer is employed as both transmitter and receiver transducer. The ultrasonic wave is transmitted and received from the same transducer. It alternately transmits and received the ultrasonic wave and hence there is only one access to the transducer when it is in operation. In comparison, the transmission mode includes a pair of identical transducers, transmitter and receiver transducers, placed in-sight position of the active surface. This mode allows the operation of continuous ultrasonic wave to the under-test object and analysing frequencies at which resonance frequencies are the importance of interest and determine the properties of the object [12].

The evaluation uses air- and water- coupled medium on transmission mode propagation, in which the transmitter and receiver transducers were in sight at a fixed distance of 3 cm. This test employed two pair transducers with each pair frequency of 200 kHz and 1 MHz, respectively. Table 1 shows the specifications of each transducer employed. A pair of transducer 1 is air coupled medium with 18 mm in diameter with the center frequency of 200 ± 8 kHz, maximum pulsed voltage 200 V_{PP} and 400 pF capacitance. This transducer is air medium-dedicated usage. As comparison, transducer 2 is a water-flow sensor type, 21 mm in diameter with the center frequency of 1.0 ± 5% MHz, maximum pulsed voltage 500 V_{PP} and 1200 pF capacitance.

The generated ultrasonic waves propagate through air from the transmitter to the receiver transducer. The receiver converts the mechanical ultrasonic waves into an electrical signal. It is very weak with a very low signal-to-noise ratio (SNR) value. A high frequency and broadband amplifier (OPA365 based operational amplifier) amplifies this signal prior to displaying and recording. The amplified signal is subsequently displayed and recorded as a digital signal by a high speed digital oscilloscope storage (GW-Instek GDS2014). The spike pulse is connected to the oscilloscope for signal triggering. It is displayed and recorded as well with number of samples of 5000, recording with frequency sampling of 50 MHz, 250 MHz, and 2 GHz. All wiring connections use high frequency coaxial cables to suppress noises may interfere the measurement.

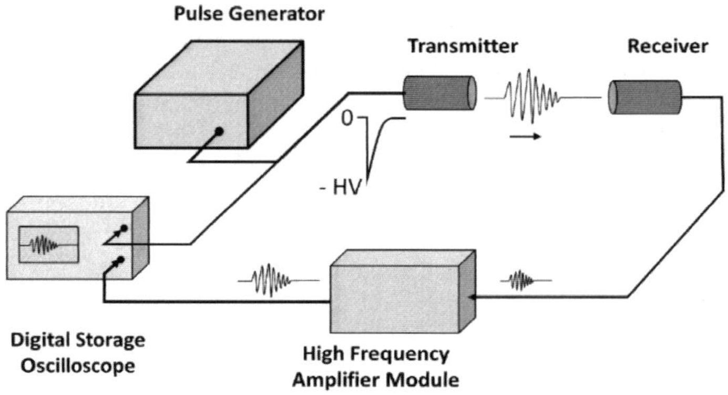

Figure 1. Schematic system for testing the fabricated pulse generator using the transmission mode propagation.

Table 1. Specifications of transducers employed for the testing.

Parameters	Transducer 1	Transducer 2
Brand	ZHIPU	XIN NUO QI
Diameter	18 mm	21 mm
Center Frequency	200 ± 8.0 kHz	1.0 MHz ± 5%
Bandwidth	16 kHz	0.2 MHz
Oscillation Excitation 0 dB	200 Vpp, 50 Bursts, 20 cm	500 Vpp, 8 Bursts, 8 cm
Min. Parallel Resistance	600 Ω + 30%	130 Ω ± 30%
Capacitance	400 pF + 20%	1200 pF ± 20%
Input Voltage Max. (Pulse)	500 Vpp	500 Vpp
Directivity at -3dB	7° ± 2°	7° ± 2°
Detection Distance	0.1 - 2 m	0.1 - 5 m

2.2 Frequency Domain Transformation

The received ultrasonic signal may contain a direct current (DC) which does not carry any information. This appears as the mean of recorded signal. The elimination of DC content is performed as described in equation (1) [5].

$$\alpha x_i = x_i - \frac{1}{N}\sum_{i=0}^{N-1} x_i \quad (1)$$

with x_i is the signal of current in the recorded digital signal, N is number of samples, and $\alpha\, x_i$ is the DC-eliminated signal.

Stromer and Ladani applied rectangular, Tukey, Hann, Hamming and Blackman-Harris window functions to a transmission-mode received digital ultrasonic waves for the detection of inclusion in an object. This reported that the resulting power spectra of received signal were smooth and computational efficiency improved. Hence, for suppressing noises due to signal discontinuity, it applies a Hanning Window prior to the Fourier Transform and Power Spectrum calculations. Power spectrum of signal is calculated using equation (2) and (3) [5].

$$\hat{P}(f) = \frac{1}{N}\left|\sum_{n=1}^{N} x_n exp(-jfn)\right|^2 \quad (2)$$

$$\bar{P}(f) = \frac{1}{A^2}\hat{P}(f) \quad (3)$$

IOP Conf. Series: Journal of Physics: Conf. Series **1080** (2018) 012034 doi:10.1088/1742-6596/1080/1/012034

with $\hat{P}$ is the power spectrum of signal. The normalized power spectrum, $\bar{P}$ by a factor A.

3. Results and Discussions

Figure 2(a) shows the excitation pulse of transducer. The pulse is negative spike -740 V with 107.1 ns at half width. This signal has data length of 5000 and sampling frequency of 50 MHz. Spike excitation pulse optimizes broadband response and near surface resolution. The fast rise times of pulse proportionally contains broadband frequency spectral excitation resulting in wideband transducer response. The optimization of transducer response can be done by selecting pulse energy and damping values, which adjust pulse rise time, width, and voltage. The pulse must have a width at which the pulse frequency content covers the bandwidth of transmitter transducer.

Frequency domain transformation of the excitation pulse is shown in figure 2(b). The frequency content spreads up to 10 MHz. The maximum peak is at a frequency of about 1 MHz. In the range of 0-5 MHz, the power spectrum exhibits higher magnitude. These results identify that the pulse contains frequency bandwidth of 5 MHz enabling transducer excitation with frequencies up to 5 MHz. This indicated that the excitation pulse adequately drives the transmitter transducer with a center frequency up to 5 MHz. This pulse is subsequently applied to excite transmitter transducers with frequencies of 200 kHz and 1 MHz as described in the following discussions.

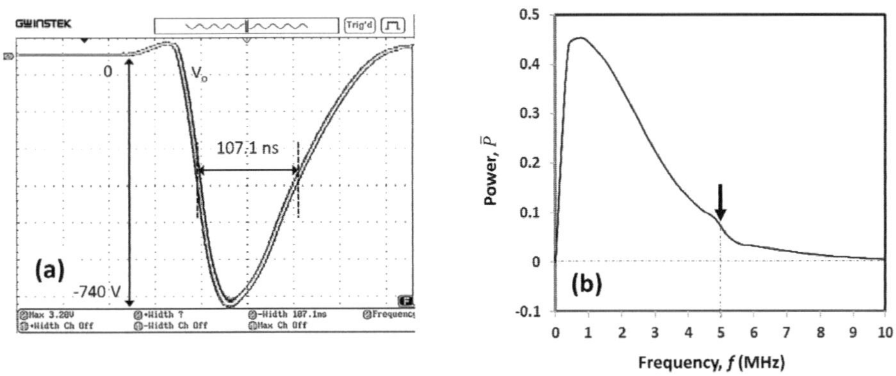

Figure 2. The excitation pulse of transmitter transducer: (a) time-domain negative spike pulse and (b) frequency domain.

The typical pulse-excited ultrasonic signal is an underdamped oscillation with a certain damping ratio as shown. A spike negative voltage -740 V with a half-width of 107.1 ns excites the transmitter transducer with frequency of 200 kHz. This ultrasonic wave propagates through the medium in which a portion of ultrasonic energy absorbed by the medium resulting in the reduction of signal amplitude of received signal [13]. Figure 3(a) shows the time-domain digital signal received at the receiver transducer, with data length of 5000 and sampled at a frequency of 250 MHz. The signal is underdamped oscillation with a frequency of 200 kHz. The oscillation amplitude diminished in accordance with damping ratio of transducer.

It is difficult to visually extract the information contains in the time-domain signal, particularly, frequency and amplitude components [14]. Prior to Fourier Transformation, a Hanning window with length of 5000 is applied to the signal (figure 3(a) and 3(b)). As a result, the points at begin and end of the windowed signal has zero values, suppressing high frequency noises due to signal discontinuities and enhances the signal-to-noise ratio (SNR) of the signal (figure 3(b)) [15].

Using equation (1-3), the windowed signal is calculated as a frequency domain signal as shown in figure 3(d) It is seen that a dominant peak appears at 0.2 MHz. No other peaks appear. This results strongly indicated that the received ultrasonic signal contains dominantly frequency of 200 kHz, in accordance with the center frequency of transducer used.

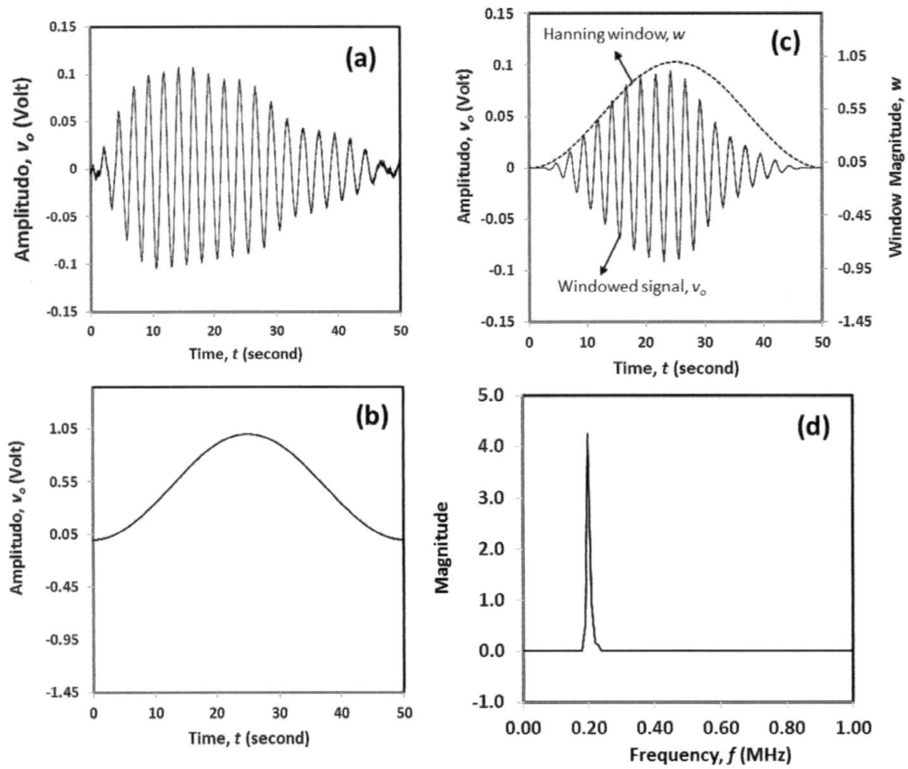

Figure 3. Ultrasonic signal received at the 200-kHz receiver transducer, (a) as-received time domain signal, (b) Hanning windowed signal, (c) Hanning-windowed signal and (d) frequency domain signal.

In addition to air coupled transducer, a pair of 1-MHz water-coupled transducer (Transducer 2) was applied using this method as well. The similar spike signal excites a water-coupled transducer transmitter having a center frequency of 1 MHz. The time domain of one packet received signal is shown in figure 4(a). The measurement shows that it is containing a frequency 961.5 kHz lower than its specified center frequency value (table 1). However, this value is acceptable because it is in the range of 5% tolerance of center frequency (range of 950-1050 Hz). At one spike pulse excitation, the amplitude diminished, decaying from maximum peak-to-peak 2.5 V to zero, exhibiting under damped oscillation characteristic. These results demonstrate that the frequency content of pulse is adequately capable to excite ultrasonic transducer in the order of 1 MHz of frequency.

The recorded signal is initially windowed to cut out zero values at the beginning, resulting signal with length of 3801 (figure 4(b)). Figure 4(c) shows the application of a 3801-data-length Hann window, resulting in zero values on begin and end points. The power spectrum of windowed signal is apparent in figure 4(d). Two adjacent peaks of 0.965 and 1.0 MHz appear, indicating that the received signal contains two dominant frequencies which they are indistinguishable in time domain.

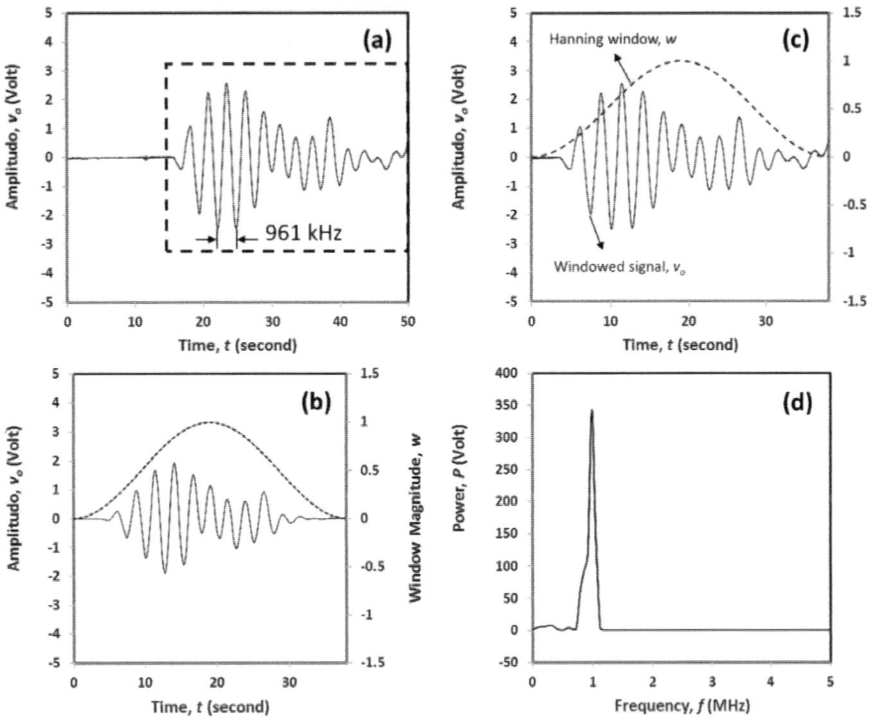

Figure 4. Ultrasonic signal received at the 1-MHz receiver transducer, (a) time domain signal and (b) Hann window function, (c) Hann windowed signal and (d) frequency domain signal.

4. Conclusion

Ultrasonic data acquisition in transmission mode has been demonstrated. Digital signal processing using Fourier Transform of received signals has been demonstrated to evaluate its frequencies content. The results concluded that calculated frequencies of Fourier Transformed signals strongly in accordance with that of measurement in the time domain. In conclusion, this technic is prospective applicable to ultrasonic signal analysis.

Acknowledgements

We acknowledged Kementerian Riset, Teknologi dan Pendidikan Tinggi (Kemenristekdikti) Republic of Indonesia for providing research scheme fund of Penelitian Unggulan Perguruan Tinggi (PUPT) Universitas Padjadjaran no. 718/UN6.3.1/PL/2017 year of 2017.

References

[1] Park J H, Yang S H, Lee H R, Yu C B, Pak S Y, Oh C S, Kang Y J and Youn J R 2017 *J. Sound Vib.* **397** 17
[2] Pal B 2017 *Ultrasonics* **73** 140
[3] Hidayat D, Widiyastuti W, Ogi T and Okuyama K 2010 *Aerosol Sci. and Technol.* **44** 692
[4] Purwanto A, Hidayat D, Terashi Y and Okuyama K 2008 *Chem. Mater.* **20** 7440
[5] Xu Y, Xu Y, Ding Z, Chen Y, Su B, Ma Z and Sun Y -N 2017 *Ultrasonics* **73** 77
[6] Queirós R, Alegria F C, Girão P S and Serra A C 2015 *Ultrasonics* **63** 86
[7] Stromer J and Ladani L 2018 *Biomed. Signal Proces.* **40** 454
[8] Granchi S, Vannacci E, Biagi E and Masotti L 2016 *Ultrasonics* **68** 89
[9] Jiao J, Liu W, Zhang J, Zhang Q, He C and Wu B 2013 *Mech. Syst. Sig. Process.* **35** 69
[10] Sen C and Tie G 2012 *Trans. Nonferrous Met. Soc. China* **22** s323
[11] Song Y, Ma J and Feng Y 2011 *Procedia Eng.* **15** 2211
[12] Maria V, Felice and Fan Z 2018 *Ultrasonics* **88** 26

[13] Hidayat D, Joni I M, Setianto, Panatarani C and Okuyama K 2013 *AIP Conf. Proc.* **1554** 87
[14] Tamrakar P, Azari H, Yuan D, Nazarian S 2017 *Transp. Geotech.* **12** 101
[15] Tiwari K A, Raisutis R, Samaitis V 2017 *Procedia Struc. Integrity* **5** 1184

IOP Conf. Series: Journal of Physics: Conf. Series **1080** (2018) 012035 doi:10.1088/1742-6596/1080/1/012035

Mapping and analysis of the physical properties of rocks for building materials in the Sunda Strait region

N Haerudin[1*], A Zaenudin[1], Y Aribowo[2] and I G B Darmawan[1]

[1]Geophysical Engineering, University of Lampung, Bandar Lampung, Indonesia 35145

[2]Geological Engineering, Diponegoro University, Semarang, Indonesia

[*]nandi.haerudin@eng.unila.ac.id

Abstract. This research connected the gravity value with the existence of andesite rock with good quality physics. Based on gravity mapping, high gravity anomalies in southern Lampung relate to igneous rocks, especially granite and granodiorite plutons, meanwhile andesite rock in medium gravity anomalies. Further geological observation and test of andesite rock physical properties become very important, in order to support the fulfilment of quantities with good quality for the acceleration of development of the Sunda Strait area. The andesite potentials in the Tarahan and Bakauheni areas are predominantly located on the medium Bouguer anomaly. The results of petrographic analysis showed three types of rocks, namely hornblende andesite, basalt amphibolite and tuff. The magma series based on the Atomic Absorption Spectrometry (AAS) test results on the three samples shows the calc-alkaline magma series. While from the testing of the physical properties of rock shows properties that meet the standard SII.0378-80 by the ultimate of the strength, force value range between 1125 to 1558 kg/cm^2, in which the material can be used for medium to heavy buildings raw materials.

1. Introduction

The Sunda Strait area is located between the southeastern tip of Sumatra island with the northwest tip of Java island separated by the Sunda Strait as strategic location for being the link between Java and Sumatra also one of the strategic central government development programs. As a strategic area under development, infrastructure development activities require the availability of sufficient and adequate raw material resources in quantity and quality. The availability of volcanic rock formations dominated by andesite rocks around the Sunda Strait area becomes an interesting option to be explored. The distribution of rock resources can be done by mapping gravity response of gravity method, while the physical quality of the rock can be known by uniaxial compressive strength test.

This study aims to map the potential location of rock material resources by identifying gravity response and ultimate strength force (qu) from some geological observation sample especially in Tarahan and Bakauheni area. The utilization of gravity data as a density distribution response (Xu *et al.*, 2015), is excellent for identifying subsurface structures (Evariste *et al.*, 2014), lithology of bedrock and near surface rocks (Kilaru *et al.*, 2013; Zahra *et al.*, 2016). This study will also illustrate the gravity anomaly relationship, especially regarding to the existence of andesite rocks that have been and are being explored for further needs in the future. In addition, physical properties tests were

IOP Conf. Series: Journal of Physics: Conf. Series **1080** (2018) 012035 doi:10.1088/1742-6596/1080/1/012035

conducted to ensure that the andesite rocks meet the SII.0378-80 standard as good quality materials, for light to heavy building.

2. Geological Setting

The schist rocks as the oldest rock in the study area consisted of mica-schist and amphibole-schist identified as the basement (Mangga *et al.*, 1993) and interpreted as a metamorphism product of deep ocean sediments (Amin *et al.*, 1994) and deformation during the accretion process at volcanic arc sediments is deposited in subduction trenches (Barber *et al.*, 2005). The tertiary andesitic lava exposed in the Tarahan and Bakauheni areas is a potential target as an area of observation and sampling while dominance of the quaternary hornblende andesite and the pyroclastic rocks are exposed on Mount Rajabasa (Darmawan *et al.*, 2015; Haerudin *et al.*, 2012).

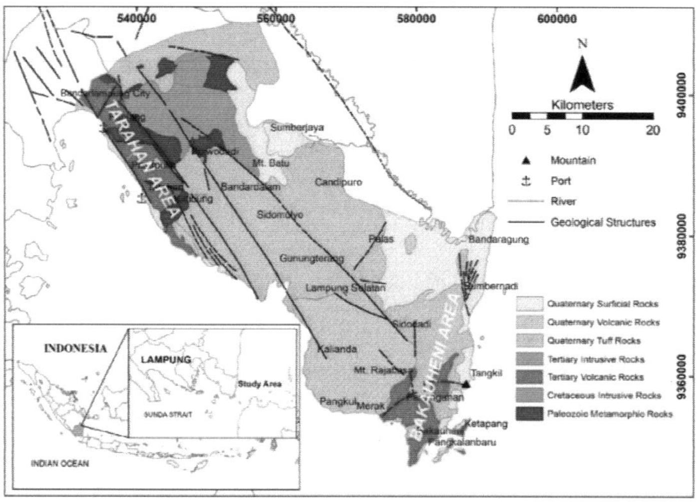

Figure 1. Geological map showing the geological conditions of the Tarahan and Rajabasa regions (modified from Mangga *et al.*, 1993).

The dominant northwest-southeast fault structure (figure 1) known as the Sumatran fault system (Pulunggono and Cameron, 1984), is interpreted as a result of a subduction system in the Sunda arc. The complex accretion zones with anticlinal ridges and elevated faults (Malod *et al.*, 1995) are thought to be the result of changes in the direction of subduction in the Sunda Strait. The process of magmatism can be observed in Anak Krakatau volcano activity up to Mount Rajabasa, even the volcanic system formed on Rajabasa Mountain is the result of volcanic cone reconstruction of pre-Rajabasa (Bronto *et al.*, 2012).

3. Gravity Anomaly

The result of rock density mapping represented by the Bouguer anomaly response in the research location is shown in figure 2. Location of the rock observation in the field is focused on two areas namely Tarahan area and Bakauheni Area. Based on the observations in the field, the andesite rock distribution is mostly located in medium gravity anomaly response. The pattern is generally the same as the pattern of andesitic rock distribution in Tarahan and Bakauheni area that is on the medium Bouguer anomaly. The low anomaly response is thought to be the dominance of surface rock and geothermal potential in the Mount of Rajabasa. This condition is related to the fracture zone due to reservoir and geothermal fluid path in Mount of Rajabasa.

An interesting pattern of Bouguer anomalies is in the area between Tarahan and Bakauheni with a high value. This condition needs to be confirmed whether the high anomaly is caused by the presence of intrusive rocks or associated with bedrocks that is close to the surface. Due to the regional geology, the area is in the Lampung Tuff Formation and has not been found any rock outcrops intrusion yet. This anomaly covers the Sidomulyo-Gunungterang-Palas area.

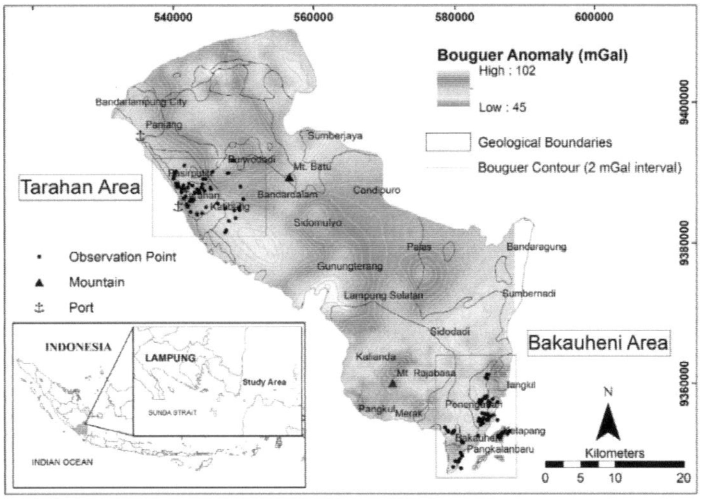

Figure 2. Complete Anomaly Bouguer Map of research area (Tarahan and Bakauheni). The red box is the location of observation and sampling. The black dots are a rock observation point.

4. Field Observation

Several geological observations in the field have shown some of Tertiary volcanic rock outcrops in the Tarahan area such as amphibole and andesite rocks (figure 3). Amphibole lithology is generally dark in color, very compact and hard, and hornfelsik texture. Based on the field observations, the main constituents are plagioclase and hornblende. While the andesitic outcrops are dark, hard and compact, there is a massive, heavily padded structure with afanitic-porphyritic textures. The results showed that the main constituent minerals in these andesite rock outcrops were composed of plagioclase and hornblende minerals.

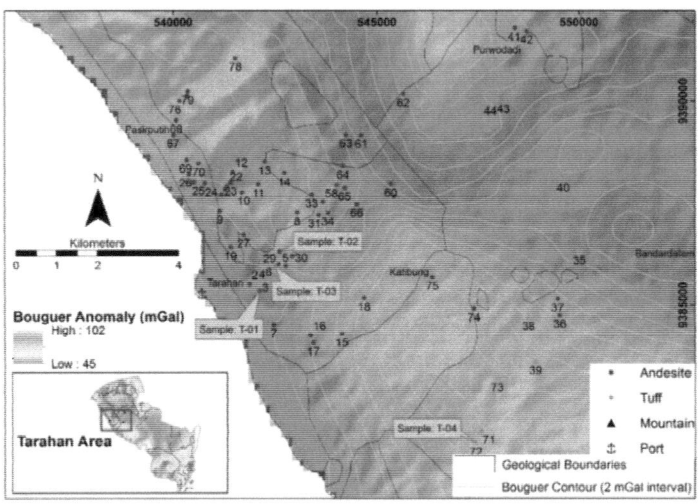

Figure 3. Geological observation and sampling point in Tarahan area. Red dots show the andesite outcrop and yellow dots is tuff.

IOP Conf. Series: Journal of Physics: Conf. Series **1080** (2018) 012035 doi:10.1088/1742-6596/1080/1/012035

Some tuff outcrops are also found in observations in the Tarahan area included in the Lampung Tuff unit. This unit is characterized by bright grey lithology, fragmental texture composed of fine-grained volcanic ash, coarse, low hard / relatively soft.

The outcrop of andesite rock in the Bakauheni area has similar characteristics with the fresh colour of dark grey and the brownish-grey colour of the weather. This rock has aphanitic to porphyritic texture with sheeting joint structure. The condition of andesite rocks exposed in the fresh state is very harsh and the observation of mineral shows the main constituents of andesite rock in Bakauheni in the form of plagioclase and hornblende. Other outcrops of pyroclastic rocks such as tuffaceous are also often found to be mixed with tuffaceous sandstone and tuffaceous breccia (figure 4).

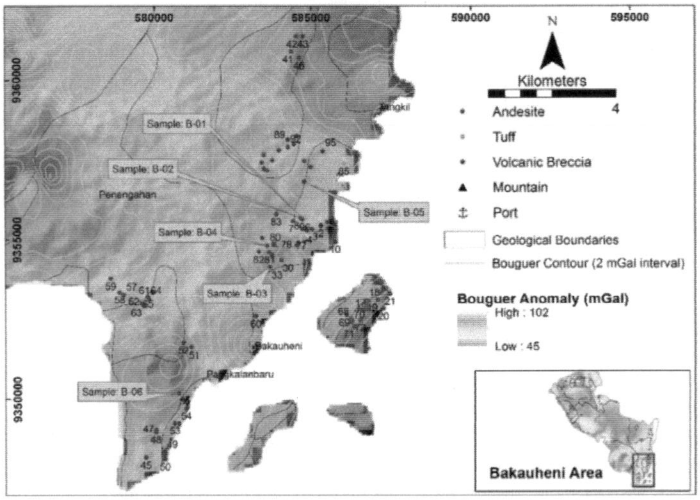

Figure 4. Geological observation and sampling point in Bakauheni area. Red dots show the andesite out-crop, purple dots are volcanic breccia and yellow dots is tuff.

5. Petrography and Strong Pressure Uniaxsial

5.1. Petrography Analysis

Petrographic analysis of microscopic rock samples was performed with the object of analysis of the thin section from rocks sample, rocks cut and honed to a thickness of 0.03 mm, and then was analyzed by using polarization microscope. Petrographic analysis is intended to identify the microscopic appearance of rocks, including textures, structures and types of minerals that make up the rocks along with their abundance. The determination of microscopic mineral types is based on differences in mineral optical properties (mineral properties associated with the ability of minerals to absorb and forward polarized light).

Petrographic analysis is needed to confirm the rock type and rock minerals more accurately, due to limited observations with the eyes without the use of tools. The analysis was performed on 10 selected samples that were considered representative of each visually visible variation of rocks and represented rock units from each location in Tarahan and Bakauheni. There are three types of rocks observed by petrography analysis: basalt-amphibole, hornblende andesite, and tuff, with andesite rock texture variations including porphyry andesite, while tuff variations include fine/ash tuff and lapilli tuff. The constituent minerals from 10 samples of petrographic analysis are shown in table 1.

5.2. Strong Test Press Uniaxial

The uniaxial pressure on samples of cylinder rocks is the most commonly used for tests of mechanical properties. The uniaxial compressive strength test is performed to determine the compressive strength of the rock (σt), Young's Modulus (E), Poisson's Ratio (v), and stress-strain curve. Examples of cylindrical rocks are pressed or burdened to collapse. The comparison between the height and

IOP Conf. Series: Journal of Physics: Conf. Series **1080** (2018) 012035 doi:10.1088/1742-6596/1080/1/012035

diameter of the commonly used cylinder example is 2 to 2.5 with a flat, smooth and parallel loading surface perpendicular to the axis of the rock sample axis. From the test results will be obtained data of compressive strength of rock (σc or qu) as the following.

Table 1. The constituent minerals from petrographic analysis of samples in both Tarahan and Bakauheni area.

Component	Sample (%):									
	T-01	T-02	T-03	T-04	B-01	B-02	B-03	B-04	B-05	B-06
Plagioclase	65	25	35	15	-	45	15	25	25	45
Hornblende	25	55	25	-	-	15	10	10	20	15
Opaque	10	10	5	-	-	10	10	10	10	30
Groundmass	-	10	10	-	35	40	65	55	45	10
Volcanic Glass	-	-	25	15	-	-	-	-	-	-
Volcanic Ash	-	-	-	70	65	-	-	-	-	-
Rock Type	HA[a]	BA[b]	HA[c]	LT[c]	AT[d]	HA[a]	HA[a]	HA[a]	HA[a]	HA[a]

[a] Hornblende Andesite [b] Basalt Amphibole [c] Lapilli Tuff [d] Ash Tuff

The voltage price at which the rock sample is destroyed is defined as the uniaxial strength of the rock and is given by the relationship:

$$\sigma c = \frac{F}{A} \tag{1}$$

by:
σc = Uniaxial rock strength (MPa)
F = Force that works when rock samples are destroyed (kN)
A = Area of the initial cross-section of rock samples perpendicular to the force (mm)

Based on the analysis of compressive strength and rock density of 3 samples from Tarahan and Bakauheni, generally it showed a high compressive strength value, andesite had the ultimate compressive strength range of 1125 to 1364 kg/cm^2 while ampibolite had a greater compressive strength value of 1558 kg/cm^2 (table 2). When referring to SII.0378-80, the minimum compressive strength for heavy building foundations is 1500 kg/cm^2 and the minimum compressive strength for medium building foundations is 1000 kg/cm^2, then ampibolite type rocks are eligible for heavy building foundation (qu > 1500 kg/cm^2), while all tested andesites meet the criteria as a medium building foundation because of the 1500> qu-value> 1000 kg/cm^2 (table 2).

Table 2. Strong Pressure and rock density analysis results.

Code	**Location**	**Rock type**	**Ultimate strength force (qu) kg/cm^2**	**Receive strength force (qa) kg/cm^2**	**Density (gr/cm^3)**
T-01	Tarahan	Andesite	1364.755	454.9184	2.89
T-02	Tarahan	Amphibole	1558.244	519.4148	3.23
T-03	Tarahan	Andesite	1254.858	418.2859	2.83
B-02	Bakauheni	Andesite	1294.154	431.3846	2.79
B-03	Bakauheni	Andesite	1125.224	375.0746	2.86
B-04	Bakauheni	Andesite	1232.621	410.8738	2.77
B-05	Bakauheni	Andesite	1253.924	417.9746	2.74
B-06	Bakauheni	Andesite	1139.289	379.7629	2.59

6. Atomic Absorption Spectrometry (AAS)

To find out the magma series of andesite rocks in Tarahan and Bakauheni, AAS test was conducted on three selected samples. This test is performed to determine the source of magma and the ratio of quartz content to other major oxides. This analysis is also used to determine the abundance of the major

IOP Conf. Series: Journal of Physics: Conf. Series **1080** (2018) 012035 doi:10.1088/1742-6596/1080/1/012035

oxides that are closely related to the formation of the main mineral in the rocks. The Abundance of certain minerals also characterizes the process of tectonic formation and setting, one of which is an alkaline compound which is represented by oxides K_2O and Na_2O. The result of AAS analysis is shown in table 3.

Based on its silica abundance (SiO_2) ranging from 59-62%, still within the range of slightly acidic intermediate magma composition, reinforced with significant Al_2O_3 concentration, approximately 15% indicates a close association with plagioclase abundance. Alkali (sodium and potassium) are present in significant amounts in which K_2O ranges from 1.9-2.5% and Na_2O 3.8-4.6% shows a slight influence of continental crust and magma formation is closely related to subduction of Indian oceanic crust with the influence of Eurasian Continental Crust.

Table 3. Analysis results of AAS test show the major oxide of rocks sample.

Molecular	T-02	B-02	B-06
SiO_2	59.3	60.1	61.9
TiO_2	0.52	0.7	0.6
Al_2O_3	15.9	14.8	16.4
Fe_2O_3	6.4	6.9	5.1
MgO	2.7	2.8	1.7
CaO	7.2	5.9	6.6
Na_2O	4.6	3.8	4.3
K_2O	2.1	2.5	1.9
LOI	1.3	1.8	1.6
	99.96	99.3	100.1

Based on the results of the major oxide plot shown in figure 5, it is known that the andesite rock magma series at the study site are calc-alkaline. The composition of magma in the sample rocks is dominated by partial melting and mixing between continental crust and oceanic crust that undergo fractional crystallization and become felsic magma with low magnesium and iron content. Calc-alkaline type rocks are commonly found in subduction zones, especially in volcanic arcs located above continental plates.

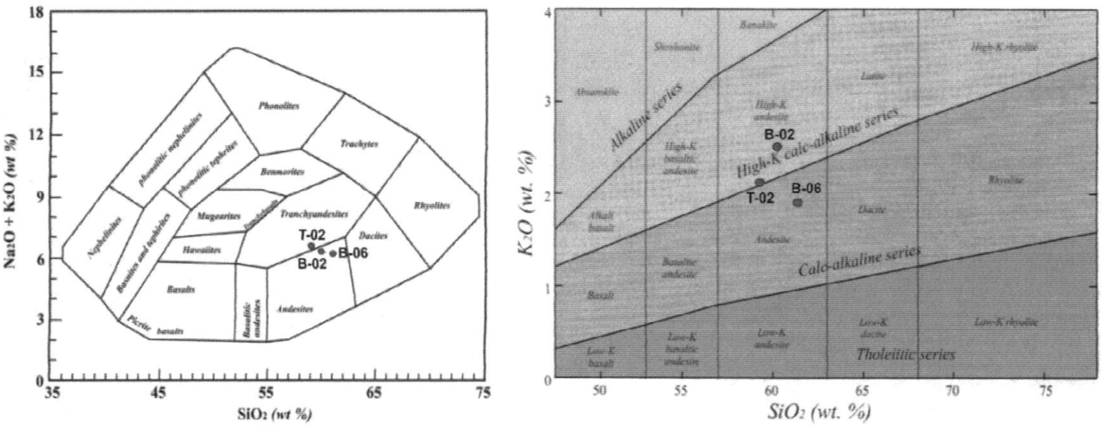

(a) (b)

Figure 5. Plot of magma classification based on (a) percentage of major oxide SiO_2 with alkaline and (b) total alkaline. The plot was based on AAS analysis data on three samples (indicated by red dots).

The plot results also support the results of petrographic analysis, which indicate that rock samples analyzed are andesite rocks with the condition of one sample is an trachyandesite (sample T-02). The composition of the sample rock texture with the appearance of the orientation of minerals / orientation direction is the direction of flow visible on the petrography section. These results confirm that the andesite rocks in the Tarahan area, especially in the T-02 sample are intermediate rocks formed on the back arc tectonic order. The trachyte texture of this sample also characterizes the development of rocks in shallow intrusions or lava.

7. Conclusion

1. The andesite rocks distribution are mostly located in medium gravity anomaly response. The pattern is generally the same as the pattern of andesitic rocks distribution in Tarahan and Bakauheni area that is on the medium Bouguer anomaly.
2. The outcrop of andesite rocks in both Tarahan-Bakauheni areas has similar characteristics with the dark color, hard and compact to dark gray and the brownish-gray color with aphanitic to porphyritic texture and are composed of plagioclase and hornblende minerals.
3. There are three types of rocks observed by petrography analysis: basalt-amphibole, hornblende andesite and tuff, with andesite rock texture variations including porphyry andesite, while tuff variations include fine/ash tuff and lapilli tuff.
4. Based on compressive strength analysis generally it shows a high enough value of compressive strength, where andesite has the ultimate compressive strength between 1125 to 1364 kg/cm^2 while ampibolite has a greater compressive strength value of 1558 kg/cm^2. Referring to SII.0378-80, the ampibolite type rocks are eligible for heavy building foundations (number of > 1500 kg/cm^2), while all tested andesites meet the criteria as medium building foundation material because 1500 > its value > 1000 kg/cm^2.
5. The andesite rocks magma series at the study site are calc-alkaline dominated by partial melting and mixing between continental crust and oceanic crust, meanwhile the trachyandesite rocks in the Tarahan area are intermediate rocks formed on the back arc tectonic order.

Acknowledgments
The authors would like thanks to KEMENRISTEK-DIKTI-Indonesia who has funded this research through the scheme of the MP3EI grant.

References

[1] Amin T C, Sidarto, Santosa S and Gunawan W 1994 *The Geology of the Kotaagung Quadrangle (1010) Sumatra Scale 1:250 000* (Bandung: Geological Research and Development Centre)
[2] Barber A J, Crow M J and Milsom J S 2005 *Sumatra: Geology, Resources and Tectonic Evolution* (UK: The Geological Society Publishing House)
[3] Bronto S, Asmoro P, Hartono G and Sulistiyono 2012 *Indonesian J. of Geology* **7** 11
[4] Darmawan I G B, Setijadji L D and Wintolo D 2015 *Proc. World Geothermal Congress IGA* (Australia)
[5] Evariste N H, Genyou L, Tabod T C, Joseph K, Severin N, Alain T and Xiaoping K E 2014 Crustal structure beneath Cameroon from EGM2008 *J. Geodesy and Geodynamics* **5** 1
[6] Haerudin N, Wahyudi, Suryanto W and Sarkowi M 2012 *Proc. PIT API XII* (Indonesia)
[7] Kilaru S, Goud B K and Rao V K 2013 *J. Geoscience Frontiers* **4** 717
[8] Malod J A, Karta K, Beslier M O and Zen M T JR 1995 *J. Southeast Asian Earth Sciences* **12** 85
[9] Mangga S A, Amirudin, Suwarti T, Gafoer S and Sidarto 1993 *Geological Map of Tanjungkarang Sumatra* (Bandung: Geological Research and Development Centre)
[10] Pulunggono A and Cameron N R 1984 *Proc. 13th Annual Convention IPA* **1** 121
[11] Xu C, Wang H, Luo Z, Ning J and Liu H 2015 *J. of Applied Geophysics* **114** 174
[12] Zahra H S and Oweis H T 2016 *J. of Astronomy and Geophysics* **5** 106

IOP Conf. Series: Journal of Physics: Conf. Series **1080** (2018) 012036 doi:10.1088/1742-6596/1080/1/012036

Fracture analysis to delineate tectonic evolution: case study at Ciherang, Bogor-Sukabumi, West Java, Indonesia

C Endyana[1,3*], Hendarmawan[1,3], E Sukiyah[1] and I A Dharmawan[2]

[1]Faculty of Geological Engineering, Universitas Padjadjaran, Indonesia
[2]Department of Geophysics, Faculty of Mathematics and Natural Science, Universitas Padjadjaran, Indonesia
[3]School of Post Graduate, Universitas Padjadjaran, Indonesia

[*]cipta.endyana@unpad.ac.id

Abstract. Indonesia, especially Java island is strongly influenced by tectonic processes and volcanic processes. The phenomenon of fractures in volcanic areas provide a significant contribution to the process of calculating the potential of its groundwater resources. Ciherang area, West Java is located among two quaternary volcanoes, namely Mt. Salak and Mt. Pangrango, composed by volcanic deposits of both volcanoes, and also affected by Pelabuhanratu fault as regional fault. Fracture pattern on Mt. Salak and Mt. Pangrango not only influenced by volcanism only but also influenced by Pelabuhanratu fault with relatively older than the volcanic processes that occur in both the mountain. By using the method of automatic lineament extraction on ASTER GDEM raster image with the resolution of 30m × 30m per pixel, then obtained the lineament of the ridge and valley. Each group has its character itself with influence by volcanic processes, tectonic processes or both of volcanic and tectonic processes. Subwatershed created to divide fracture and to describe tectonic style involved in each subwatershed, due to subwatershed become important parameter to restrict population of lineament. Delineation of fracture pattern based on subwatershed have a difference style of direction for each subwatershed.

1. Introduction

The Indonesia Archipelago is a country which highly potential of natural resources. Many natural resources in Indonesia such as oil and gas, mineralization and groundwater. Potential of its natural resources appears because the area of Indonesia Archipelago is included in active tectonic and active volcanic area. Therefore, there are many active mountains and unstable lands found as a result of tectonism and volcanism.

Volcanic area is an area containing a large groundwater resources. Volcanic areas are contained large groundwater resources. Groundwater is contained in the volcanic rocks that have a high permeability values. Volcanic rocks with many variation of deposit because difference of eruption mechanism, many of eruption period. All those processes induce more variation of volcanic deposit. Igneous rocks- lava, with highly hardness, pyroclastic rock with variation of component-breccia, and tuff. Fracture in volcanic area involved as secondary porosity. Nowadays, the fracture system become one of many parameters for calculating groundwater reserves in volcanic area [1]. Ciherang as a study area has been chosen as an area with tectonic and volcanic affected. Ciherang area is affected by

Pelabuhanratu Regional Fault and also two active volcanoes Mt. Salak and Mt. Pangrango. Two different processes may develop two fracture patterns which each pattern will depend on their process.

The fracture patterns become important to be discovered as detail to know clearly the fracture system that develops in study area. In generally, fracture pattern appear is different for each volcanic fracture patterns. There is different fracture pattern between fracture pattern in Mt. Salak and Mt. Pangrango (figure 1). Beside of two volcanic patterns, there is another fracture pattern which is Pelabuhanratu Fault as a tectonic product involved to this Ciherang area This regional fault has been formed relatively older then volcanic processes and still an active fault until present.

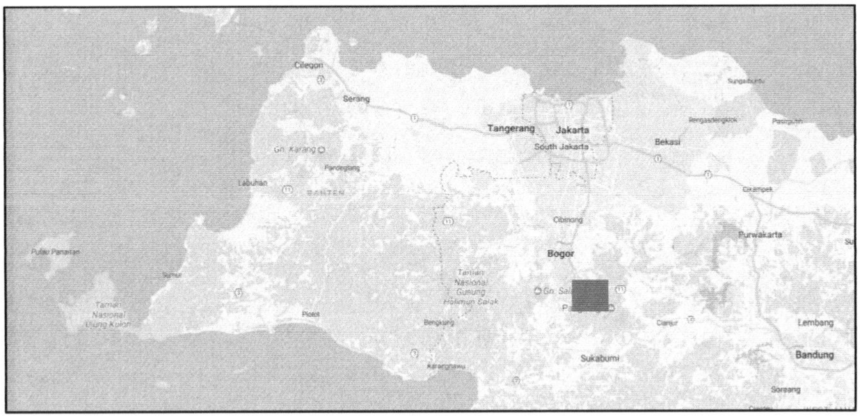

Figure 1. Location of Ciherang between Bogor - Sukabumi, West Java, Indonesia (shown in red point in map of Java Island).

2. Regional Setting

The oldest deposit in Ciherang area consist of pumice tufaceous included in Old Volcanic Rock Unit. This unit spread out at eastern, northern, and southern part of study area. The Old Volcanic Rock Unit is composed of andesitic and basaltic lava that included as Mt. Pangrango volcanic deposit. The western part of study area composed with lava, tufaceous breccia and lapilly tuff. The characterization of lava flow is andesitic basalt which included on Mt. Salak volcanic deposit. All deposits as included to quaternary Volcanic Rock Unit as shows in figure 2 [2].

Tectonic activity in Ciherang area has been started at early Tertiary, followed with Plio-Pleistocene tectonic activity which is reactivation of products of early tectonic period as a fault with direction of northeast-southwest and northwest-southeast. Fractures develop was formed become weak zone and as a media for late quaternary volcanic deposits appears in surface area. Geological setting of his area, generally consist of quaternary rock unit as a young volcanic deposit unit that consist of tuffaceous pumice, tuffaceous laharic breccia from Mt. Pangrango with thick deposit. Below of this unit, the old volcanic rock had been deposited which is consist of andesitic-balsaltic breccia, andesitic lava, tuff and agglomerate. Fine clastic sediment rock until coarse spreading out southwardly which had been folded and faulted [3].

Appearance of volcanic mountain in Java Island closely related with subduction activity between oceanic plate and continental plate of Java Island. Based on geophysics data, show that subduction zone at southern Java Island is a constantly zone, but the appearance of volcanic mountain on surface area appears desperately between the others. This phenomenon indicate that volcanic activity not appears to surface itself, but must be fracture as a media become a pathway to release magma to surface. Those Fracture are always become fault zone, such as in Sumatera Island, the position of its volcanic mountain is in volcanic arc that its appearance on surface is located in Sumatera Fault Zone. As a geological fact above, many active mountains found with paleogence until present volcanic age certainly must be related with fault zone [4].

IOP Conf. Series: Journal of Physics: Conf. Series **1080** (2018) 012036 doi:10.1088/1742-6596/1080/1/012036

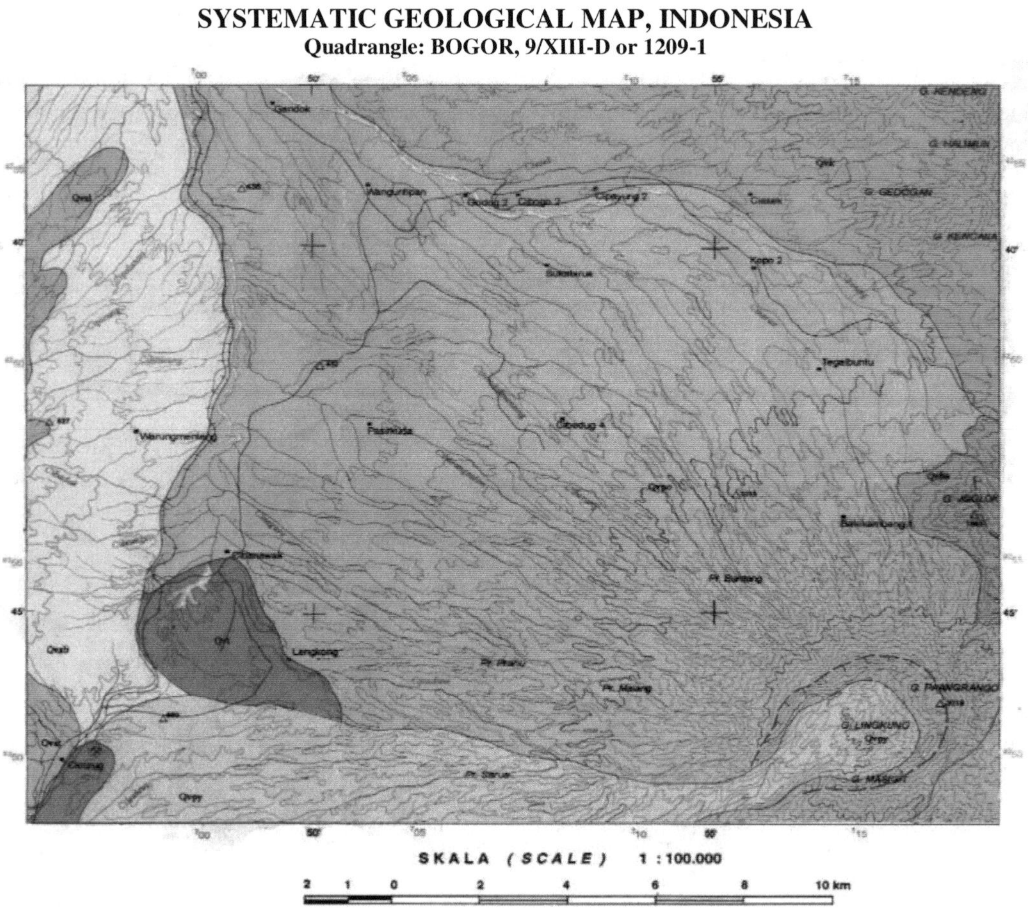

Figure 2. Geological setting of Ciherang area, West Java, Indonesia.

3. Methodology

The indication of geological structure in field almost undefined clearly at volcanic deposits. It was due to weathering rock condition. A Fault can be form as a fault plane or single fracture, but also found as fault zone which is consist of more than one fault. The fault zone or shear zone have long and wide dimension. The variation of fault zone dimension from minor until hundreds of kilometers. Joint shown the deformation and also state as minor fault. Fault with extensional mechanism such as subsidence, landslide, called fissure is not in fault definition [5].

Morphotectonic aspect like describe above can be analysis with remote sensing method and with Digital Elevation Model (DEM) image. This method can describe and visualize morphology at study area and the dominant of processes involved. The texture of DEM shows the tectonic that was developed, rock hardness, and other geological processes. Further, this method namely remote sensing. Structural analysis can be processed directly, which mean with detail observation at some outcrop or indirectly with analyzing of maps, satellite images, inter-correlation well profiles or seismic profiles based on concepts and theories. Several studies of morphotectonic are DEM enhancement with relief shading, lineament pattern, fracture domain density, and finally combining all parameters in overlaid map in thematic map with GIS technique.

Digital Elevation Model (DEM), in this case ASTER GDEM can be use and easy to get it. It is free to download. The resolution of ASTER GDEM is 30m x 30m and it is enough to use for fracture analysis with relief shading in volcanic area. Slope is incline plane with leveling from plain to steep. DEM must be convert to TIN raster in order to make a relief shading image, then slope can be calculated as an elevation point at each location [6]. Unit dimension of slope can be degree or percent. The slope aspect is slope orientation, measured clocked wise in degree 0-360° where 0° facing to north direction, 90° to east direction, 180° to south direction, and 270° to west direction. Hill-shading is the technique that used to show relief shading of rock, illuminated with hypothetic light. Illumination value for each raster cell determined by orientation of illumination based on slope direction and slope aspect [7].

Frequency and length of lineaments produced by DEM extraction depend on input of parameters value representing from LINE module in PCI Geomatica software [8]. The algorithm module extract lineaments from raster DEM and convert this lineament feature to vectors. Six parameters in LINE module are Filter Radius (RADI), Edge Gradient Threshold (GTHR), Curve Length Error Threshold (LTHR), Line Fitting Error Threshold (FTHR), Angular Difference Threshold (ATHR), and Linking Distance Threshold (DTHR) [9]. Recommended value to extract lineaments related to tectonism in study area and the optimal value for parameters in LINE module display in table 1. Relief Shading Image as a dataset used for data input in LINE module and calculation each image to predict length, orientation, number, and density.

Table 1. Threshold parameters in extraction lineaments.

Threshold Parameter	Value
RADI: Filter Radius	5
GTHR: Edge Gradient Threshold	75
LTHR: Curve Length Error Threshold	2
FTHR: Line Fitting Error Threshold	10
ATHR: Angular Difference Threshold	20
DTHR: Linking Distance Threshold	1

4. Result and Discussion

In DEM processing to determine of lineament pattern at study area need to expanded due to easier interpretation. Presenting of lineaments at topographical objects in DEM, it will be processed with hill-shade method as described section before [10]. First step is producing four relief shading images with each direction of illumination: 0°, 45°, 90°, 135°, 180° for positive image as a primary dataset. Second step is producing four relief shading images with each direction of illumination: 225°, 270°, 315° for negative image as a secondary dataset. Inclination of sun angel at 30° with contrast 0.20 to produce greater shade. Setting ambient light only for scale factor at shading in topographic program [11]. The edge enhancement is one of many methods for sharpening raster image which is detail geometric from an image can be modify or increase [12]. Lineament, drainage, and morphology feature particularly almost indicated by extreme difference of its value in radiometric respond. The edge enhancement is useful technique and efficient to increase those features. The edge enhancement should be applied to both relief shading dataset to improve frequency and contrast of both raster image [13]. Then, both dataset is already to extract lineament automatically by LINE module algorithm. Lineaments appears and convert to line vector and restricted by subwatershed shows in figure 3 and 4.

Spatial techniques used to delineate distribution of lineament pattern that has been produced by automatic extraction before. Subwatershed created to divide fracture and to describe tectonic style involved in each subwatershed, due to subwatershed become important parameter to restrict population of lineament. Delineation of fracture pattern based on subwatershed have a difference style of direction for each subwatershed. Distinction of style direction of each subwatershed has been plotted to Rose Diagram of lineament. Each direction of each subwatershed (SW) describe in table 2. (figure 5).

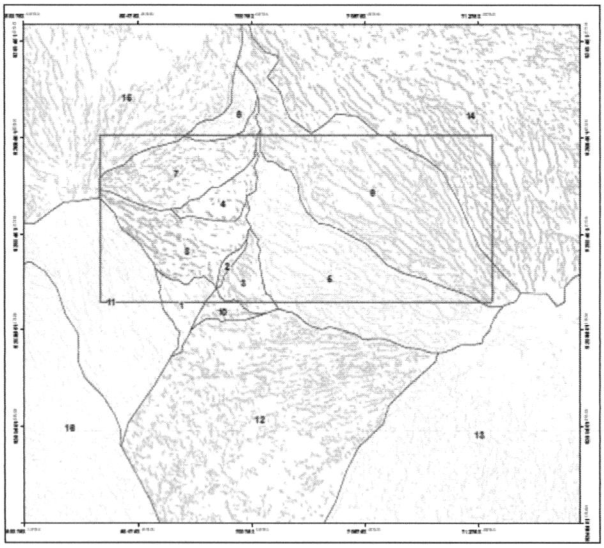

Figure 3. Positive relief shading dataset.

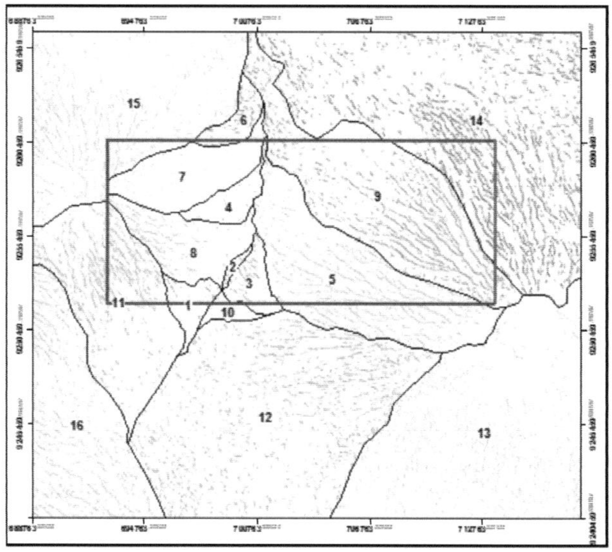

Figure 4. Negative relief shading dataset.

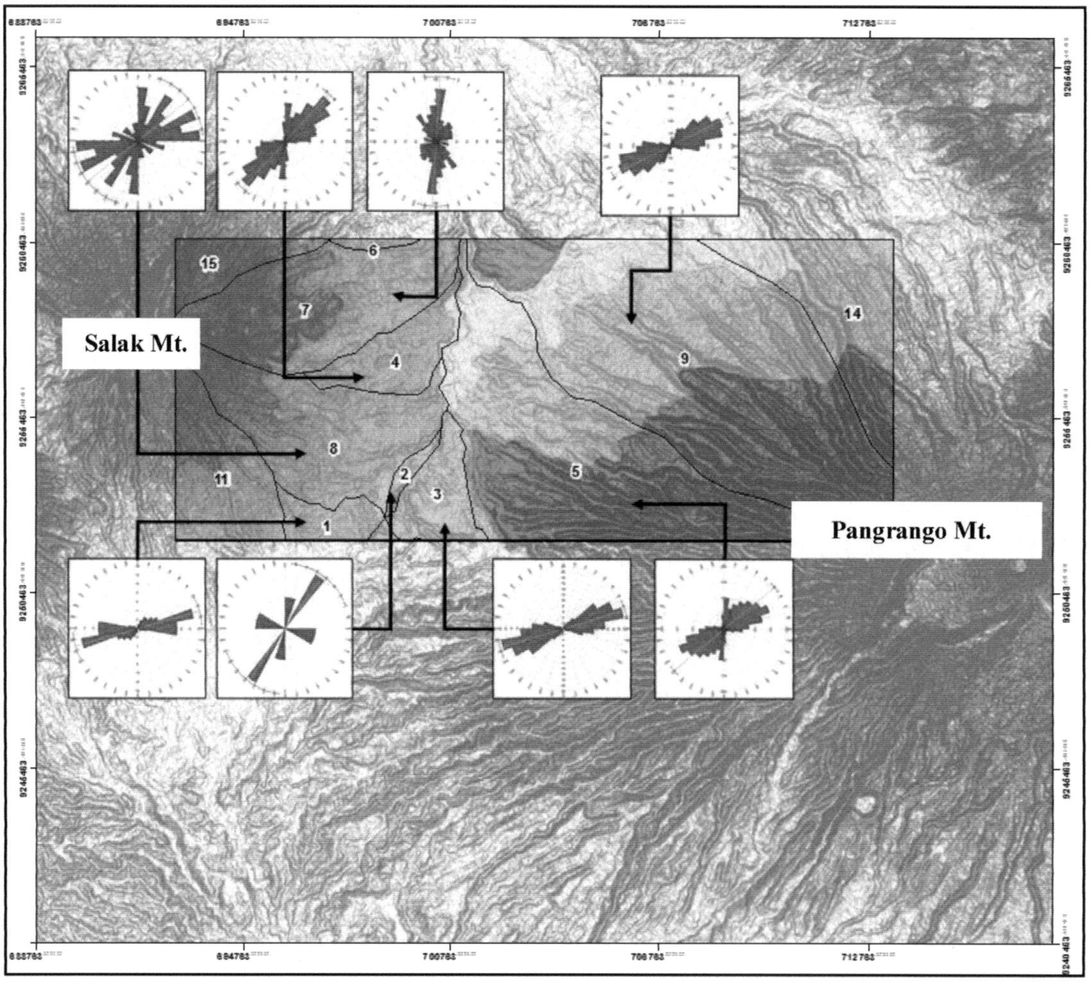

Figure 5. Style direction shown in rose diagram of each subwatershed to delineate tectonic style involved.

Table 2. Direction of each subwatershed to delineate tectonic or volcanic process in Ciherang area.

Subwatershed	Description
SW-2, SW-8	high various direction
SW-4, SW-7	low various direction
SW-4, SW-5	relative identical direction
SW-7	unique direction, different from another SW
SW-2, SW-3	Northeast direction 40° NE, equal to Pelabuhanratu fault (regional fault system)
SW-4, SW-5, SW-7, SW-8	Many truncated lineament, affected by regional fault

Subwatershed-2 dominated with northeast-southwest direction or relatively 40° NE, it may be predicted to have identical direction with regional fault system of Pelabuhanratu Fault direction with dextral strike slip fault mechanism. This regional fault gives an impact to subwatersheds near of Pelabuhanratu Fault Zone. Subwatershed 4, 5, 7, and 8 are subwatershed that highly affected by this fault. Other characterization of lineament in those subwatershed are truncated lineament that indicate as a fault zone also.

5. Conclusion

Subwatershed with northeast-southwest direction or relatively 40^0 NE, it may be predicted to have identical direction with regional fault system of Pelabuhanratu Fault direction with dextral strike slip fault mechanism. This regional fault gives an impact to subwatersheds near of Pelabuhanratu Fault Zone. Other characterization of lineament is truncated lineament that indicate as a fault zone also.
Pelabuhanratu strikeslip fault will contribute a significant impact to groundwater flow system in study area. This regional fault become a bearing plane as a boundary of groundwater flow system between Mt. Salak and Mt. Pangrango. Moreover, this regional fault will provide many fractures near fault zone and generate anomalies in groundwater flow system.

Acknowledgement

This study is a part of thesis in doctoral program with the major theme "Fracture Pattern On Volcanic Deposits and Its Implication to Groundwater Flow System at Ciherang, West Java, Indonesia" at Faculty of Geological Engineering Universitas Padjadjaran, 2016. The research facilitated under collaboration research between Faculty of Geological Engineering, Universitas Padjadjaran and PT. Danone as a mineral water industry.

References

[1] Denny S C, Allen D M & Journeay J M 2007 *Hydrogeology Journal* **15** 483
[2] Efendi A C and Hermanto B 1998 *Geological Map of Bogor, scale 1:100.000* (Indonesia: Geological Research and Development Center)
[3] Muertianto E 2006 *Hydrogeological Map of Bogor, scale 1:100.000* (Indonesia: Geological Research and Development Center)
[4] Katili J A 1975 *Tectonophysics* **26** 165
[5] Davis G H and Reynolds S J 1996 *Structural Geology of Rocks and Regions* (2nd edition: Wiley)
[6] Smith M J & Clark C D 2005 *Earth Surface Processes and Landforms* **30** 885
[7] Endyana C, Hendarmawan, Sukiyah E, Dharmawan I A And Helmi F 2014 *Proceeding International Seminar on Chemistry (ISC)* 1
[8] Ahmed S A 2014 International Journal of Computer Applications **93** 975-8887
[9] Sarp G 2005 *Lineament analysis from satellite images, north-west of Ankara* (Master of Science Dissertation, School of Natural and Applied Science of Middle East Technical University)

[10] Kolesinski M 2017 *Terrain relief shading enhancing system, device, and method*: U.S. Patent No. 9582857 (Washington, DC: U.S. Patent and Trademark Office)
[11] Abdullah A, Akhir J M and Abdullah I 2010 *Electronic Journal of Geotechnical Engineering* **15** 1
[12] Hung L, Batelaan O and De Smedt F 2005 *Proc. of SPIE* **5983** 1
[13] Endyana C, Hendarmawan, Sukiyah E and Dharmawan I A 2016 *IOP Conference Series: Earth and Environmental Science* **29** 012029

IOP Conf. Series: Journal of Physics: Conf. Series **1080** (2018) 012037 doi:10.1088/1742-6596/1080/1/012037

Evaluation and comparison of hourly and daily Tropical Rainfall Measuring Mission (TRMM) precipitation products with the rain gauge data in the Sungai Baung district, South Sumatera

M Nur[1], D O Lestari[2], W Mardiansyah[3], M Y Nurkhakim[3], D Setiabudidaya[3] and I Iskandar[3*]

[1]Master Program of Physics, Faculty of Mathematics and Natural Sciences, Sriwijaya University, Palembang 30139, South Sumatera, Indonesia
[2]Doctoral Program of Environmental Science, Sriwijaya University, Palembang 30139, South Sumatera, Indonesia
[3]Physics Major, Faculty of Mathematics and Natural Sciences, Sriwijaya University, Indralaya, Ogan Ilir 30662, South Sumatera, Indonesia

[*]iskhaq@mipa.unsri.ac.id

Abstract. Precipitation is a highly dynamic process, constantly changing in form and intensity as it passes over a given area. Nowadays, rainfall data required for a wide range of scientific applications can be achieved through meteorological satellites. However, the satellite data require a validation by comparing with in-situ. This study is designed to evaluate and to compare the hourly and daily precipitation obtained from the Tropical Rainfall Measuring Mission (TRMM) satellite and an Automatic Weather System (AWS) located at the Sungai Baung district, South Sumatera. Due to instrument failure, only short period from 1 November 2013 to 31 December 2014 of rainfall data were obtained from the AWS. The correlation and regression analysis were used to examine the TRMM data in this study. The results revealed that the TRMM data have significant correlation with the AWS data. On the hourly timescale, the coefficient correlation of the precipitation data is $r= 0.12$ (significant at 99% $r > 0.043$). Furthermore, the correlation significantly increases on the daily timescale, which $r= 0.64$ (significant at 99% of $r > 0.135$). In general, TRMM data can be used for estimating the precipitation over the rural or unpopulated areas where rain gauge is unavailable.

1. Introduction

Indonesia is the maritime continent which characterized as one of the highest precipitation areas on Earth [1]. Split by the equator, Indonesia has an extreme variation of precipitation that is linked to the monsoons. This monsoon transports moisture from the surrounding ocean directly toward the center of the Indonesia's continent. In general, the westerly monsoonal wind of Indian Ocean dominates from December to February (wet season). On the other hand, in June through August (dry season), the moisture is primarily transported from the Pacific Ocean [2-4]. Therefore, understanding and monitoring of precipitations variability are important in reducing various impacts.

There are three primary sources of precipitation observations; gauge measurement, ground-based radar, and satellite estimations. Gauge measurement provides the most accurate to measure precipitation. However, it is usually limited by their spatial coverage. Ground-based radar observations are excellent alternatives with relatively accuracy [5-6]. However, these ground instruments are relatively high costs of operation and maintenance and also leaving many unmeasured areas [7]. Thus, satellite observations may be the best solution for adequate temporal and spatial coverage of precipitation.

The launch of the Tropical Rainfall Measuring Mission (TRMM) satellite in November 1997 by the United State National Aeronautics and Space Administration (NASA) and the Japan Aerospace Exploration Agency (JAXA) provided a better understanding of precipitation over tropical and subtropical regions. One of the best retrieval algorithms is TRMM Multi-Satellite Precipitation Analysis (TMPA). Its algorithm has generated two main satellite precipitation products, that are post-real time (3B42) and near-real time (3B43). These products have been used in different applications and regions. However, it is necessary to quantify the biases of the TRMM products relative to any rain gauge measurements.

Several studies have been done to compare and to evaluate TRMM products with other rainfall measurements [8-9]. However, these studies are usually focused on monthly or seasonal scales. Shorter period, such as hourly and daily are more appropriate for flood monitoring or other hydrological applications. Comparison of precipitation products at daily scale have been carried out for the Europe, US, Australia, Asia, especially in Thailand [10-11]. The comparison over Indonesia, particularly in South Sumatera, is still lacking. Therefore, we evaluate and compare hourly and daily TRMM precipitation products with the rain gauge data in the Sungai Baung District in this study.

2. Data and methods

2.1. Data

The rain gauge data is obtained from an Automatic Weather System (AWS) which named HQ. Sei Baung. HQ. Sei Baung hosted by PT. OKI Pulp and Paper Mills HQ Sei Baung Palembang. It is lying between 2° 44' 08" N and 105° 17' 06" E with the altitude 11 m above sea level. Administratively, the AWS is located at the Sungai Baung District, northeast of South Sumatera, which closed to the Bangka Strait.

Satellite data obtained from the Global Satellite Mapping of Precipitation (GSMaP) project in the Global Rainfall Watch website (http://sharaku.eorc.jaxa.jp/GSMaP/). The Global Rainfall Watch is published by the Japan Aerospace Exploration Agency (JAXA) [12]. It is able to provide rainfall information on approximately 4-hours lag from the time of satellite monitor (quasi real-time basis) and on a global scale. The website takes rainfall and cloud cover products from several satellites that monitor these conditions. Its algorithm now produces 0.1 degrees of spatial resolution (latitude and longitude) and 1 hour of temporal resolution [13-14].

2.2. Methods

2.2.1. Data processing

The data period used in this study is from 1 November 2013 to 31 December 2014. Because of differences in spatial and temporal resolution among the two products, a direct comparison of these data is impossible in the original format of each dataset. Therefore, a data preprocessing is required because it performed a point-to-pixel analysis [15-16]. We extracted the satellite data to the point-based station location instead of interpolating the gauge data into a gridded product. Two temporal scales, that is hourly and daily, was conducted in this study.

2.2.2. Statistical analyses

There are five statistical analyses used to examine these precipitation products. Those are correlation coefficient (r) is used to evaluate the degree of agreement between satellite and gauge data, that has values ranging from -1 to 1, Mean Absolute Error (MAE). MAE is the average distance between each data sets and the mean, Root Mean Square Error (RMSE) is a frequently used to measure the average magnitude of the estimate error, Mean Bias Error (MBE) that is evaluates the systematic bias between satellite and gauge data. A positive MBE shows that the estimated precipitation is overestimated, while a negative sign shows underestimated, and Standard Deviation (STD) that is used to quantify the amount of variation of data value. Those statistical metrics were calculated as follows [17]:

$$r = \frac{1}{N-1}\sum_{i=1}^{N}\frac{\left(x-\bar{x}\right)\left(y-\bar{y}\right)}{s_x s_y} \quad (1)$$

$$\text{MAE} = \frac{1}{N}\sum_{i=1}^{N}\left|x_i - y_i\right| \quad (2)$$

$$\text{RMSE} = \sqrt{\left[\frac{1}{N}\sum_{i=1}^{N}\left(x_i - \langle \text{MBE} \rangle - y_i\right)^2\right]} \quad (3)$$

$$\text{MBE} = \frac{1}{N}\sum_{i=1}^{N}\left(x_i - y_i\right) \quad (4)$$

$$\text{STD} = \sqrt{\frac{\sum_{i=1}^{N}\left|x-\bar{x}\right|^2}{N}} \quad \text{or} \quad \text{STD} = \sqrt{\frac{\sum_{i=1}^{N}\left|y-\bar{y}\right|^2}{N}} \quad (5)$$

where x refers to the gauge measurement, y refers to satellite measurement, and N is the number of samples. S_x and S_y are the standard deviations of those two measurements. In addition, other statistical metrics were also included in this study, that is, slope coefficient of linear regression and coefficient determination (R^2). It can be described as follows [18]:

$$y_i = ax_i + b \quad (6)$$

$$R = \frac{\sum_{i=1}^{N}\left(x_i - x\right)\left(y_i - y\right)}{\sqrt{\sum_{i=1}^{N}\left(x_i - x\right)^2 \sum_{i=1}^{N}\left(y_i - y\right)^2}} \quad (7)$$

where a and b are the linear regression coefficient. If a is close to unity, a high level of consistency is expected between the satellite estimate and gauge measurement. R^2 is another widely used index for effective assessment of consistency [19]. It measures the relationship between the estimate and measurement.

3. Results and Discussions

This study examined precipitation data of TRMM satellite and AWS Sungai Baung. The evaluation was conducted at hourly and daily scales from 1 November 2013 to 31 December 2014. Comparison of hourly precipitation data is shown by figure 1. The hourly comparison is shows similar pattern. There are two rainfall peaks during this period. From figure 2, the highest rainfall intensity appears in the evening until early morning (22.00-05.00 Local Time/LT). Meanwhile, moderate rainfall intensity is occurring in the daytime (13.00-16.00 LT). High rainfall variability over the study area influenced by land-sea interaction [20-22], that is also known as land-sea breezes. The two breezes occur along coastal areas or areas with adjacent large water bodies. Sea takes more time to warm up and is able to retain the heat longer than that land. In the daytime, the land heats up quickly and the air above it warms up more than the air over the sea. The warm air over the land is less dense and begins to rise and it causes low pressure is created. The air pressure over the sea is higher with cold dense air, which moves to occupy the space created over the land. The cool air that comes along is called a sea breeze. Meanwhile, the land quickly releases the heat while the water retains its warmth in the night time. It causes the air over the water is warmer, less dense and begins to rise. Low pressure is created over the water. Cold and dense air over the land begins to move into the sea surface to replace the warmer rising air. The cool breeze from the land is called a land breeze. These phenomena causing high precipitation over the land in the nighttime. On the other hand, low or moderate precipitation occur over the land in the daytime. Generally, most of the TRMM satellite rainfall products shows overestimation, except in the daytime, which indicates AWS rainfall intensity higher than that satellite rainfall intensity (slight underestimation).

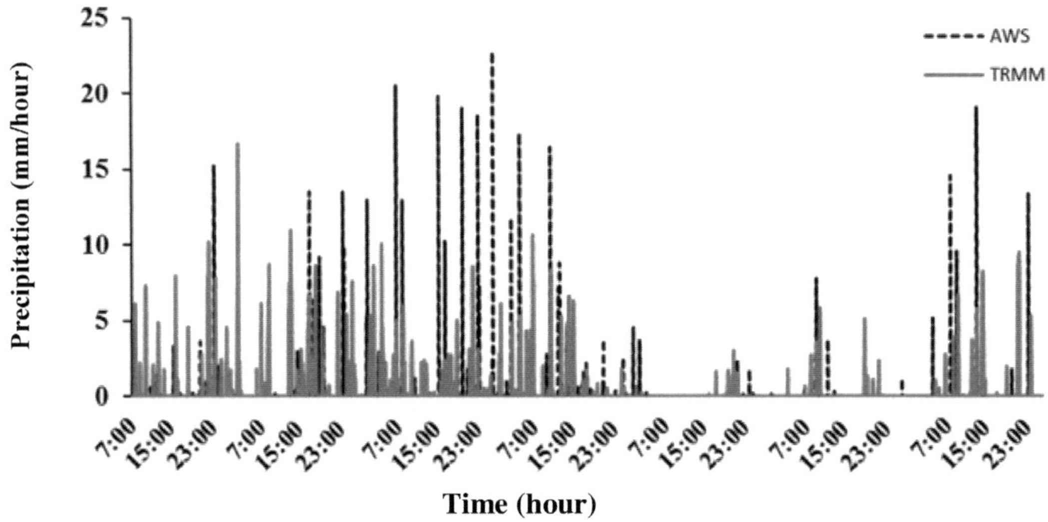

Figure 1. Hourly comparison of precipitation intensity between TRMM satellite and AWS Sungai Baung.

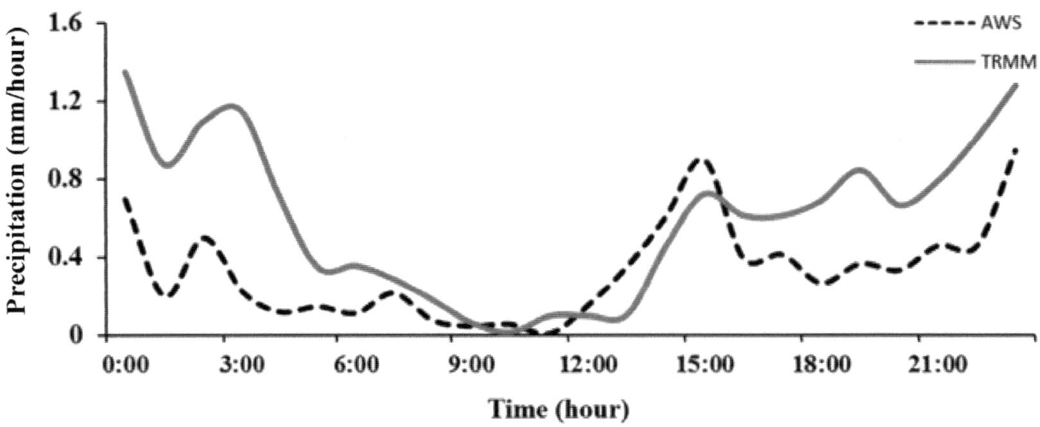

Figure 2. Diurnal climatology of precipitation intensity between TRMM satellite and AWS Sungai Baung.

Figure 3 shows the time series of daily comparison between TRMM satellite and AWS precipitation products. Daily precipitation estimates from those rain measurement process also indicates similar pattern. Although the double peaks in the rainy seasons are discernible, the variations at the daily scale are still quite prominent. All products captured high daily precipitation during December until January (first peak) and March until early May (secondary peak). By contrast, low daily precipitation showed during end of May until October. Moreover, the TRMM satellite rainfall products tend to overestimate daily precipitation almost in the whole periods.

Table 1 summarizes the results of statistical analysis that compare satellite rainfall estimates and rain gauge data at hourly and daily resolution. The best performance was observed by daily comparison. This comparison shows moderate correlation between TRMM satellite and AWS with coefficient correlation $r = 0.64$ (significant at 99%, $r > 0.135$). This result is consistent with the findings reported in Blue Nile Basin [23-24], Singapore [25], Thailand, Japan, and Korea [26]. On the other hand, low agreement between two data sets is demonstrated by hourly comparison, which $r = 0.12$. This correlation is also significant at 99% because of its coefficient, $r > 0.043$. Although the coefficient correlation of daily evaluation is better than that hourly evaluation, but statistical error between that comparison tend to increase in daily comparison. The MAE values of the hourly and daily comparisons are 0.56 mm/hour and 3.61 mm/day, respectively. The RMSE values of those comparison is 2.36 mm/hour (hourly comparison) and 7.16 mm/day (daily comparison). All comparison shows TRMM satellite products over estimated precipitation. That is indicated by a positive value of MBE, 0.14 mm/hour and 0.84 mm/day, hourly and daily comparison, respectively.

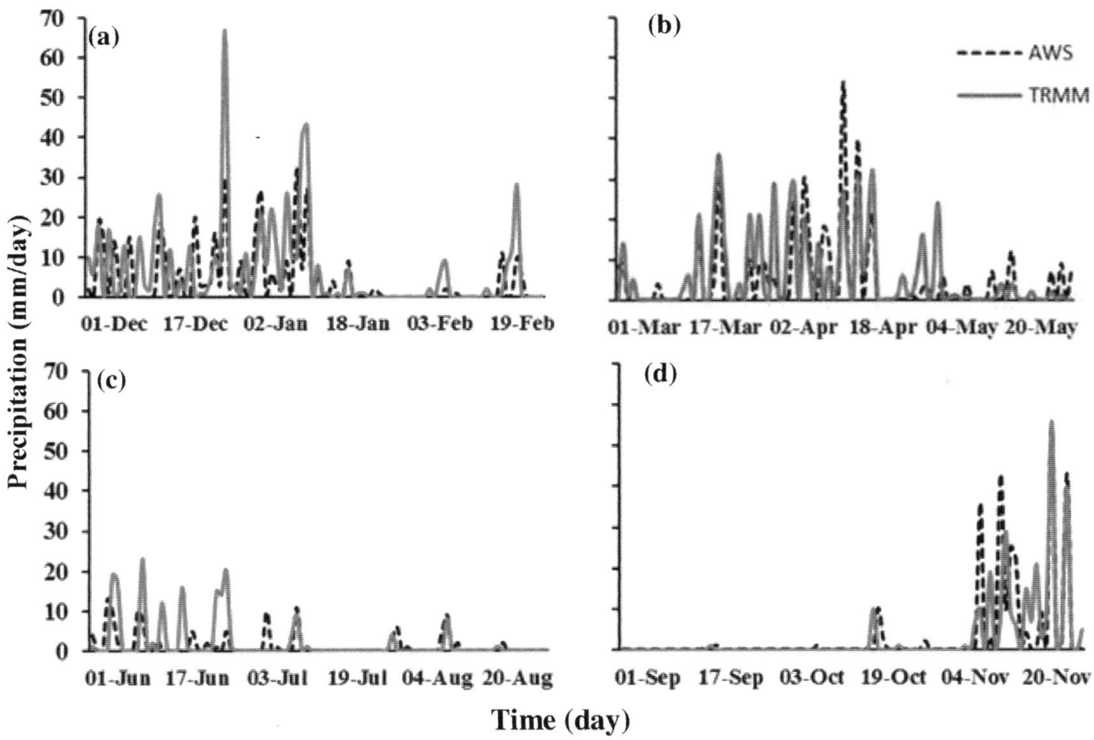

Figure 3. Daily comparison of precipitation intensity between TRMM satellite and AWS Sungai Baung: (a) December-February, (b) March-May, (c) June-August and (d) September-November.

The scatter plot of data from rain-gauge station against satellite-based rainfall estimates is shown in figure 4. Figure 4 (a) shows the relationship between TRMM satellite and AWS at hourly scale. The linear regression (a = 0.11) and the determination (R^2 = 0.013) coefficients demonstrated a low agreement between two datasets. But, the correspondence increases at daily scale (figure 4(b)). The TRMM satellite had a value of 0.71 with the rain gauge. The coefficient of determination in this period is R^2 = 0.41. These coefficients indicate a moderate correlation between TRMM satellite and AWS rainfall products. The correlation between evaluated satellite and measurement from gauge is

IOP Conf. Series: Journal of Physics: Conf. Series **1080** (2018) 012037 doi:10.1088/1742-6596/1080/1/012037

still low and also moderate in hourly and daily precipitation estimation. A possible explanation is the limited number of gauge used by this study. In order to improve the satellite products, a more gauge station in study area should be developed and applied in the satellite bias correction. In addition, it also needs to investigate the correlation between both data for a monthly timescale.

Table 1. Statistical metrics of hourly and daily precipitation estimates from TRMM satellite and AWS.

Parameter		Hourly	Daily
r		0.12*	0.64*
MAE		0.56	3.61
RMSE		2.36	7.16
MBE		0.14	0.84
STD	TRMM Satellite	1.36	8.82
	AWS	2.06	6.64

*Significant level of correlation coefficient at 99%, which are $r > 0.043$ and $r > 0.135$, hourly comparison and daily comparison, respectively.

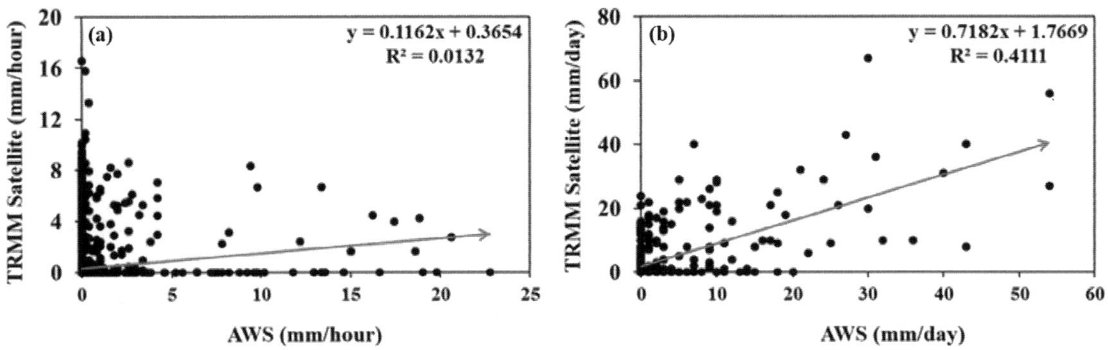

Figure 4. Scatter plot of hourly and daily comparisons between TRMM satellite and AWS products: (a) Hourly comparison and (b) Daily comparison.

4. Conclusion

The comparison of TRMM satellite precipitation product using rain gauge in the Sungai Baung District, South Sumatera during 1 November 2013 until 31 December 2014 is presented in this study. Hourly and daily comparisons were conducted by applying several statistical methods.

The results show low and moderate correlation between TRMM satellite and AWS precipitation products, with the coefficient correlation $r = 0.12$ (hourly timescale) and $r = 0.64$ (daily timescale), respectively. The relationship of those precipitation products was calculated with confidence intervals estimated at 99% confidence level. This agreement is also supported by regression and determination coefficient, with $a = 0.11$ and $R^2 = 0.013$ and $a = 0.71$ and $R^2 = 0.41$, in the hourly and daily timescales, respectively. Although the correlation (r, a, and R^2) of daily evaluation is better than that hourly evaluation, but the statistical error (MAE, RMSE, and MBE) in the daily scale increased. Overall, satellite-based precipitation product generally showed overestimation rainfall events. That is indicated by the time series of hourly and daily comparison and positive value of MBE (0.14 mm/hour and 0.84 mm/day) at both scales.

Finally, since one rain-gauge data used in this study, it is suggested that future studies to use more rain-gauge stations, so that the correlation between satellite-based rainfall product could show strong relationship and a pixel-to-pixel analysis can be applied to conduct in-depth analysis. Moreover, we recommend that an additional statistical correction, such as Probability of Detection (PoD), False

IOP Conf. Series: Journal of Physics: Conf. Series **1080** (2018) 012037 doi:10.1088/1742-6596/1080/1/012037

Alarm Ratio (FAR), Critical Success Index (CSI), and Probability Density Function (PDF). These statistical analyses should be conducted to hourly and daily of TRMM satellite precipitation estimates before used in other interests.

References

[1] Ramage C S 1968 *Mon. Wea. Rev.* **96** 365

[2] Aldrian E and Susanto R D 2003 *Int. J. Climatol.* **23** 1435

[3] Chang C P, Wang Z, McBride J, and C H Liu 2005 *J. Clim.* **18** 287

[4] Sekaranom A B and Masunaga H 2017 *US Meteo. Soc.* **98** 1867

[5] Krajewski W F 2007 *Spring.* **8** 403

[6] Guo R and Liu Y 2016 *Water* **8** 281

[7] Durkee J 2010 *Geogr. Compass* **4** 956

[8] Chiu L, Liu Z, Rui H, and Teng W 2005a *E. Sci. Sat. Rem. Sens.* J Qu, *et al.* (Editors) Springer-Verlag (in press)

[9] Nur M and Iskandar I 2013 *Proc. Int. Sym. for Sust. Hum* Bengkulu ISSN 2008-9127 125

[10] Ebert E E 2002 *1st Intl. Prec. Work. Gr. (IPWG) Workshop* Madrid 23-27 September 2002

[11] Chokngamwong R and Chiu L 2005 *Fairfax* VA 22030-4444 USA

[12] Aonashi K *et al.* 2009 *J. of the Meteo. Soc. of Japan* **87A** 119

[13] Ushio T, Sasashige K, Kubota T, Shige S, Okamoto K, Aonashi K, Inoue T, Takahashi N and Iguchi 2009 *Journal of the Meteorological Society of Japan* **87A** 137

[14] T, Kachi M, Oki R, Morimoto T, and Kawasaki Z I 2009 *J. of the Meteo. Soc. of Japan* **87A** 203

[15] Cohen L, Matos T J, Boillat J L, and Schleiss A 2012 *Hyd. And E. Sys. Sci.* **16** 489

[16] Thiemig V R, Rojas, Zambrano-Bigiarini M, Levizzani V, and De Roo A 2012 *J. of Hyd.* **13** 1760

[17] Tan M L, Tan K C, Chua V P, and chan N W 2017 *Water* **9** 57

[18] Behrangi A, Khakbaz B, Jaw T C, Aghakouchak A, Hsu K, and Sorooshian S 2011 *J. Hydrol.* **397** 837

[19] Prakash S, Sathiyamoorthy V, Mahesh C, and Gairola R M 2014 *Int. J. Remote Sens.* **35** 3018

[20] Mori S 2004 *Month. W. Rev.* **132** 2021

[21] As-syakur A R, Tanaka T, Osawa T, and Mahendra M S 2013 *Int. J. Remote Sens.* **34** 7723

[22] McPherson R D 1970 *J. of Appl. Meteo.* **9** 767

[23] Sahlu D, Nikolopoulus E, Moges S, Anagnostou E N, and Hailu D 2016 *J. Hydrometeo.* **17** 2875

[24] Dembele M and Sander J Z 2016 *Remote Sens.* **37** 3995

[25] Tan M L and Duan Z 2017 *Remore Sens.* **9** 720

[26] Kim K, Park J, Baik J, and Choi M 2017 *Atmos. Res.* **187** 95

IOP Conf. Series: Journal of Physics: Conf. Series **1080** (2018) 012038 doi:10.1088/1742-6596/1080/1/012038

Characterization of *ethyl cellulose* (EC) microcapsules for lime oil encapsulation

E Julaeha[1*], R Nugeraha[1], M Nurzaman[2], D Kurnia[1], T Wahyudi[3] and Y Rosandi[4]

[1]Department of Chemistry, Faculty of Mathematics and Natural Sciences, Universitas Padjadjaran, Jatinangor, Sumedang 45363, Indonesia
[2]Department of Biology, Faculty of Mathematics and Natural Sciences, Universitas Padjadjaran, Jatinangor, Sumedang 45363, Indonesia
[3]Indonesia Center for Textile, Jl. Jendaral Achmad Yani No. 390 Bandung, 40272, Indonesia
[4]Department of Geophysics, Faculty of Mathematics and Natural Sciences, Universitas Padjadjaran, Jatinangor, Sumedang 45363, Indonesia

[*]euis.julaeha@unpad.ac.id

Abstract. We report on the preparation of microcapsules using ethyl cellulose as a shell and lime oil (*Citrus aurantifolia* oil) as the core. In this work the lime oil was extracted from lime syrup production residue by hydrodestilation method. Microencapsulation is prepared using a simple coaservation method. The size distribution of the microcapsules is observed using the Particle Size Analyzer (PSA). The effectiveness of the encapsulation process was determined by evaluating the images captured by the optical microscope. Whereas, the morphology of the produced microcapsules was identified using the scanning electron microscopy. Our result shows the size distribution ranged from 39 to 104 μm, with oil containment up to 14%. This size distribution is suitable for functional textile products application.

1. Introduction

Microcapsules technology is used to control the containment and release time of oil substances. The technology has a wide spread applications in many sectors, for instance, health, medical, cosmetics, textile, and agriculture. These utilization of the microcapsules is continually developing with broader field of application. Research on the encapsulation process of substances by microcapsules started very recently. The capsules are used to isolate and prevent the mobility of liquids, especially essential and fragrant oil by polymer shell or membrane. By tuning the properties of the membrane, it is possible to control the mechanism to release the encapsulated substances. By encapsulating fragrant liquid in the microcapsule, the release of the fragrant can be made gradual. This mechanism is very suitable to be applied for instance, in the functional textile products [1].

Microcapsules can be prepared from ethyl cellulose (EC) using various methods such as phase separation, coacervation, solvent evaporation [2]. EC is also frequently used as a hydrophobic polymeric coating material for extended drug release applications. In this work the coacervation method is applied to encapsulate essential lime oil (LO). The essential oil, processed from residue of lime home industry, is used as the core liquid wrapped by capsule shells made from ethyl cellulose.

2. Material and method

2.1. The preparation of lime oil

In this work, we prepared lime oil extracted by hydrodistillation process of lime peel, using the standard distillation equipment (Stahl), by hydrodistillation process of lime peel, using the standard distillation equipment (Stahl). This processing time takes about three hours. In order to remove water content in the oil we add sodium sulfate ($NaS0_4$) to the yield. In our experiment, 1.281 g of lime peels produced 12.9 mL of lime oil, with yellowish color and a characteristic lime odor. In order to maintain the quality of the lime oil, the obtained oil transferred into an airtight amber-colored vial and stored in a refrigerator at 4°C for further analysis. The quality of the essential oil was determined using the method proposed in [3], which analyses the specific mass, the refraction index, the acid number, and solubility in alcohol.

2.2. Microcapsule preparation procedure

We applied a simple encapsulation procedure of LO proposed by Mirabedini *et al.* [4]. The amount of LO (0.5 g) is kept constant with the variation of the EC, ranged from 0.6, 0.8, 1.0, and 1.2 g. This process was performed in two stages using the solvent evaporation method. In the first stage (the oil phase), the EC powder was gradually solved in 10% ethyl acetate. The process was performed in 10 minutes and followed by LO addition which take 30 minutes, gradually. Throughout the process the system condition was kept in room temperature and the magnetic stirrer was always activated with the speed of 600 rpm.

In the second stage (the water phase) the oil phase drops to the 1% sodium dodecyl sulfate (SDS) solution, under the same condition for 15 minutes. In this stage the process was performed inside a closed container. The resulted dispersion mixture is removed to the 60°C water bath and the stirrer was running for one hour. The mixture was deposited in the same container and let cooled for one day until the microcapsules was formed. The microcapsules were collected using the millipore filter and dried in a desiccator for one day.

2.3. Microcapsule characterization

The particle size distribution was identified using the Particle Size Analyzer (PSA) apparatus. The morphology of the microcapsules was observed using an optical microscope with magnifications of 40, 100 and 200 times. The on-screen magnification was 128, 320 and 640 times. We also evaluated the morphology using Scanning Electron Microscope (SEM) apparatus. A UV-spectrophotometer was used to analyze the efficiency of the encapsulation process. The spectrophotometer operated at 340 nm wavelength, according to the standard curve. The oil containment in the microcapsules is calculated using linear regression, where the encapsulation efficiency *EE* (%) is determined by equation (1).

$$EE = \frac{w_o}{w_c} \times 100\% \tag{1}$$

where w_o and w_c is the weight of oil content and microcapsules, respectively.

3. Result and discussion

3.1. Preparation of the lime oil (LO)

Lime oil (LO) was obtained from *C. aurantifolia* (lime) peels by hydrodistillation. The essential oil is extracted by the water steam pressure from oil cells of the lime peel. The oil is mixed with the water steam, and flows through a condenser device. This mixture was condensed and the separation between oil and water phase occurred. Due to smaller specific mass of oil, the LO will be deposited on top of water [5]. We obtained LO with the yield of 0.89%, and it's the quality is shown in table 1. Our values fit well with the literature [4].

Table 1. The quality of lime oil produced by hydrodistillation. The reference value is taken from Guenther, 1987 [3].

Parameter	Value	Ref [3]
density	0.854	0.862–0.868
refraction index	1.4756	1.4750–1.4770
acid number (%)	0.5148	0.5–1.5
solubility in 90% ethanol	soluble	soluble

3.2. Encapsulation of the lime oil

Essential oils are volatile and chemically unstable in the presence of air, light, moisture and high temperatures. Hence, it is beneficial to microencapsulate volatile ingredients to limit aroma degradation or loss during processing and storage [6]. Microencapsulation has the ability to enhance the oxidative stability, thermostability, shelf-life, and biological activity of oils. In addition, encapsulation can control the release time of the essential oil, hence the volatility and usage time can also be controlled [7]. In the study, we used a simple coacervation method, based on the addition of a poor solvent to a hydrophilic colloidal solution which results in the formation of two phases: one is rich in colloid molecules (coacervate), and the other is almost coacervate free [5].

From this simple method, the yield of the microcapsule is taken from the ratio of the ethyl cellulose (shell) and the essential lime oil (core). In this work, we put 0.5 g of LO to the system with the varied weight of EC from 0.6, 0.8, 1.0, and 1.2 g. The result is shown in Table 2, where the yield of 69% is reached using the membrane shell weight of 1.0 g.

Table 2. The yield of the microcapsule with the variation of the membrane shell. The essential oil weight is 0.5 g.

Sample no.	Membrane shell	Weight	%yield
I	0.6037	0.4687	45%
II	0.8014	0.4966	60%
III	1.0067	0.9847	69%
IV	1.2034	1.0871	69%

3.3. Microcapsule morphology

The surface morphology of microcapsules depends on the core material properties and the micro-encapsulation process of the shell. Microcapsule particle size and the thickness of shell are believed to be dependent on the ratio of core to shell material, and it greatly influences the encapsulation efficiency [8]. In the study, four different weight of shell material 0.6, 0.8, 1.0, and 1.2 g were selected while core material fixed at 0.5 g to evaluate the effectiveness on the micro-encapsulation process. The influences of weight of shell to the size of particle size is shown in figure 1, which shows the result of size distribution measurement using PSA apparatus.

The data shows that the average particle size decreases with the increase of EC weight from 0.6 to 1.0 g. However, it increases at the core weight of 1.2 g. At the core/shell ratio of 0.5:1.0 g (1:2) we found the smallest microcapsule particle size. The average particle size distribution is decreased by increasing oil to EC ratio, however, the decrease in most cases was not statistically significant.

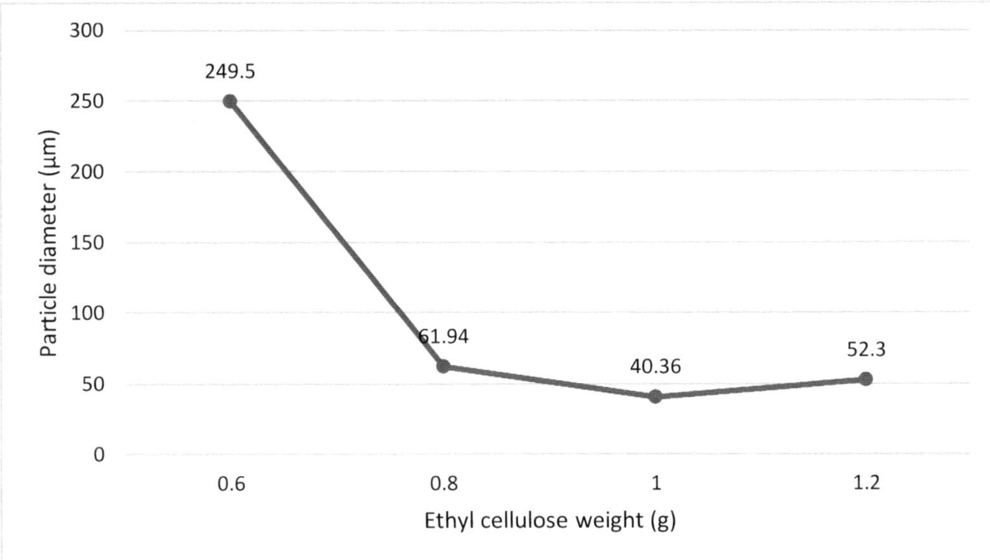

Figure 1. The PSA measurement of the particle size, demonstrates the dependence of microcapsule size to the EC shell concentration.

The fabricated microcapsule was characterized using optical microscope and scanning electron microscope (SEM) in order to observe the morphology. The result is shown in figure 2 and 3. In figure 2, it is shown the difference between the empty and filled microcapsules in the various magnification. The lime oil appears as bubbles inside the capsule shells. It indicates that on micro-encapsulation, the lime oil (core) has been coated by the ethyl cellulose coating (shell).

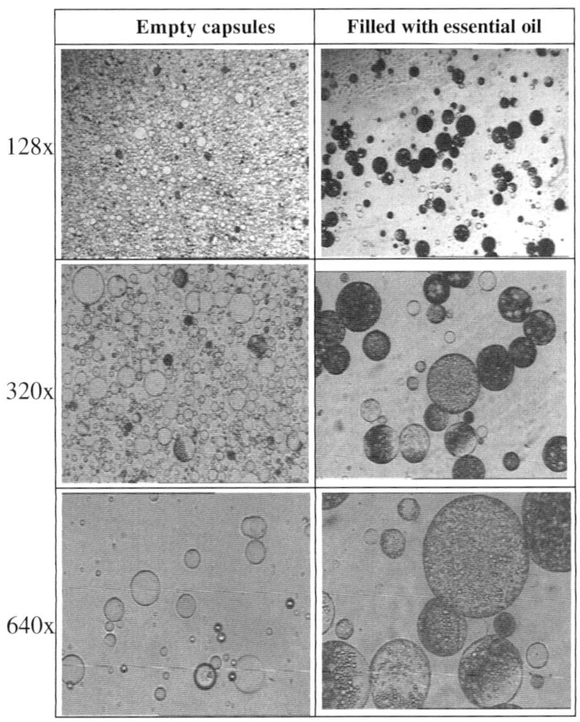

Figure 2. The morphology of empty microcapsules (left column) and microcapsules filled by the essential oil (right column), as seen by the optical microscope. The magnification is shown on the left side of each image, 128, 320, and 640 times magnification.

Figure 3 shows electron microscope imaging result. By direct observation, the microcapsules seem approximately spheres and the surface morphology is smooth without coacervation. From the image, we

IOP Conf. Series: Journal of Physics: Conf. Series **1080** (2018) 012038 doi:10.1088/1742-6596/1080/1/012038

found that the diameter is distributed between 39 to 104 μm. The average size of the fabricated capsules increase by increasing the oil phase. No free oil material is seen in the optical micrographs, indicating that the oil is well contained inside the shell and no air bubbles is present inside the the microcapsule.

We used the digital UV-vis device to determine the content of essential oil inside the capsule shells. From the standard curve of lime oil, we obtained obtain 2.57 g/L oil content with ε value of 14%. We consider that the oil content is low, which might be caused by several factors, such as not optimal ratio between the shell and core (the oil), incomplete encapsulation of the oil, incomplete forming of microcapsules, and the leaking of the shell and the oil during the process.

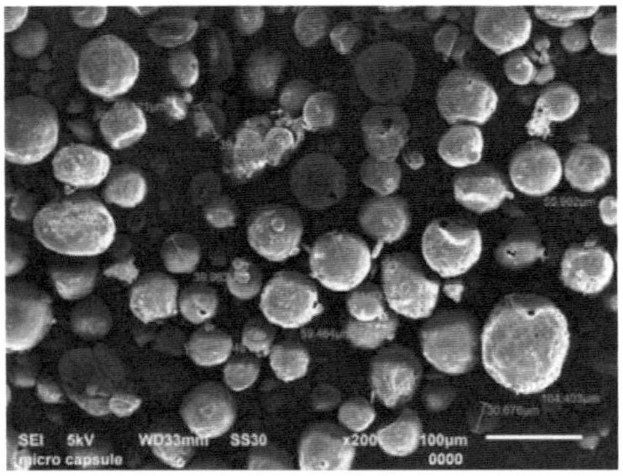

Figure 3. Scanning electron microscope (SEM) image, shows the morphology of the microcapsules. The measured diameter is indicated by yellow bars. The scale (100µm) is shown by the white bar at right-bottom position.

4. Conclusion

Lime oil microcapsules have been prepared using *C. aurantifolia* oil (lime oil, LO) as core material and *ethyl cellulose* (EC) as shell material by coacervation technique. By analysis of the particle size, the encapsulation efficiency was achieved at the ratio of core material to shell material of 1:2, and stirring speed is 600 rpm, with the yield is 69%. The particle size of microcapsule is around 39 to 104 μm, the encapsulation efficiency, *s*, of the microcapsule is 14%. The microcapsules are in sphere shape without coacervation. The quality criteria of the produced lime oil agree well with the literature.

Acknowledgment

We are grateful for internal research funding of Universitas Padjadjaran, contract No. 718/UN6.3.1/PL/2017.

References

[1] Wahyudi T, Mulyawan A S, Kasipah C, Prayudie U and Julaeha E 2017 *Arena Tekstil* **32** 1

[2] Badulescu R, Vivod V, Jausovec D and Voncina B 2008 *Carbohydrate Polymers* **71** 85

[3] Guenter E 1987 *Essential Oils* (New York : Robert E. Krieger Publishing Co.)

[4] Mirabedini S M, Dutil I and Farnood R R 2012 *Colloids and Surfaces A: Physicochemical and Engineering Aspects* **394** 74

[5] Asbahani A E, Miladi K, Badri W, Sala M, Addi E H A, Casabianca H, Mousadik A E, Hartmann D, Jilale A and Renaud F N R 2015 *International Journal of Pharmaceutics* **483** 220

[6] Jun-xia X, Hai-yan Y and Jian Y 2011 *Food Chemistry* **125** 1267

[7] Bakry A M, Abbas S, Ali B, Majeed H, Abouelwafa M Y, Mousa A and Liang L 2016 *Comprehensive Reviews in Food Science and Food Safety* **12** 143

[8] Wang J M, Zheng W, Song Q W, Zhu H and Zhou Y 2009 *Journal of Fiber Bioengineering and Informatics* **1** 293

IOP Conf. Series: Journal of Physics: Conf. Series **1080** (2018) 012039 doi:10.1088/1742-6596/1080/1/012039

Electrode capacity and voltage performance of lithium iron phosphate – polyaniline coin cell battery

I Rahayu[1*], A Wijayati[1], A R Noviyanti[1], S Hidayat[2] and Risdiana[2]

[1]Department of Chemistry, Faculty of Mathematics and Natural Sciences, Universitas Padjadjaran Jl. Raya Bandung Sumedang km 21 Jatinangor, West Java 45363, Indonesia

[2]Department of Physics, Faculty of Mathematics and Natural Sciences, Universitas Padjadjaran Jl. Raya Bandung Sumedang km 21 Jatinangor, West Java 45363, Indonesia

[*]iman.rahayu@unpad.ac.id

Abstract. Battery is the main component in many portable electronic devices that serves as energy storage. Lithium cobalt oxide ($LiCoO_2$) is one of the most popular types of lithium-ion battery due to its low self-discharge and high energy density. However, $LiCoO_2$ is toxic to the environment, thus a safer material is important to be studied. A non-toxic and thermally stable lithium iron phosphate ($LiFePO_4$) could be a good alternative for battery material. Nevertheless, the low electrical conductivity of $LiFePO_4$ (10^{-9} $S.cm^{-1}$) should be enhanced by using a conductive polymer, such as polyaniline (PANI), to form a composite. This study aims to investigate the electrode capacity of the coin cell battery composite using galvanostatic cycle and to examine its voltage performance using the cyclic voltammetry methods. The cathode was prepared in powder form. The composite was assembled in the form of coin cell battery. It consists of lithium, lithium hexafluorophosphate-ethylene carbonate-dimethyl carbonate, and lithium iron phosphate-polyaniline-polyvinylidene fluoride as the anode, electrode, and cathode, respectively. It is found that the electrode capacity of [Li|1 M $LiPF_6$/EC DMC|$LiFePO_4$-PANI-PVDF] coin cell battery was 38 mA h g^{-1} with a working voltage of 3.6 – 4.0 V.

1. Introduction

In latest decade, eco-friendly technology and energy storage are the opportunities for technology advancement. Those opportunities have been received great attention from various parties involving research groups, industry, and government. The desire for the better portable energy source increases every year. The consumer needs energy storage that have higher capacity and working voltage to integrate multiple devices into a single product, like digital camera, radio, music player, and GPS in one smart phone. The device requires a source of massive and flexible energy to support these functions [1-4].

From all of the existing battery technologies, lithium-ion batteries have enormous energy storage capacity, for example, $LiCoO_2$ battery as a cathode which has a storage capacity of 900 $mAhg^{-1}$ [4-7]. Despite its enormous capacity, $LiCoO_2$ is not environmentally friendly and unstable at high temperatures. Therefore, battery manufacturers are looking for alternative replacement cathode materials such as $LiNiO_2$, $LiMn_2O_4$, and $LiFePO_4$ [7-10].

Theoretically, specific capacity of $LiFePO_4$ is 170 $mAhg^{-1}$ at a voltage 3.45 V, which is well within the window of the polymer as well as the carbonate electrolytes [6-10]. It is, therefore, an inexpensive, environmentally benign, safe cathode capable of thousands of charge/discharge cycles without degradation. Therefore, $LiFePO_4$ is considered to be a leading candidate material for the cathode of the large-capacity batteries.

Polyaniline (PANI) is one of the most studied conductive polymers because of its high stability in the environment, readily synthesized, and has reversible electrical changes via protonation. PANI is conductive in the form of emeraldine salts (ES) [5-7]. In recent studies, many cathode materials combined with polymers, for example, LiV_3O_8 / polypyrrole have been successfully synthesized and show better cycling performance in lithium cells [5-10]. Among the various types of conductive polymers, PANI has been used extensively because it is easily produced with the desired morphology and structure. PANI is superior to polypyrrole and polythiophene in energy density and durability. PANI is electrochemically active in the range 2.0 - 3.8 V. Furthermore, PANI is not only useful as a conductive adhesive such as carbon in the electrode, but also participate as an active material [5-7].

From our previous study, it is found that the process of electrode coating with conductive polymers is an effective method to improve the electrical performance of cathode and anode materials in lithium-ion batteries [5-10]. The conductive polymer is an additional material in lithium ion batteries. It has several properties that could facilitate redox reactions in batteries involving lithium ions or anion electrolyte. Conductive polymers also could be isolated crystalline particles, preventing agglomeration on the surface of the electrode. Furthermore, conductive polymers may prevent the dissolution of the active material into the $LiPF_6$ electrolyte. In this work, we attempt to use PANI in $LiFePO_4$ - PANI cell battery and investigate its charge capacity and working voltage with structure [Li | 1 M $LiPF_6$/EC DMC | $LiFePO_4$-PANI-PVDF].

2. Materials and Method

2.1. Materials

The material used are aniline (pro analyst, Merck), lithium iron phosphate, ammonium hydroxide (pro analyst, Merck), ammonium peroxydisulfate (pro analyst, Merck), distilled water, hydrochloric acid (37%, Merck), methanol, 1 methylpyrrolidinone (99.5%, Sigma Aldrich), toluene (pro analyst, JT Baker), polyvinylidene fluoride (pro analyst, Merck), lithium (Kyokuto Metal Co., Japan), Celgard® 25021 polypropylene (Celgard LLC, USA), lithium hexafluoro phosphate (pro analyst, Merck), ethylene carbonate (pro analyst, Merck), and dimethyl carbonate (pro analyst, Merck).

2.2. Method

PANI was synthesized by interfacial method [5-7]. The electrode was prepared by mixing 70% by weight of $LiFePO_4$, 15% by weight of PANI ES, and 15% by weight of polyvinylidene fluoride (PVDF) in a mortar. The powder mixture was then added with N-methylpyrrolidinone (NMP) to become slurry, then molded with a thickness of 20 m and a diameter of 2 cm above the aluminum sheet. Afterwards the sample was dried in the oven for 24 hours at a temperature of 80 °C. The dried sample were compacted using the roll to roll coater. The electrode was re-dried in a vacuum oven for 12 hours at 80 °C. Battery manufacture refers to R2032 Standard coin cell battery. All electrodes which have been vacuumed for 12 hours at 80 °C were included in a pure Argon Humidity Controlled Glove Box and H_2O gas concentration <1 ppm and O_2 <1 ppm gas. The lithium sheet was used as a negative electrode (anode) and a reference electrode. Celgard® 25021 polypropylene is used as a separator, and 1 M lithium hexafluoro phosphate ($LiPF_6$) in ethylene carbonate (EC) and dimethyl carbonate (DMC) mixture 1:1 (v/v) were used as an electrolyte [5-7]. The battery was characterized using cyclic voltammetry, and Galvanostatic cycle to measure its capacity. The coin-cell making scheme is shown in figure 1.

There are two categories of solvent that can be used: (i) an organic solvent-based system and (ii) a water-based system. The organic solvent is more commonly used to obtain good suspension. The examples of the organic solvent are NMP. If using a water-based system, there will be an agglomeration of the powder in the suspension caused by strong hydrogen bonding and electrostatic forces. One of the keys in the coating process is to control agglomeration, which can improve dispersion and slurry stability.

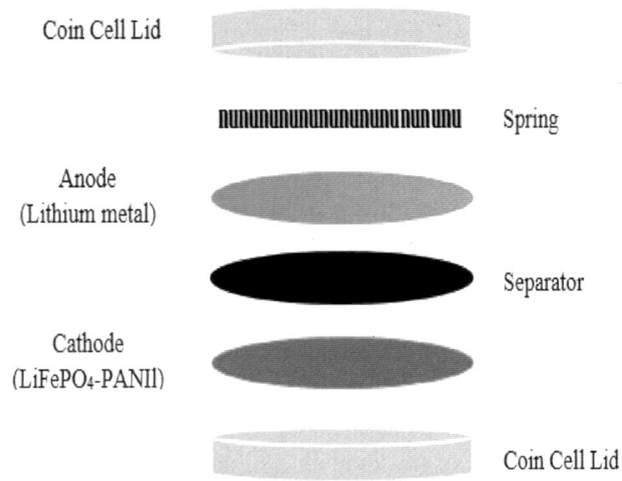

Figure 1. Coin Cell Battery Illustration.

The method used for the preparation of cathode slurry coatings was the doctor blade method. The cathode slurry was placed in a doctor blade reservoir and printed on an aluminum sheet. The thickness of the cathode is the gap between the doctor blade and the aluminum sheet. The coating is very influential on the electrode results. The addition of excessive NMP causes the slurry to become very dilute (low viscosity) so there is much cracking in the electrode that results in less optimal battery performance. After the cathode was printed, the next step was done to evaporate the solvent in the slurry.

When the coin cell is an assembly at Glove Box, it is important to keep O_2 and H_2 pressures less than 1 ppm to minimize the bursting of lithium metal. In addition, coin cells must be sealed completely during the sealing process to prevent reactions between lithium metal and air. Figure 2 shows an example of a coin cell sealing.

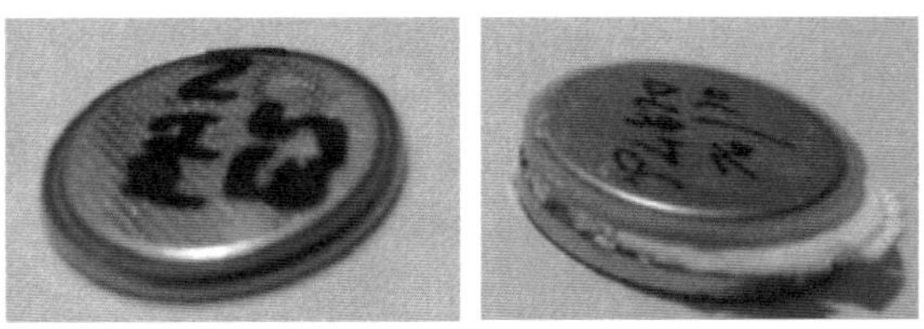

(a) (b)

Figure 2. Sealing of (a) coin cells in well-sealed and (b) not well-sealed.

The assembled battery cells were then electrochemically tested with cyclic voltammetry and galvanostatic cycles to identify redox potential and storage capacity.

3. Results and Discussion

Cyclic voltammetry of $LiFePO_4$ - PANI cell battery is presented in figure 3. The oxidation and reduction peak of the cell is at 2 V - 4 V which indicates good reversible property. The increasing of applied voltage increase current, but when the current is reduced the coin cell battery's voltage is decreased gradually. It shows the coin cell battery has an internal redox.

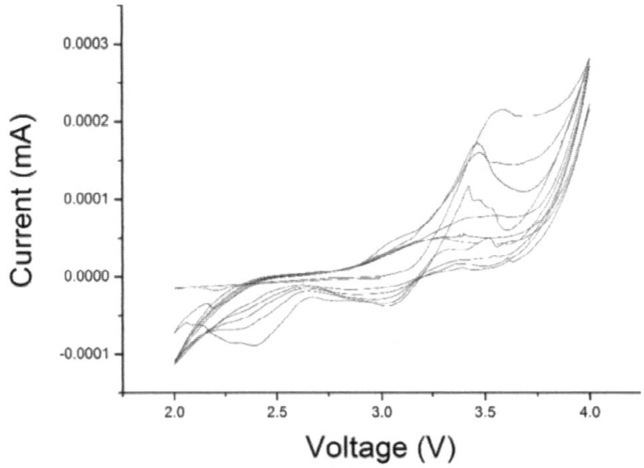

Figure 3. Cyclic voltammetry of coin cell battery [Li|1 M $LiPF_6$/EC DMC|$LiFePO_4$-PANI-PVDF].

The PANI cyclic voltammogram shows two oxidation peaks at 3.5 and 4.0 V and three reduction peaks at 2.0, 3.2 and 3.6 V. Redox peaks (oxidation/reduction) at 3.5/3.2 V shows benzoquinone, and the last peak at 4.0 / 3.6 V indicates presence of head-to-tail dimer oxidation and conversion of emeraldine structure to pernigraniline. A reduction peak at 2.0 V indicates leukomeraldine formation.

The galvanostatic cycle is used to identify the charging and discharging capacity in a battery system over a given cycle. Figure 4 shows graphs of charging and discharging coin cells [Li | 1 M $LiPF_6$ / EC DMC | $LiFePO_4$-PANI-PVDF] at 298 K for 25 cycles. The measurement of the coin cell is carried out at a current density of 30 A cm^{-2}. $LiFePO_4$-PANI-PVDF electrode produces maximum voltage at 4.0 V during filling and discharging process. The charging capacity of the electrode was 38 $mAhg^{-1}$ and was decreased to 3 $mAhg^{-1}$ for 25 cycles.

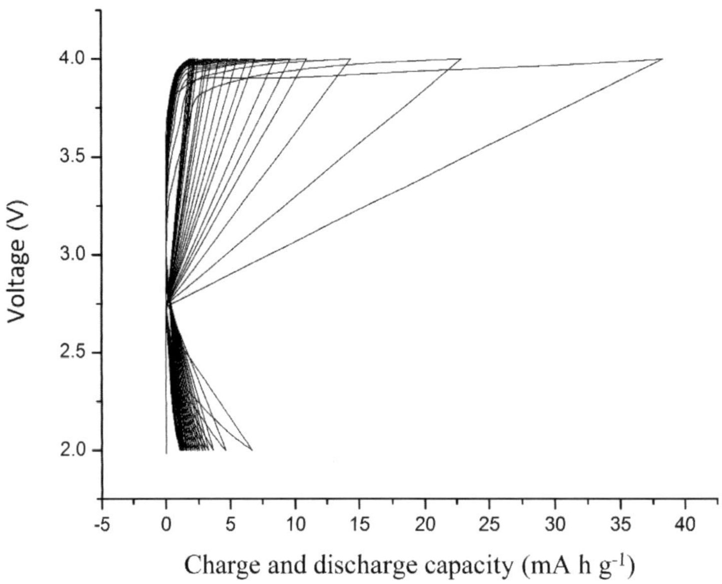

Figure 4. Charging and Discharging graphic of coin cell battery [Li|1 M $LiPF_6$/EC DMC|$LiFePO_4$-PANI-PVDF].

The processes that occurred during the charging and discharge process take place are [9, 10]:

$$LiFePO_4 \rightarrow xFePO_4 + (1\text{-}x)LiFePO_4 + xLi^+ + xe^- \quad (1)$$

$$FePO_4 + xLi^+ + xe^- \rightarrow xLiFePO_4 + (1\text{-}x)FePO_4 \quad (2)$$

The reasons why practical capacity is lower than theoretical capacity is, firstly not all the Lithium can be removed from the lattice of the host material. The rest of the Li is removed above the cut off potential. Therefore, it is not accessible. Secondly, the lithium metal reacts with the electrolyte during the filling process, producing a film. During lithium deposition, the film temperature will be high due to ion transfer in the film itself. The result of heating the film is the film into an electronic conductor, and lithium metal is deposited in the film. If heating does not occur, then isolated lithium becomes electrochemically inactivated [11]. Thirdly, the slurry is too dilute, resulting in the cracked coating and hence decreasing the amount of composite printed on the aluminum sheet.

Besides that, the existence of PVDF is also affect the capacity of cell battery. Excessive percentage of PVDF lowers the value of composite conductivity, and results in low capacity values. The reason was due to PVDF only serves as an adhesive between the composite and aluminum sheet and it has no conductivity value. The presence of PVDF will inhibit the intercalation of lithium ions and the extension of PANI conjugation.

4. Conclusion

The electrode charging capacity of the [Li | 1 M $LiPF_6$ / EC DMC | $LiFePO_4$-PANI-PVDF] coin cell battery is 38 $mAhg^{-1}$ and the working voltage at 3.6 – 4.0 V in 25 cycles. The practical capacity (38 $mAhg^{-1}$) is lower than theoretical capacity (170 $mAhg^{-1}$) due to not all the Lithium can be removed from the lattice of the host material.

Acknowledgement

We are grateful to General Directorate of Higher Education (DIKTI) through PUPT program No. 718/UN6.3.1/PL/2017 as a financial support. We also thank Prof. Dr. M.V.V. Reddy, senior research fellow and Mr. Senthil Kumar, a post graduate student of National University of Singapore (Advanced Battery Laboratory) for the support during the coin cell assembly and characterization. This work also supported partially by Academic Leadership Grant, Universitas Padjadjaran 2017, No. 872/UN6.3.1/LT/2017.

References

[1] Nishi Y, Minami M, Tatsumisago M, Wakihara C, Iwakura S, Kohjiya I and Tanaka 2005 *Solid State Ionics for Batteries* (Springer-Verlag Tokyo)
[2] Zhang X, Toprakci O, Toprakci H A K and Ji L 2010 *KONA Powder and Particle Journal* **28**
[3] Yuan L X, Wang Z H, Zhang W X, Hu X L, Chen J T, Huang Y H and Goodenough J B 2011 *The Royal Society of Chemistry* (Energy & Environmental Science)
[4] Detsri E and Dubas T 2009 Journal of Metals, Materials and Minerals **19** 39
[5] Wijayati A, Susanti S , Rahayu I and Hidayat S 2016 *AIP Conference Proceeding* **1712** 050021
[6] Rahayu I, Hidayat S, Noviyanti A R, Rakhmawaty D and Ernawati E 2017 *J. Phys.: Conf Ser.* **812** 012094
[7] Rahayu I, Hidayat S and Aryadi L 2016 *AIP Conference Proceeding* **1712** 050026
[8] Noviyanti A R, Hastiawan I, Yuliyati Y B, Rahayu I, Rosyani D and Syarif D G 2017 *AIP Conference Proceeding* **1848** 040001
[9] Hidayat S, Cahyono T, Mindara J Y, Alamsyah W and Rahayu I 2017 *IOP Conference Series: Materials Science and Engineering* **196** 012035
[10] Hidayat S, Amelia O, Rahayu I and Fitrilawati 2015 *Material Science Forum* **827** 125
[11] Chew S Y, Feng C, Ng S H, Wang J, Guo Z and Liu H 2007 *J. Electrochem. Soc.* **154** A633

IOP Conf. Series: Journal of Physics: Conf. Series **1080** (2018) 012040 doi:10.1088/1742-6596/1080/1/012040

Zero cross algorithm performance on raspberry pi machine for ECG QRS detection

M I Rizqyawan*, M F Amri, A I Simbolon, M A Suhendra, T Haiyunnisa and D E Kusumandari

UPT BPI, Indonesian Institute of Sciences, Bandung 40135, Indonesia

*muha176@lipi.go.id

Abstract. The electrocardiogram (ECG) is one of the most popular instruments in the medical field. Have been developed a smart IoT system for ECG Arrhythmia classification using Android device as the IoT Hub. In order to minimize the computing load on the Android device, some parts of the signal processing are moved to the data acquisition module by using Raspberry Pi 3 as the processor. In the Raspberry machine, a QRS detection system with Zero Cross Algorithm was implemented and its performance was measured. To measure the performance of the algorithm, a timer was placed before and after the execution of the process. The result shows that the algorithm can be run in real-time use in Raspberry Pi 3, although it is a bit slow and produced a high computational load of a single core processor in the machine.

1. Introduction

The electrocardiogram (ECG) is the electrical activity signal of the heart which widely used to diagnose different heart abnormalities [1]. Any difference in the main rhythm of the ECG is called "Arrhythmia". Automatic detection of arrhythmia [2, 3, 10] is one of the most popular research fields in biomedical signal processing because heart arrhythmias can be used to diagnose heart disease, which is considered one of the deadliest disease. To avoid these deaths, it is necessary to diagnose these arrhythmias accurately and fast [4]. The detection of the QRS complex is the most important task in automatic ECG signal analysis to interpret and decide whether the heartbeat belongs to the normal state or the arrhythmia state [5].

Pan-Tompkins is an algorithm developed by Pan and Willis J. Tompkins that most commonly used to detect the QRS wave of the ECG in real time. This algorithm uses an adaptive threshold after the signal is processed with a bandpass filter, derivative, squaring, and moving average [6]. In order to design an algorithm which is more effective to compute and easily applied to mobile applications, the zero-cross method for QRS detection is proposed. This method uses an addition of high-frequency signal and zero crossing count to detect QRS complex which has a low-frequency oscillation. By employing the zero-crossing count, this method is robust against noise and performs well in the noisy ECG signal [7].

Nowadays, the smartphone has been widely used for heart monitoring system. Stefan Gradl implemented the real-time application for monitoring ECG and providing automated arrhythmia detection on mobile devices [8]. However, due to QRS detection and arrhythmia classification performed by the smartphone, this system caused high computational loads and reduced overall sensitivity for abnormal beat detection. To overcome high computational load, in this paper QRS detection will be carried by Raspberry Pi 3 as the processor. Raspberry Pi 3 performance in QRS detection system with Zero Cross Algorithm will be evaluated and will be analysed whether it is suitable for real-time use or not.

2. Methodology

We use MIT-BIH arrhythmia database [9] as our testing data for this experiment, using specifically the MLII channel as our raw data. Selected records of the database, named dataset 1 (ds1) in protocol proposed by Chazal *et al.* [10] is used for the experiment. The algorithm and testing is writes in Python 3 programming language. The processing time of QRS detection of each processing windows is measure in order to examine the computation load required by the algorithm. The processing time is measure by setting up a timer just before and after the process of each steps of QRS detection.

The testing program then loaded and executed on the Raspberry Pi device. We use Raspberry Pi 3 model B as the device for this experiment. It has Quad Core 1.2GHz 64bit CPU with 1GB RAM. The device equipped with some features to support the IoT system such as wireless LAN and Bluetooth Low Energy (BLE) on board, 4 USB 2 ports and HDMI port. It is also contained Micro SD port for device storage the operating system and data. The operating system used in this experiment is Raspbian version 9 with Python version 3.5.3 to run the testing program. The Raspberry Pi 3 can be seen in figure 1.

Figure 1. Raspberry Pi 3 used in the experiment.

3. Zero Cross Algorithm

The processing unit on Raspberry that we used on this paper has low computing ability compared to the CPU with big computation processor or GPU. The major factor we use raspberry because we need to build it a small device for a mobile monitoring system that affordable but has a big capability. Because of the limited resources from the Raspberry, we need an algorithm that can meet the requirements, such as low computational load and less time consuming but not compromising on the performance. We used Zero Crossing Count [7] for QRS detection method on our system. The method can be separated into three steps of computation, there are feature extraction, event detection, and temporal localization.

The feature extraction aims to generate feature signal, which would be used as the input of event detection process. The event detection block searches the possible location of QRS complex by comparing the feature signal with the threshold. Temporal localization searches the actual peak location in the block of event detected in the previous process.

IOP Conf. Series: Journal of Physics: Conf. Series **1080** (2018) 012040 doi:10.1088/1742-6596/1080/1/012040

3.1 Feature Extraction

The first process this step is filtering the raw signal. The first treatment is a Bandpass filter that filled with a cut-off frequency of 18 Hz and 35 Hz. Then the output will be processed with the nonlinear transform, by squaring the filtered signal, multiplied by its signum. This process follows the equation (1).

$$y(n) = sign(x_f(n)) \cdot x f^2(n) \tag{1}$$

where $y(n)$ is the transformed signal and $x_f(n)$ denotes the bandpass filtered signal. This process will separate the important features of the signal, and the rest of other features will be transformed close to the x-axis.

The second step is adding the high-frequency sequence to the output signal. But for this process, we must calculate the amplitude of the high-frequency sequence before adding it. The calculated amplitude of the high-frequency sequence can be obtained by equation (2).

$$K(n) = \lambda_K K(n-1) + (1-\lambda_K)|y(n).c| \tag{2}$$

where the estimated amplitude symbolled with $K(n)$, $\lambda_K$is the forgetting factor, $y(n)$ is the nonlinear transformed signal, and c implied the constant gain. After the amplitude is calculated, then we derived the high-frequency sequence by equation (3).

$$b(n) = (-1)^n \cdot K(n) \tag{3}$$

where $b(n)$ is the high-frequency sequence and $K(n)$ is the amplitude. The next step is adding the high-frequency sequence to the nonlinear transformed signal and make the zero-cross in equation (4).

$$z(n) = y(n) + b(n) \tag{4}$$

where $z(n)$ is the signal after the high-frequency addition, $y(n)$ is the nonlinear transformed signal, and $b(n)$ is the high-frequency sequence. The whole process of the high-frequency addition can be seen in figure 2.

The last step in the feature extraction is zero crossing count. The zero crossing is counted by using equation (5).

$$D(n) = \lambda_D D(n-1) + (1-\lambda_D)d(n) \tag{5}$$

where $\lambda_D$is forgetting factor and $d(n)$ is the count of zero-crosses at the segment.

3.2 Event Detection

The main purpose of this process is to provide the predicted area of QRS complex by comparing the extracted featured signal with the threshold. The mathematic calculation of threshold is shown by equation (6).

$$\Theta(n) = \lambda_\Theta \Theta(n-1) + (1-\lambda_\Theta)D(n) \tag{6}$$

where is λ_Θ forgetting factor, $D(n)$ is the feature signal and $\Theta(n)$ is the adaptive threshold. Using this method, the feature signal that lower than the calculated threshold, the event is marked as the start point of the process. The algorithm would continue the process and when getting the feature signal that higher than the threshold, it is counted as the end of the process.

3.3 Temporal Localization

The R peak is the highest peak or lowest peak on the QRS complex. In this section, R-peak would be detected. To locate the R-peak, the temporal Localization will calculate the maximum and minimum of the signal in the predicted area of QRS complex that obtained before. If the maximum value is higher than the absolute value of the minimum, then the temporal location of the maximum value is considered as the R-wave location. But if otherwise, the temporal location of the minimum value is considered as the R-wave location.

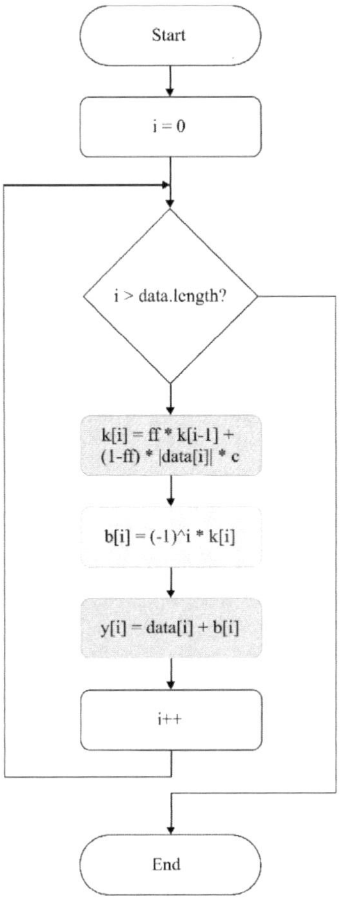

Figure 2. Flowchart of the High-Frequency Addition. The blue box indicates the amplitude estimation. The yellow box indicates the high-frequency sequence. And the green box is the high-frequency addition.

4. Result

The testing is conducted using Linux terminal from Raspbian OS. The result of the testing is shown in table 1.

Table 1. Processing time test on each sub-process.

	Min (ms)	Max (ms)	Average (ms)
Bandpass Filter	51.4	90	52.5
Nonlinear	0.67	0.96	0.79
High-frequency Addition	298.1	315.7	307.1
Zero Crossing Count	466.5	487.3	478.3
Event Detection	219.3	252.7	230.8
Total Process of Algorithm	1,044.7	1,095.6	1,069.5

The result from table 1 then converted to real-time factor proposed by Sufi *et al.* [1] to measure the real-time capability of the method using equation (7).

$$R_f = \frac{P_t}{MeasurementWindow} \quad (7)$$

IOP Conf. Series: Journal of Physics: Conf. Series **1080** (2018) 012040 doi:10.1088/1742-6596/1080/1/012040

where R_f is real-time factor value, and P_t is processing time acquired from table 1. The real-time factor can be seen in table 2. Figure 3 shows the processing time of each sub-process of the Zero Cross Algorithm.

Table 2. The real-time factor of the Total Process of Zero Cross Algorithm.

	Zero Cross Algorithm (Real-time factor)
Min	0.0174
Max	0.0182
Average	0.0178

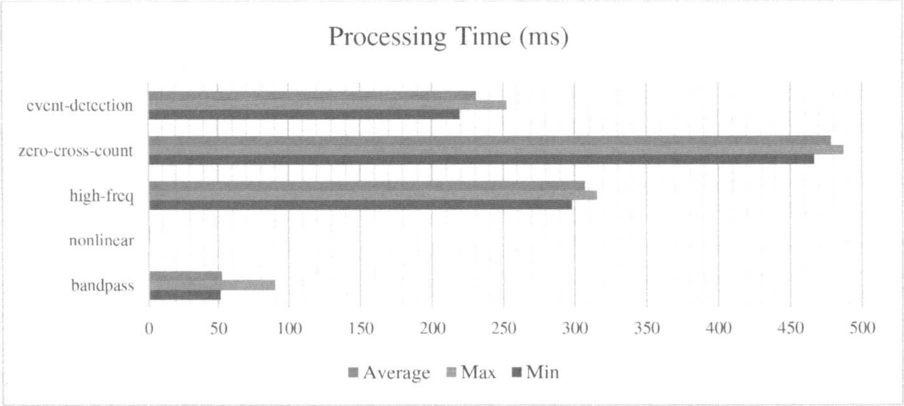

Figure 3. The processing time of each sub-process of the Zero Cross Algorithm. As can be seen, the zero-cross count step takes the most amount of time.

As can be seen, the zero-crossing count sub-process takes most of the processing time in the algorithm with the minimum time of 466.5ms. It is still longer than the maximum value of the second longest processing sub-process, which is High-frequency Addition with 315.7ms.

Please note that the implementation of the algorithm is far from being optimized, as Köhler mentioned in the original paper that the most computationally expensive stages are the bandpass filter and event detection [7]. Therefore, there is still an improvement room on the overall processing time of the algorithm by optimizing the implementation of high-frequency addition and zero-crossing count stages.

As the whole algorithm itself, the algorithm takes 1,069.5ms to process the 60s of data in processing window. While the algorithm generally takes a lot of time (around 1s), it is still considered suitable for real-time use. At maximum, the real-time factor of the algorithm is 0.0182, which is still much less than 1. According to [1], whenever Rf is much less than 1, the algorithm is classified as real-time.

5. Conclusion

This paper presents Zero Cross algorithm performance on Raspberry Pi 3 device for ECG QRS detection. The result shows that this method is suitable for real-time use of QRS detection in the raspberry device. Although the algorithm generates a high computing load on the device and seems to take a lot of processing time, the real-time factor shows that the algorithm fulfils the requirement of real-time use by 0.0182 at maximum which is less than 1. Furthermore, there is a lot of improvement room to make by optimizing the implementation of the algorithm, especially in the zero-crossing count and high-frequency addition stages. This result also shows that it is possible to build the part of ECG signal processing unit into the embedded system using Raspberry Pi and send the feature vector to the

IOP Conf. Series: Journal of Physics: Conf. Series **1080** (2018) 012040 doi:10.1088/1742-6596/1080/1/012040

cloud where the classification is performed. This will eliminate the needs of the smartphone as stated in our initial aims.

Acknowledgements
This research was supported by the Insinas Ristek funded by Minister of Research, Technology, and Higher Education, Indonesia.

References
[1] Sufi F, Fang Q and Cosic I 2007 *29th Annual International Conference of the IEEE Engineering in Medicine and Biology Society* p 3697
[2] Luz E, Schwartz W, Cámara-Chávez G and Menotti D 2016 *Computer Methods and Programs in Biomedicine* **127** 144
[3] Escalona-Moran M, Soriano M, Fischer I and Mirasso C 2015 *IEEE Journal of Biomedical and Health Informatics* **19** 892
[4] Karimifard S, Ahmadian A, Khoshnevisan M and Nambakhsh M S 2006 *28th IEE EMBS Annual International Conference* p 1367
[5] Park K S, Cho B H, Lee D H, Song S H, Lee J S, Chee Y J, Kim I Y and Kim S I 2008 *Computers in Cardiology* **35** 229
[6] Pan J and Tompkins W 1985 *IEEE Transactions on Biomedical Engineering* **32** 230
[7] Köhler B, Hennig C and Orglmeister R 2003 *Progress in Biomedical Research* **8** 138
[8] Gradl S, Kugler P, Lohmuller C and Eskofier B 2012 *Annual International Conference of the IEEE Engineering in Medicine and Biology Society*
[9] Moody G and Mark R 2001 *IEEE Engineering in Medicine and Biology Magazine* **20** 45
[10] deChazal P, O'Dwyer M and Reilly R 2004 *IEEE Transactions on Biomedical Engineering* **51** 1196

Identification of magnetic anomalies in subsurface area of Sinabung mountain

T Tampubolon*, Rahmatsyah and M A T Harahap

Department of Physics, University of Medan, Jl. Willem V Medan Estate Iskandar Pasar ZIP code 20222

*topartam@gmail.com

Abstract. This research was conducted by using geomagnetic methods. It aims to identify cross section anomalies and subsurface models. Measurements using a PPM (Proton Precession Magnetometers) Elsec 770 types data collection is done randomly with 27 measurement points. Data analysis was performed using 11 surfers to get a contour map and Mag2DC software to get the magnetic anomaly cross section. The results showed that the anomalous cross section with the geomagnetic method has the lowest value of 18.45 nT and the highest value reached 64.92 nT. From the anomaly, it was found that the value of vulnerability obtained is 0.0004, 0.00018, 0.0002 emu where subsurface layer model consists of dolomite and limestone.

1. Introduction

Mount Sinabung is a type of volcano located in Karo, North Sumatra, Indonesia. It is classified into types strato volcano not far from the residence of the population. This is the second highest peak in North Sumatra. It is nearby with Sibayak Mountain and has a height of 2,451 meters. The mountain is not erupted since 1600, suddenly, it was reactivated by the eruption in 2010. The last eruption occurred since 2013 until now. It causing major problems. people had to evacuate and leave the shelter and land in dangerous zone eruption.

The eruption of Mount Sinabung history occurred on August 29, 2010 at approximately 12:15, Mount Sinabung produced lava mixed with stones. This mountain status was raised to be careful. At least 12,000 people have been evacuated and accommodated in 8 locations. Sinabung mountain ash tends to glide from southwest to northeast. These bursts of volcanic ash up to 5,000 meters in the air. Then, slowly return to subside [1]. No casualties were reported, but thousands of people were forced to evacuate to a safe area. As a result of this event, the status of Mount Sinabung is raised to level 3 into standby. After a fairly high activity for a few days, another problem is when the eruption occurred, agricultural land crop failure due to damaged canal caused by volcanic ash struck. By looking at the effects of the eruption of Mount Sinabung, researchers interested in conducting research to find out what kind of rocks that form the surface of the mountain with geomagnetic method.

Geophysics magnetic methods in the application depends on accurate measurement of the local geomagnetic field anomalies generated by variations in the intensity of magnetization in the rock formations. The intensity of geomagnetic induction depends on the magnetic susceptibility of rocks, magnetic force, and intensity of the permanent. The value of magnetic susceptibility depends on the type of material, positive to negative for the paramagnetic and diamagnetic property. Magnetic susceptibility is the most basic parameters in the study of magnetic rocks. The magnetic response of rocks and minerals determined by the number and susceptibility of magnetic material in them. The equation for the total field is proportional to H, where the symbol is called the permeability of the material in a vacuum, because there is no magnetic material, then = 0, so it is called the vacuum permeability of 4π x 10-7 Wb / am [2]. The value of magnetic susceptibility in a vacuum is equal to zero because the only real things that can be magnetized. magnetic susceptibility can be interpreted as the degree of magnetism of a material [3].

Data anomalies of the total magnetic field from the continuity result would be reduced to the poles to localize the area with maximum anomalies above anomaly body causes the object, it makes interpretation easier. pole reduction is done by making the angle of the object to 900 and its declination 00 [4]. This is because the magnetic poles, the Earth's magnetic field and the magnetic induction point to down. Data reduction of these poles, qualitative interpretation can be done. This reduction is done by using Magpickc program, IGRF International abbreviation Geomagnetic Reference Field. This is a reference geomagnetic field. Basically, the value IGRF is a strong value of the main magnetic field of the earth (Ho). One type is a sedimentary rock, has very little magnetic susceptibility compared with igneous or metamorphic rock tend to have higher magnetic content. The most magnetic survey is used to map geological structures on or within the bedrock (crystalline rock beneath layers of sediment) or to detect the presence of magnetic minerals directly. In a survey of geothermal (geothermal), magnetic methods used to identify the type of igneous or hot rock that acts as a heat source. In broad outline, there are two systems of heat sources, they are a system of active volcanoes and volcanic systems other than [5] geothermal area with the type of active volcanoes have a high temperature during 1800. High temperatures will cause the low value of magnetic anomalies. The most magnetic survey is used to map geological structures on or within the bedrock (crystalline rock beneath layers of sediment) or to detect the presence of magnetic minerals directly. In a survey of geothermal (geothermal), magnetic methods used to identify the type of igneous or hot rock that acts as a heat source. In broad outline, there are two systems of heat sources, they are a system of active volcanoes and volcanic systems other than. Geothermal area with the type of active volcanoes have a high temperature during 1800. High temperatures will cause the low value of magnetic anomalies [4]. The most magnetic survey is used to map geological structures on or within the bedrock (crystalline rock beneath layers of sediment) or to detect the presence of magnetic minerals directly. In a survey of geothermal (geothermal), magnetic methods used to identify the type of igneous or hot rock that acts as a heat source

Research on magnetic anomalies that have been made for modeling using magnetic data from the measurement results in the area Arjuno Welirang Mountain. It aims to identify magnetic anomalies produced in geothermal resources allows prospects subsurface area can be predicted.

2. Research Methodology

2.1. Time and Location

Geographically, lies between 2°50' Karo-3°19 'North Latitude and 97°55'-98°38' East Longitude with an area of 2,127.25 km2 or 2.97 percent of the area of North Sumatra Province. The geographic location of research lies between 3008'45.61 North latitude and 980 east longitude 20'12.54 on 60 points from implant village, sub district Tiganderket. This study was carried out about 2 months from February to March 2017.

2.2. Equipment and Materials

2.2.1. Research Tools

PPM (Proton protezione magnetometer), Casio stopwatch, compass, Meter, Battery, DNS GARMIN, Stationary, Transmitter.

2.2.2. Materials Research

The research material was the location of the sampling point.

2.3. Procedures for Research

1. Viewing locations and shooting area will be an area of research.
2. Determine the coordinates of the reference point in the study area using GPS (global positioning system)
3. Determine basis points (base) in the area to be surveyed by determining anomalies
4. Make measurements using geomagnetic 770 PPM elsec type.
5. Process the data found in the geomagnetic elsec 770 PPM Type.
6. Analyzing the data by mag2DC software.
7. Distinguish the type of resistance value based on the anomalies and color to see the vulnerability.

3. Results and Discussion

Field data obtained during geomagnetic field measurements using the method in the form of magnetic values of each point. magnetic data collection aims to observe the total magnetic field (H) of the Earth at some point. The data was obtained by the induced magnetic objects where the value of the magnetic field (H) shall be reduced by the value of the magnetic field induces so will result in the value of the magnetic field called a magnetic anomaly (ΔH) .the data time Earth's magnetic field at the base station and time of data the earth's magnetic field is not the same, so adapted to the times in the data base that is adjacent. Field measurement data in the form of total Earth's magnetic field is still mixed with the International Geomagnetic Reference Field magnetic main earth (IGRF) and magnetic every day.

To get the total value of the magnetic anomaly, the following correction is made, such as the daily correction, correction IGRF and topographic correction.

3.1. Result

3.1.1. Earth's Magnetic Anomaly distribution pattern

Anomalies in the magnetic field of the calculated results can be displayed in the form of a contour map using software anomaly surfer 11. It can be seen in the following figure 1.

Figure 1 shows the survey area with low anomaly at the point (D3), which is an anomaly 18,45nT and high (D25), i.e. 64.97 nT. The low value of magnetic anomalies obtained in the survey area are interpreted as areas closely related to the formation of geothermal manifestations in the area.

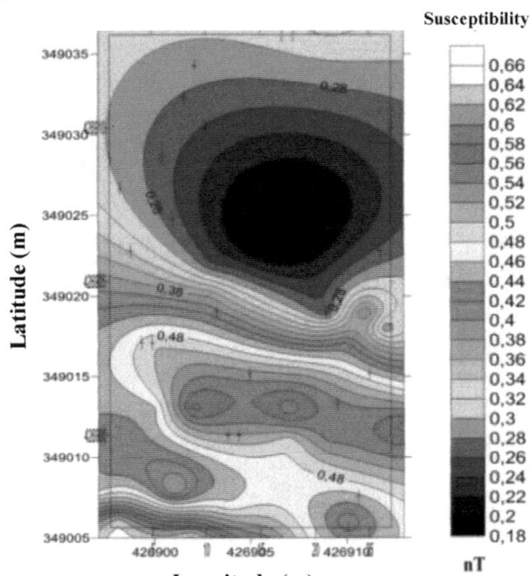

Figure 1. The distribution pattern anomalies in Earth Magnet Survey Location.

IOP Conf. Series: Journal of Physics: Conf. Series **1080** (2018) 012041 doi:10.1088/1742-6596/1080/1/012041

3.1.2 Rocks Susceptibility (K)

Susceptibility of calculation results can be displayed in the form of a map of vulnerability contour using software surfer 11 that shown in figure 2.

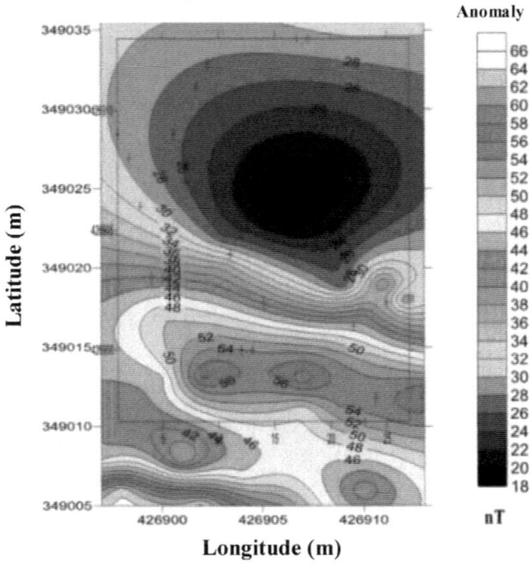

Figure 2. Contour susceptibility Map in implant Village.

From the picture above shows a survey of geothermal area in implant Village with vulnerability from lowest to highest, is: 0.18498 x 103 to 0.64922 x 103. Values obtained in the area of vulnerability of this survey will be used to determine the type of rock in the area subsurface survey.

3.2. Discussion

3.2.1. Geomagnetic Data Interpretation

The first step in modelling is to make an incision or anomalous section of the anomalous magnetic stripe low to high anomaly, it can be seen from the bottom figure 3.

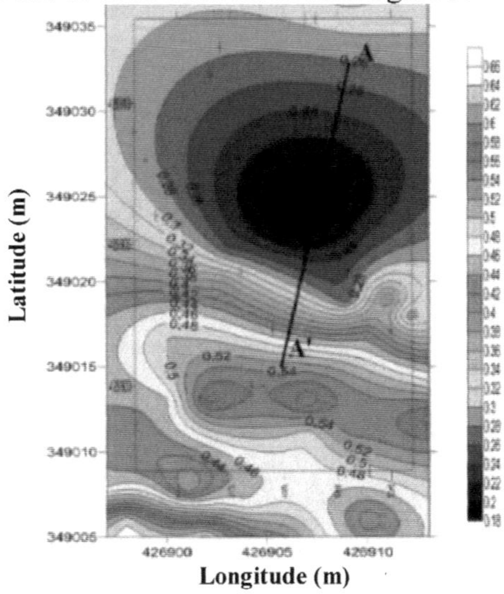

Figure 3. Contour Anomaly Map With incision AA'.

The picture above shows the incision AA susceptibility decided by geological area of research and qualitative interpretation. AA 'incision goes from the northeast to the southwest. Quantitative interpretation aims to determine the lithology of the study area. Lotology can be determined by the value of the vulnerability being modelled.

3.2.2. Part Model geomagnetic use Mag2DC

Lotology can be determined by the value of the vulnerability being modelled. This interpretation is done by modelling using software Mag2DC, resulting in the following drawings in figure 4.

Figure 4 shows a model cross-section AA 'where there are x - axis and y - axis. X - axis path distance value, negative y - axis depth value, a dotted line on the curve is observed anomalous value, while the line was squeezing the dotted line is an anomaly modelling. Modelling results obtained vulnerability value indicates the type of sedimentary rock with susceptibility value (0.0004; 0.00018; 0.0002) emu.

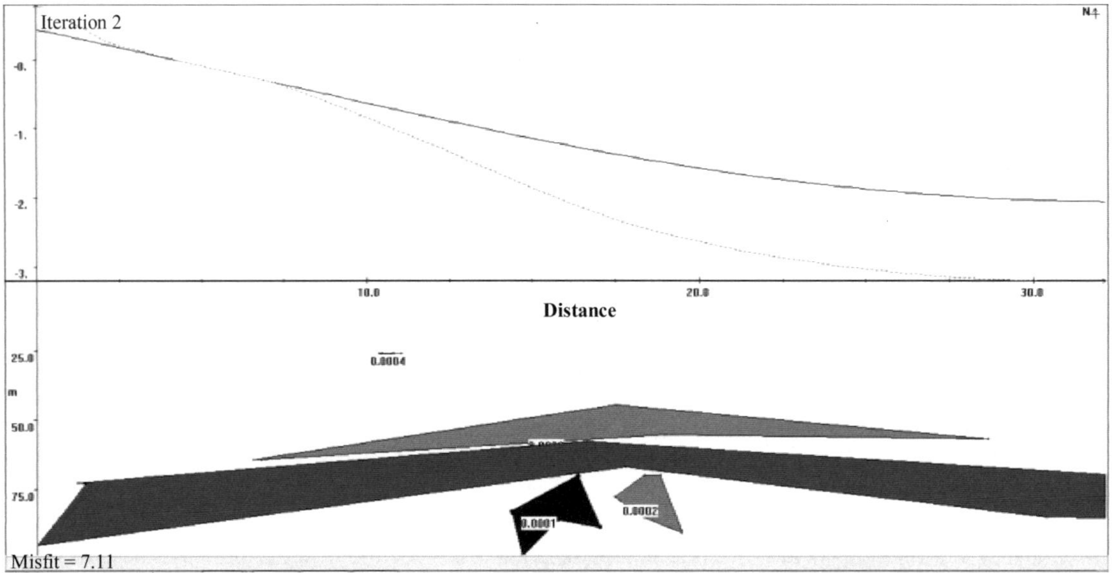

Figure 4. Geomagnetic Section Model using Mag2DC.

4. Conclusion

The values of the magnetic anomalies in the study area ranged between 18.45nT at coordinates 426 908 349 025 N and E to 64.75 nT at coordinates 426 902 349 005 N and E. Based on the value of vulnerability in the region of the village of implant, the type of rock found in the area of reserch is 0, 0004, 0.00018, 0.0002 nT where subsurface models consist of dolomite, limestone or part of sedimentary rocks.

References

[1] Bemmelen R W V 1937 *The Volcano – Techtonic Structure of The Residency of Karo, North Sumatera De Ingenieur in Ned. Indie* **IV** 9
[2] Garel F 2013 *The Influence of Wind on The Estimation of Lava effusion Rate From Thermal Remote-Sensing* (Paris: Institut de Physique)
[3] Nudianta 2011 *Calculation of Lava Effusion Rates* (Australia: Landsat TM Data)
[4] Telford W M, Geldart L P, Sheriff R C 2004 *Applied Geophysics* 2nd *ed.* (USA: Cambridge University Press)
[5] Syamsu R 2008 *Geomagnetic Method Lecture Note* (Depok: Physic Department, UI)

IOP Conf. Series: Journal of Physics: Conf. Series **1080** (2018) 012042 doi:10.1088/1742-6596/1080/1/012042

Characteristics of TiO_2 particles prepared by simple solution method using $TiCl_3$ precursor

I M Joni[1,2*], L Nulhakim[2] and C Panatarani[1,2]

[1]Department of Physics, Faculty of Mathematics and Natural Sciences, Universitas Padjadjaran, Jl. Raya Bandung-Sumedang km 21, Jatinangor 45363, West Java, Indonesia
[2]Nanotechnology and Graphene Research Centre, Universitas Padjadjaran, Jl. Raya Bandung-Sumedang km 21, Jatinangor 45363, West Java, Indonesia

*imadejoni@phys.unpad.ac.id

Abstract. In this work, TiO_2 particle has been prepared by simple solution method using $TiCl_3$ precursor. The XRD analysis confirmed the presence of rutile TiO_2 which is indicated by the peaks corresponding to (110), (101), (111), (211) and (002) plane. The crystallite size was about 7.25 nm as obtained by Scherer equation. The SEM image revealed that the rutile TiO_2 exhibits agglomeration with spherical morphology. The composition is confirmed by EDS measurement which shows 17.5 at % of titanium and 58.2 at % of oxygen, respectively. Based on SEM result, the size of the primary particle was about 100 nm. The PSA measurement showed that the average particle diameter was about 759 nm with a median d_{50} (<50%) of about 200 nm. These results show that TiO_2 sample was nanocrystalline with primary particle diameter of about 100 nm and 759 nm, respectively. The particle diameter was still large probably due to the existence of agglomeration which is acquired by PSA. However, these properties can be used as preliminary information for applications that requires nanocrystalline TiO_2 with rutile phase by using simple and low-cost preparation method.

1. Introduction

In the last few years, there has been growing interest in metal oxide particle such as TiO_2. A semiconductor is used in wide range of applications such as for lithium-ion batteries, photosensors, photocatalysis, dyes sensitized solar cells, anti-reflective (AR) coatings, electrochromic displays, optical filters, planar waveguides, photovoltaic devices, UV light sensors, biomedical applications, etc. [1-4]. Nanoparticle TiO_2 has also been used as a pigment, a thickener and sunscreen in cosmetics and skin care products [5]. Properties and applications of TiO_2 are controlled by its morphology, particle size, purity, crystal structure and phase, porosity, surface area, etc. [1, 6]. TiO_2 is an n-type wide bandgap semiconductor and has three crystalline phases, i.e. anatase (tetragonal), rutile (tetragonal), and brookite (orthorhombic) with band gap 3.02 eV (410 nm), 3.20 eV (387 nm) and 2.96 eV (419 nm), respectively [1, 6, 7]. Among these three types, the rutile phase is most stable form while anatase and brookite types are metastable and possible to transform to rutile phase under high temperature of about 750°C [8]. Synthesis of anatase, rutile and brookite of TiO_2 particle and thin film from $TiCl_3$ were reported by Leal J L *et al.* [9], Yu Q *et al.* [10], Haider A J [11] and Dusan A *et al.* [4]. The characteristics of TiO_2 such as chemical purity and crystalline phase depend on the synthesis method

[12]. The effect of various methods for preparation to obtain different particle size and morphology was reported [13]. The TiO_2 particle can be prepared by several methods such as hydrothermal, solvothermal, microwave assisted method, electrospinning and simple solution method. In the near future, the preparation of TiO_2 using low cost and simple method will be necessary. The preparation of rutile TiO_2 particle using simple solution method still needs to be developed and explored further. Therefore, in this work, the rutile TiO_2 sample will be prepared by using simple solution method. The characteristics of this sample will be analyzed and discussed.

2. Experiments

The preparation of TiO_2 particle was done by simple solution method with $TiCl_3$ 15% (Merck) as a precursor. 7 ml of $TiCl_3$ 15% was mixed with 1.5 ml HCl and 9 ml of distilled water, stirred for about 15 minutes. Then the solution was heated at 80°C temperature for about 10 minutes by gently adding 6 ml of concentrated HNO_3. The process is then continued with mixing and reheating at 80°C followed by washing and drying at about 100°C. Subsequently, the sample was calcined at 200°C for 5 hours.

The structural property of TiO_2 particle was examined by X-ray Diffraction (XRD) Rigaku using Cu-Kα (λ = 0.15419 nm) radiation source, (X'pert Pro PANalytical) over a 2θ scan range of 20°-80°. Surface morphology and element composition were studied using Scanning Electron Microscopy (SEM) equipped with energy dispersive X-ray (EDX), FEI-Inspect 550, EDAX Ametek. The diameter of particles was examined by Particle Size Analyzer (PSA) Beckman Coulter LS 13 320, while the purity of particles was examined by XRF (X-ray Fluorescence), Rigaku Nex CG.

3. Results and Discussion

The XRD pattern of sample is shown in figure 1. The peaks corresponding to rutile phase, matches well to JCPDS Card No: 21-1276 as reported by A.J. Haider *et al.* [11]. The results were also comparable with the other reports of rutile TiO_2 [10]. The position of 2θ at 27.42°, 36.08°, 41.25°, 54.33° and 63.44° corresponds to Miller indices of (110), (101), (111), (211) and (002) plane, respectively [1]. The peak position of 2θ and its correlation with Miller indices can be seen in table 1. The crystallite size (D) of sample was calculated using Debye-Scherrer's equation (1).

$$D = \frac{0.9\lambda}{\beta \cos\theta} \tag{1}$$

where λ is the wavelength of x-ray beam (Cu Kα, with λ = 0.15418 nm), β (in rad) is the full width at half maximum (FWHM) of the intense peak and θ (in rad) is the Braggs angle. The average of crystallite size for this TiO_2 sample was about 7.25 nm. This size was comparable with the result reported by J.H. Leal *et al.* [9], the crystallite size for rutile phase synthesized from $TiCl_3$ by simple low-temperature method was examined to be below 10 nm.

Table 1. The peak position of 2θ and its correlation with miller indices.

2 Theta	Miller indices	Phase
27.42	110	Rutile
36.08	200	Rutile
41.25	111	Rutile
44.20	211	Rutile
56.51	220	Rutile
63.44	002	Rutile
69.11	301	Rutile

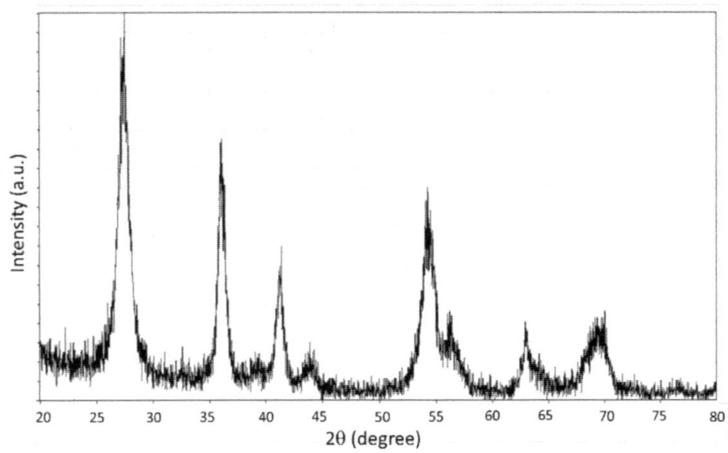

Figure 1. X-ray diffraction patterns of rutile TiO_2 particle prepared by simple solution method.

SEM images of TiO_2 particle is shown in figure 2. Spherical particles with agglomeration were observed and the diameter of the primary particle was about 100 nm as shown in the inset of figure 2. On the other hand, the elemental analysis was examined by EDX measurement and given in figure 3. The green box in the inset of figure 3 depicts the area of EDX testing. The elemental compositions of sample consist of oxygen and titanium of about 80.5 and 19.5 at%, respectively. The presence of oxygen indicates that the sample is in the form of titanium dioxide.

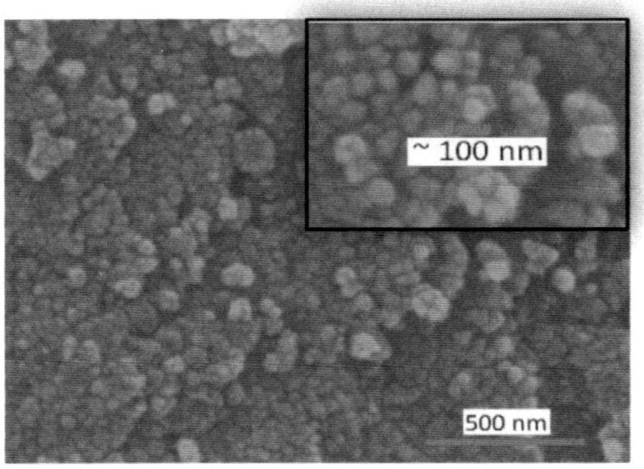

Figure 2. SEM image of rutile TiO_2 particle with two different magnification.

The PSA result showed that the average particle diameter was about 759 nm with d_{50} (<50%) = 200 nm (figure 4). This result indicated that the TiO_2 particles occur as agglomerates in suspension and hence a proper dispersing agent is necessary to obtain a well-dispersed suspension of TiO_2 nanoparticles.

IOP Conf. Series: Journal of Physics: Conf. Series **1080** (2018) 012042 doi:10.1088/1742-6596/1080/1/012042

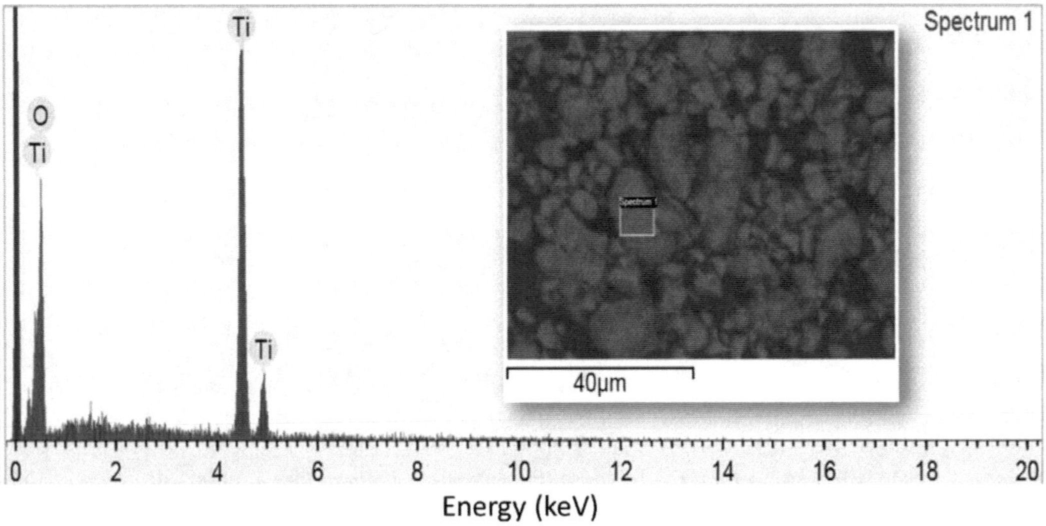

Figure 3. SEM images of rutile TiO_2 particle at various magnification.

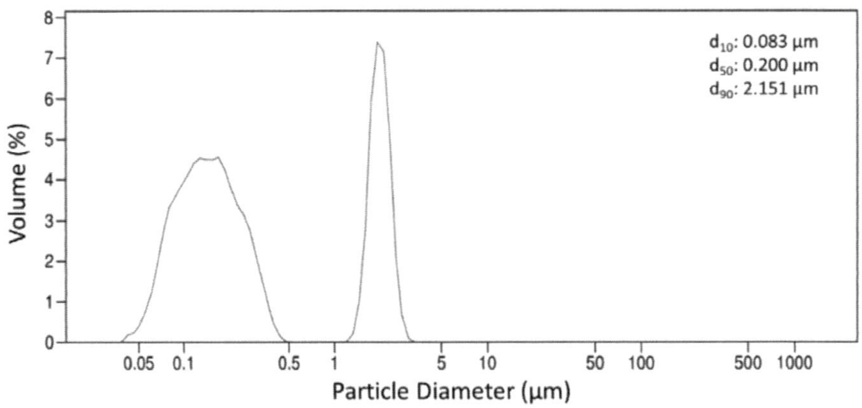

Figure 4. The particle diameter versus volume (%) of PSA measurement for TiO_2 sample.

The purity of sample was about 84 Mass % of TiO_2 as confirmed by XRF measurement and the impurities showed very low Mass percentage. The complete data can be seen in table 2.

Table 2. The percentage (Mass%) of sample composition measured by XRF.

Component	Mass%
TiO_2	83.9
K_2O	1.1
CaO	0.6
SiO_2	0.3
Zn	0.2
Al_2O_3	0.2
S	0.1
MgO	0.05
P_2O_5	0.03
Cu	0.004

4. Conclusions

TiO_2 particle was successfully synthesized by simple solution method using $TiCl_3$ precursor. XRD characterization confirmed that the particle has rutile TiO_2 phase. The average crystallite size was about 7.25 nm. SEM measurement revealed that the surface morphology of sample was spherical with agglomeration. The diameter of the primary particle based on SEM image was about 100 nm. While the particle diameter measured by PSA was about 200 nm for d_{50} (<50%). It is proposed that the TiO_2 particle in this work showed nanocrystalline nature (crystallite size < 10 nm) with particle diameter ~ 200 nm that is most probably due to the existence of agglomeration. In addition, this work described a simple solution method to prepare rutile phase with a relatively small crystallite size (below 10 nm). Since the preparation of TiO_2 is using a simple low-cost method, it will be suitable in mass production for industrial applications.

Acknowledgements

The authors would like to acknowledge Academic Leadership Grant, Universitas Padjadjaran, contract no. 855/UN6.3.1/PL/2017 for funding this research work.

References

[1] Dhandayuthapani T, Sivakumar R and Ilangovan R, 2016 *Surf. Interf.* **4** 59
[2] Chen X, Mao S S 2007 *Chem. Rev.* **107** 2891
[3] Hasan M M, Haseeb M A, Saidur R and Masjuki H H 2008 *Int. J. Chem. Bio. Eng.* **1** 92
[4] Dussan A, Bohorquez A, Quiroz H P 2017 *Appl. Surf. Sci.* **424** 111
[5] Verleysen E, De Temmerman P J, van Doren E, Francisco M A D and Mast J 2014 *Powd. Tech.* **258** 180
[6] Wang Y, He Y, Lai Q and Fan M 2014 *J. Environ. Sci.* **26** 2139
[7] Linsebigler A L, Lu G, Yates J T 1995 *Chem. Rev.* **95** 735
[8] Liao Y L, Que W X, Jia Q Y, He Y C, Zhang J and Zhong P 2012 *Mater. Chem.* **22** 7937
[9] Leal J H, Cantu Y, Gonzales D F and Parsons J G 2017 *Inorganic Chem.Comm.* **84** 28
[10] Yu Q, Wang H, Peng Q, Li Y, Liu Z and Li M 2017 *J. Hazard. Mater.* **335** 125
[11] Haider A J, Al-Anbari R H, Kadhim G R and Salame C T 2017 *Energy Proc.* **119** 332
[12] Diebold U 2003 *Surf. Sci. Rep.* **48** 53
[13] Panatarani C, Lenggoro I W and Okuyama K 2003 *J. Nanopart. Res.* **5** 47

IOP Conf. Series: Journal of Physics: Conf. Series **1080** (2018) 012043 doi:10.1088/1742-6596/1080/1/012043

Land subsidence analysis in Bandar Lampung City based on InSAR

A Zaenudin[1*], I G B Darmawan[1], Armijon[2], S Minardi[3] and N Haerudin[1]

[1]Geophysical Engineering, University of Lampung, Bandar Lampung 35145, Indonesia
[2] Geodesy Engineering, University of Lampung, Bandar Lampung 35145, Indonesia
[3] Physics Department, University of Mataram, Indonesia

[*]ahmad.zaenudin@eng.unila.ac.id

Abstract. Bandar Lampung is one of the cities in Indonesia, which has the potential to land subsidence due to human activities or natural phenomena. However, it is still lack of information and the study of land subsidence that was ever made. This research used 15 SARs data in the intervening years from 2006 to 2011 and was combined to produce the interferogram, and then, it was inverted by using SBAS algorithm. Based on data analysis, SBAS technique was reliable enough to determine the deformation symptoms quickly and accurately. However, some locations had indications of subsidence more than 5 mm/year, even the highest reaches more than 30 mm/year which is identified in the Campang Raya, Camang, Kangkung, Sukaraja, Bakung, and Beringin Raya. Subsidence that occurred is suspected caused by tectonic and human activities and is more likely to occur in the new area suffered by land conversion. Subsidence that occurred implies on damage to the building structure, the occurrence of tidal flooding in coastal areas, and landslides in hilly areas.

1. Introduction

Land subsidence is the decline in land surface that is experienced by many cities in Indonesia and in the world. This phenomenon is the decrease of the soil surface due to several reasons: excessive groundwater exploitation, ground surface loading by construction and infrastructure, natural consolidation of alluvium, and tectonic activity [1]. In western Indonesia, it was analysed for several cities experiencing an average subsidence above 22.5 cm/year. In Medan, Jakarta, Bandung and Semarang are caused by groundwater extraction for industry. In Arun, Lhoksumawe it is due to natural gas extraction and in Sidoarjo due to the discharge of the mud from the soil [2]. Land subsidence in urban areas is usually caused by a combination of excessive extraction of groundwater, the natural consolidation of alluvium soils, construction costs (i.e. high compressibility soil settlement), and sometimes tectonic activity.

The same phenomenon can occur in Bandar Lampung, due to the rapid changes of land use, the increasing exploitation of ground-water and the activity of the Panjang-Lampung Fault that crosses the city. To view the phenomenon, we examine it using DInSAR with the SBAS algorithm.

IOP Conf. Series: Journal of Physics: Conf. Series **1080** (2018) 012043 doi:10.1088/1742-6596/1080/1/012043

2. DInSAR-SBAS

In the study of land subsidence in an area, the DInSAR - SBAS method can be selected to study the symptoms of ground subsidence because it has a wide range of spatially and temporally and has high accuracy.

InSAR is a very powerful technique for detecting deformations at a point in the LOS direction occurring between two SAR acquisitions that lead to a phase displacement ($\delta\phi^{disp}$). DInSAR technique has a deficiency that is still having the influence of the atmosphere and still having the influence of phase components due to topography and due to flattening imperfect [3]. To eliminate the influence of atmospheric propagation and topographic phase components, several algorithms have been developed, one of which is the Small Baseline Subset (SBAS) technique.

3. Data and Processing

The research data used in the implementation of this research are: (a) ALOS-PALSAR dataset in horizontal aligned polarization (HH) in descending acquisition mode in Bandar Lampung City region obtained from ASF ALASKA (table 1). Digital Elevation Model DEM) SRTM 1-Arc Second area of Bandar Lampung and surrounding areas obtained from Earth Explorer - USGS.

Errors in the unwrapping process, the noise on the interferogram, as well as errors in data processing can lead to inconsistent deformation patterns. The quality index of the deformation data is also called the temporal coherence factor (C_{temp}). So by analyzing the value of C_{temp} (figure 1(a)), it can be known that the pixel has reliable information, which is generally shown in the threshold value (generally > 0.7) [3]).

Table 1. SAR dataset.

No.	Acquisition Date	Acquisition Type	No.	Acquisition Date	Acquisition Type
1	21-12-2006	FBS	11	10-02-2009	FBS
2	23-06-2007	FBD	12	01-07-2010	FBD
3	08-08-2007	FBD	13	16-08-2010	FBD
4	23-09-2007	FBD	14	01-10-2010	FBD
5	08-02-2008	FBS	15	16-02-2011	FBS
6	10-05-2008	FBD			
7	25-06-2008	FBD			
8	25-09-2008	FBD			
9	10-11-2008	FBS			
10	26-12-2008	FBS			

Based on figure 1(a), the C_{temp} value is strongly influenced by the type of land cover in the study area, where in the dynamic areas (agricultural and forest) tend to have low C_{temp} values, whereas in statutory population areas the C_{temp} is high.

Accuracy of the result of deformation measurement in study area is obtained from several parameters such as coherence and wavelength which then calculated so that will get estimation of measurement accuracy (standard deviation), where the bigger value will be lower accuracy of measurement and vice versa. Estimation of measurement accuracy is calculated by the following equation [3]:

$$\sigma = \sqrt{\frac{1-\gamma^2}{2\gamma^2}}\frac{\lambda}{4\pi} \tag{1}$$

where σ is the measurement accuracy, γ is interferomatric coherence, and λ is the wavelength used.

Based on the measurement accuracy estimation (σ), the value of σ (figure 1(b)) tends to be inversely proportional to the C_{temp} value (figure 1(a)), since the value of σ is calculated based on the value of C_{temp}.

The measurement accuracy (σ) is quite high (20-50 mm/year) in a dynamic area (high-vegetation area). Since the value of σ is not sufficiently reliable in terms of measurements of ground subsidence in the millimetre order, no relevance that can be obtained in these areas.

As for urban areas dominated by buildings and highways, the value of σ ranges from 0 to 0.7 mm/year. Accuracy is quite high considering the urban areas tend to be static (not experiencing significant changes) from time to time. Therefore, the result of SBAS measurement having high reliability is mostly obtained in urban area.

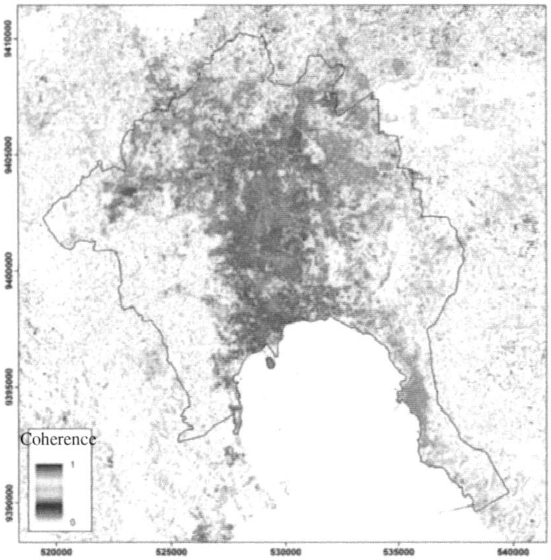

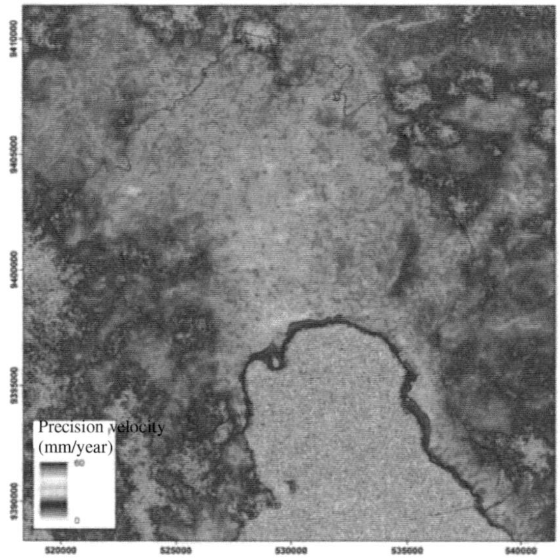

Figure 1. (a) Temporal coherence map and (b) Precision velocity map.

In the inversion process to get the value that represents the deformation velocity model in the research area, linear regression is done by using Least Square method on time-series deformation data. To know the quality of the fitting of the inversion model used for the time-series data, then a nonparametric comparative test using Chi Square method (kai square) is used. The low value of Chi Square shows good fitting to the regression line, and vice versa (figure 2(b)). The result of calculation of Chi Square test on the result of this research is shown in chi squre map in figure 2.

The value of Chi Square of the linear inversion model (figure 2(a)) is quite low in average value indicating that the fitting between the linear regression model chosen in the inversion process with the time-series data is quite good, although in some regions especially the high subsidence rate tends to have a high value of Chi Square because the trend of time-series data in the area tends to be non-linear (parabola). However, since the area with high Chi Square is not very dominant, the linear inversion model is still considered to be relevance enough to represent the deformation velocity in this study.

4. Analysis and Interpretation

The result of SAR data processing using SBAS technique in this research produces information about deformation speed in Line of Sight (LOS) direction in Bandar Lampung City (figure 3) with data resolution of 20 × 20 meter. The deformation velocity data in this study was limited to areas with an accuracy of <7 mm/year and temporal coherence> 0.7. The deformation speed in figure 3 refers to the direction of LOS (Line of Sight) and in the vertical direction (z-axis) referenced at a reference point which is a stable region (not deformed in 2006-2011).

The deformation velocity in the LOS direction has a positive value (green colour scale) to represent the relative up end deformation and the negative LOS velocity (red colour scale) to represent relative deformation away from the satellite direction (subsidence). Indication of land subsidence in Bandar Lampung City spatially has two main patterns, namely patchy (<10 km^2) in coastal area of Lampung Bay and Kemiling District, and local (10 - 100 km^2) in Eastern part of Bandar Lampung City, that is in area East Tanjung Karang.

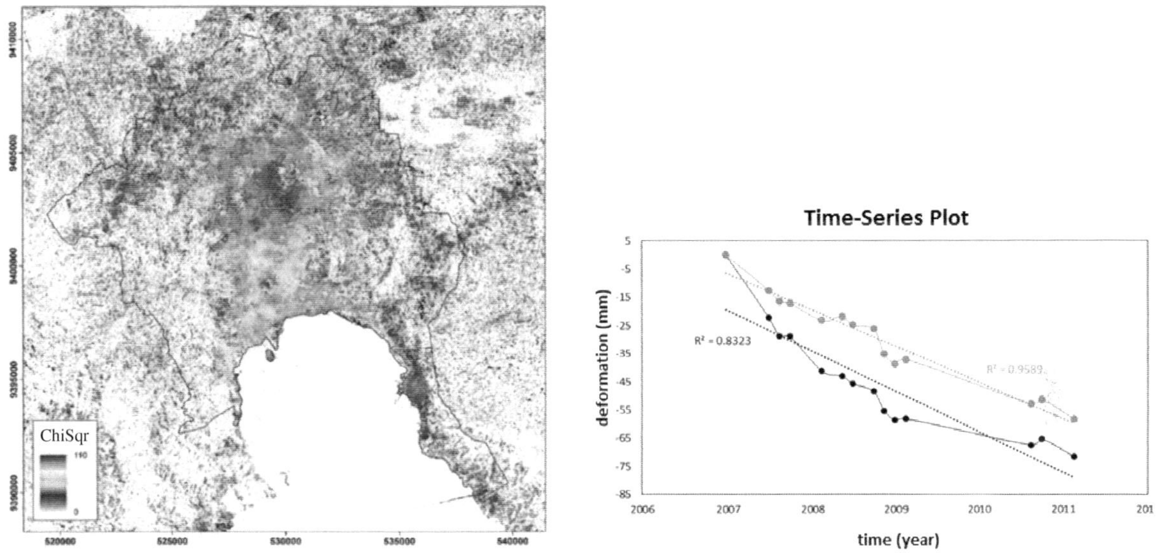

Figure 2. Chi Square map (a) Example of time-series data from two point adjacent to identical subsidence rate, but with very different Chi Square values. The black line is a time-series with a large Chi Square value (b).

For the analysis process in this study, two approaches are used, the first to investigate the indication of land subsidence on a regional scale, by identifying areas that have high indication of land degradation based on the measurement of SBAS. Further analytics were conducted on a smaller scale (local) aimed at limiting the discussion to only certain areas to reduce the complexity of the subsurface geological conditions as well as the varied causes of ground subsidence. Therefore, in this study, only six observations were done to indicate land subsidence (mean > 5 mm/year in the area 200 × 200 metres) which is expected to represent the cause factors and the pattern of land subsidence according to Chaussard [2] as depicted in figure 3.

The result of observation in the A area (figure 3) shows the area indicated to land subsidence with an average speed of 9 mm/year, in Campang Raya Village. Graph of land subsidence is shown in figure 4(a).

The process of land subsidence in the A area correlates to land use in this area, where in this area it is an industrial area and densely populated residential. The pattern of soil surface decline velocity tends to be higher in the South, that is, around the industrial estate. The development of industrial estate and settlements in this area caused additional load to the soil layer and also increased groundwater exploitation. The process of land subsidence in B area which is the area of Camang Mount (figure 4(b)), which on average undergoes a decrease of 7.4 mm/year. In addition to this, there is also intensive land conversion at the peak of Camang Mount into a luxury residential. Decreased ground level is possible although in fact Camang Mount is composed of tuff-tuffit, because the decrease of land surface due to mining activities and land conversion can intensively affect all kinds of rocks [2].

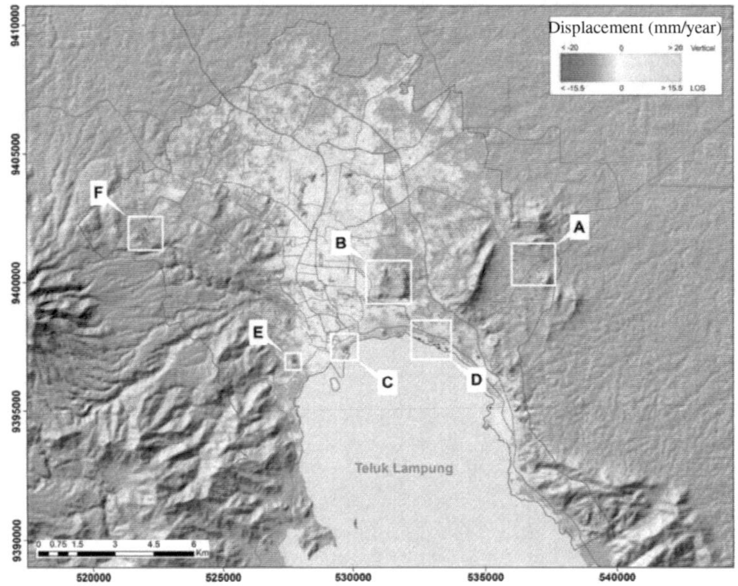

Figure 3. The location on the indicated area has subsidence more than 5 mm/year in Bandar Lampung City.

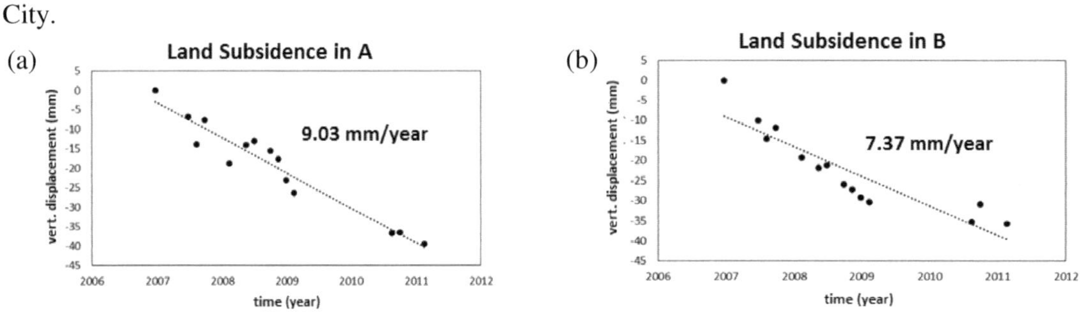

Figure 4. The land subsidence in (a) Campang Raya Village (industrial area) and (b) B area (Camang Mount and surrounding area).

In C area, the land subsidence is detected in densely populated areas in Pasar Kangkung, South of Teluk Betung (figure 5(a)), with a characterized spatial pattern of less than 10 km^2 (patchy). This area indicates a decrease in ground level with an average speed of 14.7 mm/yr. The D area is one of the beach reclamation projects on Sukaraja, South of Teluk Betung. In the west, the reclaimed area is currently used as a coal deposit site, but the East is still an untapped land. The land of reclaimed land is still largely unconsolidated and well-compacted, so that the physical properties and mechanics of the embankment material may potentially subsided, and based on the results of SBAS measurements, the land subsidence in this area averaged 30.5 mm/yr, as shown in figure 5(b).

Indication of land subsidence in the southwest of Bandar Lampung City shows an interesting pattern of land subsidence in a residential area in Bakung Village, West Teluk Betung District. In this area, it was found anomalous land subsidence that is very contrast with the surrounding area that tends to be stable. This area has decreased ground level with an average speed of 15.23 mm/year (figure 6(a)). Based on observations in the field, this area has morphology of syncline (valley) which is located between three hills and the northern part is a swamp, while in the south stands a residential area. This then leads to the occurrence of underground layer compaction in the form of organic materials that can experience decomposition and also compression due to building loads on it.

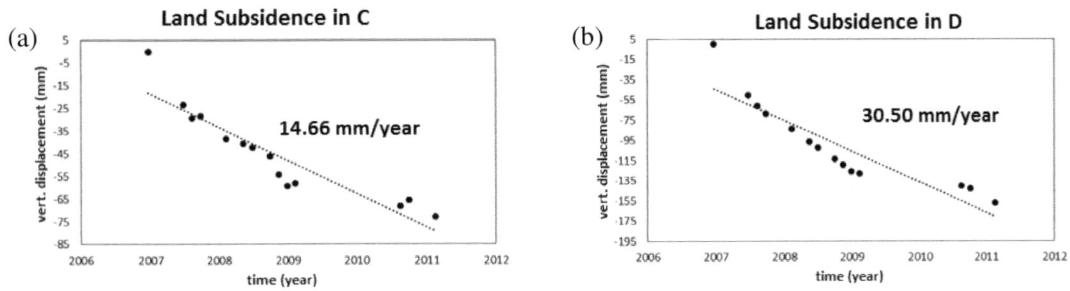

Figure 5. The land subsidence in (a) C area (Pasar Kangkung, South of Teluk Betung) and (b) D area (Sukaraja reclamation area).

Decreased ground level is indicated in Kelurahan Beringin Raya, Kemiling Sub-district (F area), which is dominated by clay-rich layer. The average decrease in this area based on the SBAS measurement is 16.04 mm/yr which is shown in figure 6(b). Due to the area dominated by the clay-rich layer, so the mechanics of the soil is what is suspected as the cause of the decrease of ground level in this area. Land that is cohesive (clay-rich soil) is very likely to decrease the face of the soil, because this layer is very sensitive to changes in water content (soil moisture). This makes it expanding when moist/wet, and then shrinking when dries (volumetric change).

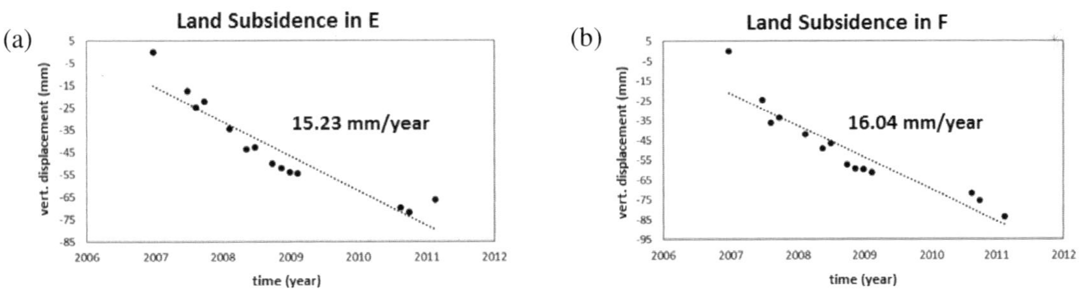

Figure 6. The land subsidence in (a) E area (Bakung area) and (b) F area (Beringin Raya, Kemiling).

Table 2. Compilation of the observations made in the six locations experiencing subsidence.

Location	Average vert. rate (mm/year)	Spatial pattern	Surface geology	Correl. with surface geology	Land use	Correl. with land use	Interpreted cause of rapid subsidence
Campang Raya (A)	9	local	Surficial deposits	×	Industrial and residential	√	Industrial water extraction and loading structure
Camang Mount (B)	7,4	patchy	Surficial deposits (tuff – tuffit)	×	Residential and mining	√	Water extraction and mining
Pasar Gudang Lelang (C)	14,6	patchy	Surficial deposits (sandstone)	√	Residential and trade and services	√	Water extraction by residence and industry
Reclamation Area, Sukaraja (D)	30,5	patchy	Surficial deposits (alluvial)	√	Industrial and stockpiling of coal	√	consolidation of embankment material
Bakung (E)	15,5	patchy	Surficial deposits and peat	√	Residential and swamp	√	decomposition of organic matter and loading structure
Beringin Raya (F)	16	patchy	Surficial deposits (clay)	√	Residential and vacant land	×	volume shrinkage in the clay layer

The results of the analytics for the six regions of indicated land subsidence are summarized in this section which are then represented in table 2. The observed area is correlated to the characteristics and factors of lands subsidence in table 2, so as to obtain a provisional estimate of the factors causing land subsidence in Bandar Lampung City based on spatial patterns, land use, and geological conditions.

Land subsidence is likely to occur in areas of alteration or land conversion that occurred less than 20 years earlier. This is why the Central Tanjung Karang and Kedaton areas with high population densities are not indicated to decrease the land surface. Based on the previous statement and if we look at the pattern in figure 5 then it is suspected that the Central of Tanjung Karang area and Kedaton may have already experienced consolidation as it were developed earlier than other regions, and in the past 10 years has tended to be stagnant in terms of development and conversion of land, this is certainly similar to the statement Sestini (1996) in [1], which states that land subsidence due to reclamation, construction and agricultural activities is limited to the first twenty years or directly after land conversion.

5. Conclusion

1. The type of land use greatly affects the reliability of the data, the more dynamic the area, the measured deformation speed information by InSAR-SBAS method becomes less precise for the measurement of deformation symptoms in millimetres. SBAS technique is reliable enough to know the symptoms of deformation in Bandar Lampung City quickly and accurately.
2. Indication of land subsidence is not evenly distributed in all places, but on average Bandar Lampung city decreased the average of 0.06 mm/year with trend which tends to linear. In some areas in Campang Raya, Camang, Kangkung, Sukaraja, Bakung, and Beringin Raya have subsidence more than 5 mm/year, where the highest one is in Sukaraja (> 30 mm/year).
3. Subsidence with a speed of more than 5 mm/year spatially has two patterns, namely patchy (<10 km^2) on the coastal area of Lampung Bay and Kemiling District, and local (10 - 100 km^2) in the East Bandar Lampung, that is in East Tanjung Karang area.
4. Land subsidence in Bandar Lampung City occurred due to tectonic activity and also due to human activities (industry, loading, mining, ground water extraction, and land conversion). The above factors affect the speed of land subsidence individually or simultaneously.
5. Subsidence is not highly correlated to population density and rock lithology, but tends to occur in areas where land conversion is less than 20 years ago.

Acknowledgments
Acknowledgements were delivered to DRPM Ristekdikti who has funded this research through the grant PPT 2017 and those who have contributed to the success of this research.

References

[1] Teatini P, Tosi L and Strozzi T 2011 *Journal of Geophysical Research* **116** B08407
[2] Chaussard E, Amelung F, Abidin H and Sang-Hong 2013 *Elsevier Remote Sensing of Environment* **128** 150
[3] Casu F 2009 *The Small Baseline Subset Technique: Performance Assessment and New Development for Surface Deformation Analysis of Very Extended Areas* (Cagliari: University of Cagliari).

Performance of supercapacitor device model based on double layer reduced graphene oxides films as electrodes in KCl electrolyte

N Syakir[1], H Addini[1], V Marcelina[1], A Bahtiar[1], Y W Hartati[2], S Wyantuti[2] and Fitrilawati[1*]

[1]Department of Physics, Faculty of Mathematics and Natural Sciences, Universitas Padjadjaran, Jl. Raya Bandung-Sumedang km 21 Jatinangor, West Java, Indonesia
[2]Department of Chemistry, Faculty of Mathematics and Natural Sciences, Universitas Padjadjaran, Jl. Raya Bandung-Sumedang km 21 Jatinangor, West Java, Indonesia

*fitrilawati@phys.unpad.ac.id

Abstract. We report the performance of supercapacitor device model based on double layer reduced graphene oxide (rGO) films as electrode in 1 M KCl. Graphene oxide (GO) films were deposited on ITO-glass use an electrochemical deposition method using 1 mg/ml GO dispersed in water. The film's thicknesses were controlled by number deposition cycles. rGO films were obtained by thermally reduced the GO films at 200 °C for 1 hour under argon flowing. A pair of 1 cm^2 rGO/ITO electrodes was employed as double layer electrodes of supercapacitor device model using a filter paper which was soaked in 1 M KCl as a separator. Performance of the supercapacitor model was investigated using cyclic voltammetry (CV) in a voltage range of 0.0 volts to +0.9 volts with varied scan rate range of 25mV/s to 125mV/s. The highest specific capacitance of 20 F/kg and specific power of 0.825 W/kg was obtained using device with rGO films that deposited for three voltage cycles.

1. Introduction

Electrical energy is a primary need of modern life. It drives significantly research activities on energy sources and energy storages. The best energy storage must be had high specific power and high specific energy. Electrochemical capacitor that also known as supercapacitor has high ratio between its capacitance and its mass. Supercapacitor has widely studied as a storage device since it has higher specific capacitance compare to the conventional capacitor. It can overcome the limit of another energy storage such battery and traditional capacitor [1]. A rechargeable battery has high specific power and low specific energy. In contrast, a capacitor has low specific power and high specific energy [2-4]. Capacitor stored energy in between 2 electrical layer as electrodes seperated by an electrolyte [2].

To overcome the disadvantage of capacitor, the material which used as electrodes has been widely investigate. Two dimensions material like graphene is a proper material as electrode, beside it has porous structure, high surface area, it is also has high electrical conductivity. Moreover, graphene has porosity and 1ettability that is suitable for supercapacitor storage system [3-5]. Availability of perfect graphene is still limited and could not be made easily, but it can be approached using a material which has close characteristic as graphene. The characteristics of reduced graphene oxide (rGO) can approach characteristics of graphene and it is easily made compare to graphene. rGO can be prepared from graphene oxide (GO) solution as a precursor and then thermally reduced to remove its oxygen functional group of GO [5].

The configuration of supercapacitor device in this experiment consist of two (2) symmetric rGO/ITO as electrode that separated by a filter paper and 1 M KCl as an electrolyte. In this experiment for investigation of device best performance it is determined the capacitance and its energy storage values based on the cyclic voltammetry (CV) data.

2. Method

2.1. Fabrication of rGO films on ITO-glass

1 mg/ml GO solution in water (Graphenea SA ES A75022608) was sonicated for an hour in room temperature. Then it was electrochemically deposited using MetrOhm Autolab on ITO-glass substrates (1 cm × 1.25 cm) using three-electrodes system. Those electrodes were Ag/AgCl as a reference, Pt as a counter electrode and ITO as a working electrode. All three electrodes are arranged in cells containing 12 ml GO solution. Deposition has been done in various voltage cycles (1 to 5 cycles) with voltage applied in the range of -1.6 V to 0 V and scan rate of 75 mV/s The samples then were dried on hotplate at temperature of 25 ^{o}C. In order to obtain rGO films, the GO thin films were thermally reduced using furnace at 200 ^{0}C for 2 hours under argon flowing

2.2. Fabrication of rGO supercapacitor

The rGO supercapacitor device model was built using two symmetric rGO films on ITO as active electrodes and a piece of filter paper as a separator. The filter paper that soaked in 1M KCl solution then was inserted between two symmetrical rGO films. Finally, the device was covered by parafilm. The schematic model of supercapacitor device is shows in figure 1. In this experiment, we prepared five (5) devices samples that consist of varied rGO film thickness. The sample was labelled as S1, S2, S3, S4 and S5 for device using 1, 2, 3, 4 and 5 voltage cycles deposition of GO films, respectively.

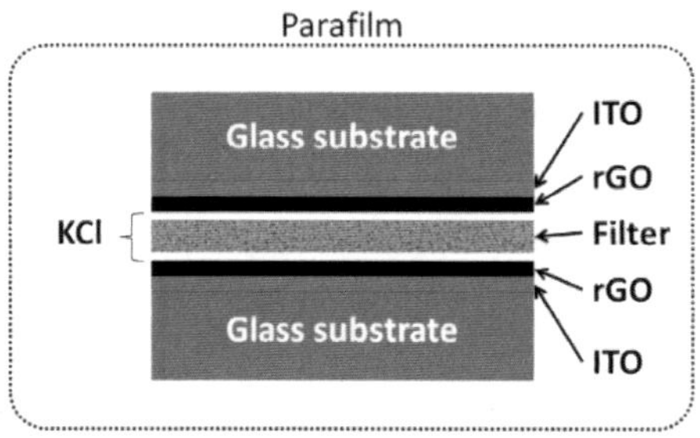

Figure 1. Schematic configuration of supercapacitor device model.

2.3. Measurements

The electrochemical capacitance measurements of the rGO supercapacitors device model were carried out using CV measurement (MetrOhm Autolab). It was done by applying voltage to the device and at the same time measured its current response of the device. The CV characteristic of the device was measured at voltage range of 0 V – 0,9 V. The device's performance was observed from their CV curve shapes for various scanrate from 25 mV/s to 125 mV/s to check its stability. The stored power in the device is calculated using equation (1) and the capacitance value is calculated using equation (2).

$$P = \int_{0}^{0,9} i \, dV + \int_{0,9}^{0} i \, dV \tag{1}$$

$$C = \frac{1}{2m \, .\Delta V \, .s}\left(\int_{-0,9}^{0} idV + \int_{0}^{0,9} idV\right) \tag{2}$$

P in the equation (1) is a power storage of device for a certain scan rate, I is a current response of device for a given applied voltages. C in the equation (2) is a specific capacitance of device, m is mass of electroactive materials, ΔV is the potential window (which is 0.9 volts), s is scanrate using in the CV measurement.

3. Result and Discussion

Samples in figure 2 that labelled as E1, E2, E3, E4 and E5 are five pair of rGO films on ITO-glass substrates. The samples of E1, E2, E3, E4 and E5 were obtained from GO films that thermally reduced at 200 ^{0}C for 2 hours. These GO films were electrochemically deposited using five different number of voltage cycles i.e. 1, 2, 3, 4 and 5 cycles, respectively. Increasing number of voltage cycles during the deposition causes increasing of GO films thickness. As can be seen on figure 2, the samples are homogeneous with a darkness level gradually increased for samples that deposited at higher number of cycles. The film darkness is related with the films thickness, thicker film appeared more dark compare to thinner GO film. Its darkness color indicates that rGO films was dominated by carbon atom and reducing oxygen atom. In this experiment we use the rGO films that prepared from GO film as "graphene-like" material.

Figure 2. The pairs of rGO films on ITO-glass substrates that electrochemically prepared using various voltage cycle during deposition, one cycle (E1), two cycles (E2), three cycles (E3), four cycles (E4), and five cycles (E5).

Then the rGO samples were employed as a pair of electrodes for supercapacitor devices model that its device structure shown in figure 1. We built five supercapacitor devices model named as S1, S2, S3, S4 and S5 by employed pair of rGO film of E1, E2, E3, E4 and E5, respectively as electrodes.

The CV characteristic of device samples S1 to S5 measured using scan rate of 25 mV/s and 50 mV/s are shown in figure 3. From figure 3 it is obvious that device of S3 has the biggest CV. It indicates that the S3 device has better power capacity than the other sample devices.

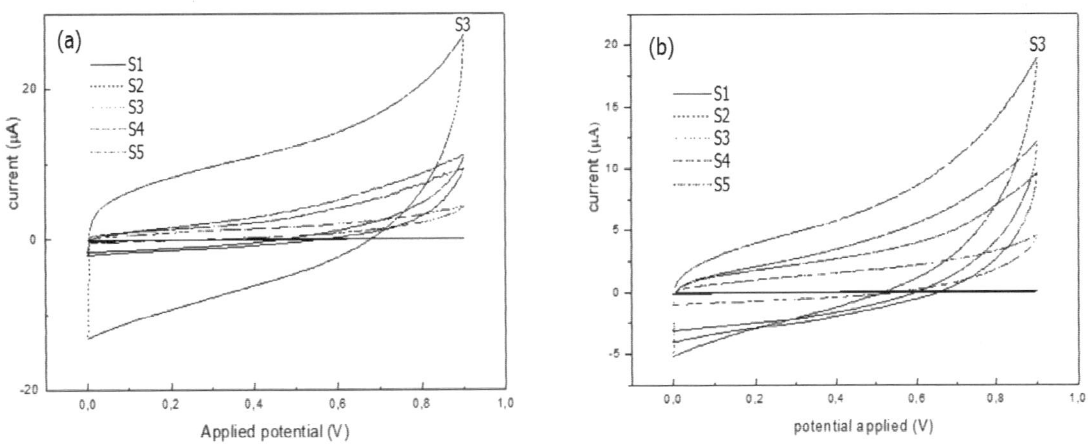

Figure 3. Cyclic voltamogram of S1 to S5 supercapacitor devices at scanrate of (a) 25 mV/s and (b) 50 mV/s.

Furthermore, figure 4 shows characteristic of CV curves of S3 device that measured at varied scan rate i.e. 25 mV/s, 50 mV/s, 75 mV/s, 100 mV/s and 125 mV/s. The curve displays the effect of scan rate on the device capacity. All of CV curves has a similar shape but diffirent size. It implies that the device model was stable enough in the voltage range of applied scan rate. The CV area increases as the scan rate was increased and it indicates that power capacity of S3 device also increased.

The values of power capacity and specific capacitance of the S3device for various scan rates were calculated using equation (1) and (2) and those results were plotted in two different graphs as shown in figure 5. The highest power density of the S3 device is 0,825 W/kg that was achieved at scan rate of 125 mV/s. At this scan rate, however, its specific capacitance is reached the lowest value i.e. 6 F/kg. The highest specific capacitance is 20 F/kg that achieved at scan rate of 25 mV/s. At that scan rate, however, its power density is the lowest i.e. 0.375 W/kg.

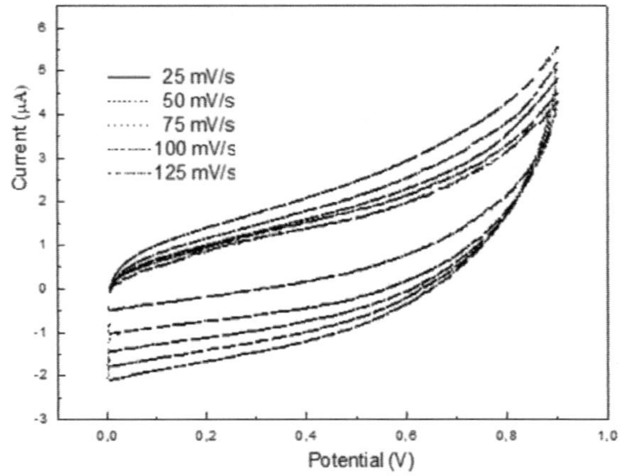

Figure 4. CV curves of S3 supercapacitor device model at various scan rate.

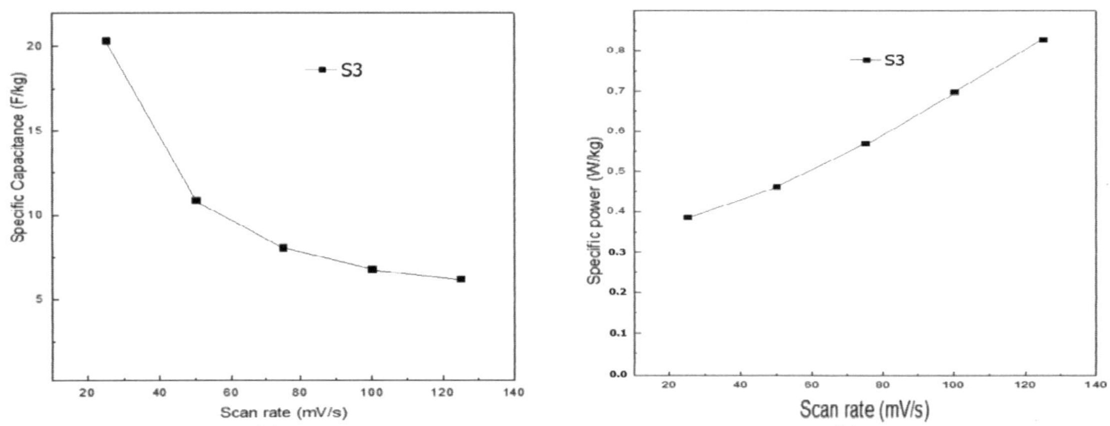

(a) (b)

Figure 5. Characteristic curve of S3supercapacitor devices at various scan rate, (a) specific capacitance and (b) power density.

As shown in the figure 5, the value of specific capacitance of S3 device decrease significantly at low scan rate, but in contrast, the value of its power density increased. For a voltage window of 0.9 V and scan rate of 125 mV/s, the time required to complete one loop of cyclic voltammogram is 14.4 s. Therefore, a power density of 0.825 W/kg at scan rate of 125 mV/s is equivalent to energy density of 0.003 Wh/kg.

Based on Ragone plot [1], a capacitor device is specified as a supercapacitor if its specific energy is above 0.05 Wh/kg and its power density is above 10 W/kg. From this point of view, the S3 device model could not classified as supercapacitor device. However, the device still could specify as supercapacitor from point of view specific capacitance since it has quite large specific capacitance in the range of 6 F/kg to 20 F/kg.

4. Conclusion

We have built supercapacitor devices model using rGO films as electrodes. The S3 supercapacitor device that consist of rGO electrodes, obtained from GO film that electrochemically prepared using three voltage cycles, has higher performance. The S3 supercapacitor has power density of 0.825 W/kg and specific capacitance of 20 F/kg. Although its specific energy still lower than the standard criteria of supercapacitor but its specific capacitance is large enough.

Acknowledgements

This work was funded by Penelitian Unggulan Perguruan Tinggi (PUPT) contract no. 718/UN6.3.1/PL/2017, dated April 17th 2017.

References

[1] R Kötz and M Carlen 2000 *Electrochim Acta* **45** 2483
[2] S R C Vivekchand, C S Rout, K S Subrahmanyam, A Govindaraj and C N R Rao 2008 *J. Chem. Sci.* **120** 9
[3] Y Chen, X Zhang, D Zhang, P Yu and Y Ma 2011 *Carbon N.* **49** 573
[4] L L Zhang, R Zhou and X S Zhao 2009 *J. Mater. Chem.* **38** 2520
[5] B Zhao *et al.* 2012 *J. Power Sources* **198** 423

Variability of surface chlorophyll-a distributions in the northwestern coast of Sumatra revealed by MODIS

Q W Sari[1], P A Utari[1], D Setiabudidaya[2], I Yustian[3], E Siswanto[4] and I Iskandar[2*]

[1]Graduate School of Environmental Science, University of Sriwijaya Palembang, South Sumatra, Indonesia- 30139.
[2]Department of Physics, Faculty of Mathematics and Natural Sciences, University of Sriwijaya, Inderalaya, South Sumatra, Indonesia- 30662.
[3]Department of Biology, Faculty of Mathematics and Natural Sciences, University of Sriwijaya, Inderalaya, South Sumatra, Indonesia- 30662.
[4]Department of Environmental Geochemical Cycle Research Research and Development Center for Global Change (concurrent) Japan Agency for Marine-Earth Science and Technology 3173-25, Showa-machi, Kanazawa-ku, Yokohama, Kanagawa, 236-0001, JAPAN.

[*]iskhaq@mipa.unsri.ac.id

Abstract. The variability of surface chlorophyll-a (chl-a) distributions in the northwestern coast of Sumatra are important components associated with the phytoplankton biomass that supports and enrich the fishery resources of the region. This study analyzes satellite data-ocean color derived from the Moderate Resolution Imaging Spectroradiometer (MODIS) Aqua for a period of January 2003 to December 2015. The results show that the surface chl-a distribution is strongly associated with the large-scale forcing (e.g. monsoonal winds, EL Niño–Southern Oscillation and the Indian Ocean Dipole). On the seasonal time scale, it is shown that the satellite observed surface chl-a concentration during the boreal fall and winter (September-January) is higher than that observed during the boreal summer (June-August). The highest concentration was observed in September which is about 0.2 mg m^{-3}. In contrast to the western coast, the surface chl-a in the eastern coast of Sumatra has excessive concentration around 0,5-10 mg m^{-3} all year around. In addition, the study also revealed that the anomaly of surface chl-a concentration was observed along the western coast of Sumatra is much higher during the positive IOD (pIOD) and EL Niño events.

1. Introduction

Dynamics of the ocean circulation in the northern Sumatra has the same importance as in the southern Sumatra. Many studies have been conducted to analyzed the dynamics of ocean circulation in the southern Sumatra. However, the dynamics in the northern Sumatra have not been studied in details, especially the dynamical distributions of surface chlorophyll-a (chl-a). The northwestern Indian Ocean is less productive compared to the southeastern Indian Ocean [1, 2].

The northern Sumatra region is part of the northern Indian Ocean that is influence by a semi-annually reversing monsoonal wind system. During the summer monsoons in June-August, the summer monsoon current (SMC) carries saltier and high salinity water mass from Arabian Sea into the northeastern of the Indian Ocean waters (Bay of Bengal). On the other hand, the North Equatorial Current (NEC) lead out warm and low salinity waters from the Bay of Bengal during winter monsoon (December–February) [3].

On the inter-annual time scale, the Indian Ocean Dipole (IOD) and the El Niño Southern Oscillation (ENSO) affect significantly inter-annual changing within the Indian Ocean. Previous study using a coupled biophysical ocean general circulation model showed that two climate modes could form unique pattern of surface chl-a distribution in the tropical Indian Ocean [4].

In addition to the temporal variation, the northeastern of Indian Ocean (especially Bay of Bengal) also experiences mesoscale dynamics namely the eddy current. The eddy variability has important role in enhancing the biological productivity up 2 to 8 times compared to the non-eddy region. The upper ocean region also has high stratified during summer monsoon [1]. The previous study also revealed that the northern Indian Ocean has shown strong intra-seasonal sea surface height (SSH) variability [5]. Meanwhile, meridional currents variability is affected by the Kelvin wave reflection due to the inclined coastline of the eastern boundary [6]. As a result, those conditions are hypothesized to influence the surface chl-a distributions in the northern of the Sumatra Island. Therefore, the main objective of this study is to identify the variability of surface chl-a distribution in that region and to examine the processes that lead the observed chl-a pattern using the important physical and biological-oceanographic properties of the marine ecosystem (the sea surface temperature (SST), the surface winds, the Ekman pumping value, and the surface chl-a concentration). The rest of the paper is organized as follow. Section 2 describes the data and methods. Section 3 explains the results and discussion. Then, the summary of the findings is present in section 4.

2. Data and Methods

The study area is located in the northern of Sumatra bounded by 12°N- 3°S and 90°E -102°E (figure 1). The surface chl-a concentration for the northern of Sumatra region was obtained from the Moderate Resolution Imaging Spectroradiometer (MODIS) aboard the Aqua satellite. The data were provided by the NASA Goddard Distributed Active Archive Center for a period of January 2003 to December 2015 at 9 km horizontal resolution.

The monthly surface wind data were obtained from the ERA-Interim of the European Center for Medium Range Weather Forecasts (ECMWF). The monthly gridded data at 0.25° × 0.25° resolution from January 2003 to December 2015 were used in this study.

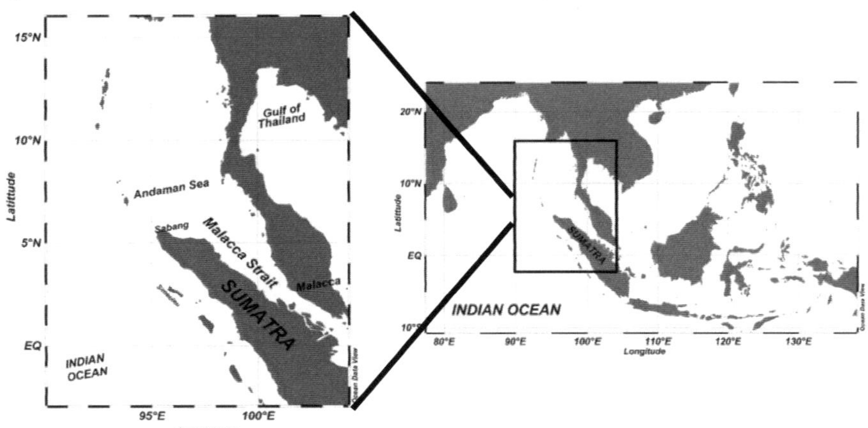

Figure 1. Maps of the northern Sumatra, the study area is bounded by 12°N-3°S and 90°E-102°E.

The monthly SST image was obtained from National Oceanic and Atmospheric Administration (NOAA) Optimum Interpolation (OI) [7].

In order to evaluate the mechanism of surface chl-a distribution, the Ekman pumping as an indicator of upwelling/downwelling processes was computed using the formula $(1/\rho f)\nabla \times \vec{\tau}$, where ρ

is the density of seawater (kg m^{-3}), f is the Coriolis parameter (rad s^{-1}) and $\vec{\tau}$ is wind stress (N m^{-2}). Note that, the $\vec{\tau}$ (wind stress) was calculated using formula $\vec{\tau} = C_D\ \rho_{air}\ U^2$, following Weisberg [8]. Where C_D, ρ_{air}, and U are constant drag coefficient (C_D=1.43 × 10^{-3}), air density (ρ_{air} =1.225 kg m^{-3}) and the surface wind monthly data (10 m above sea surface), respectively. The monthly climatology of all parameters (surface chl-a, surface wind, SST, and the Ekman pumping) were obtained from monthly time series from January 2003 to December 2015. Then, the anomaly of all parameters was calculated by deviation from the monthly data and monthly climatology.

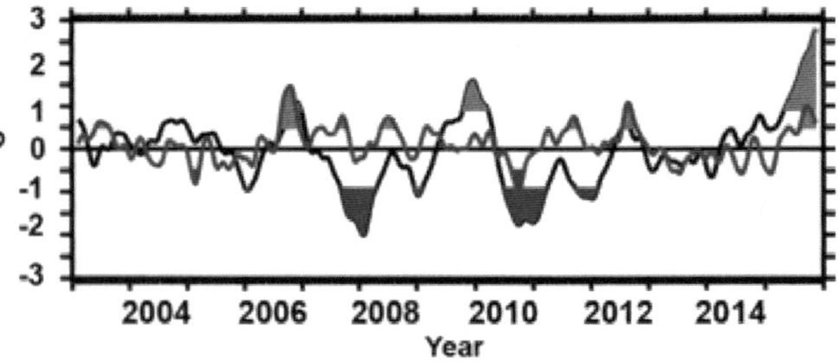

Figure 2. Time series of the Niño3.4 index (*black*) that and the DMI (*brown*) from January 2003 to December 2015. Significant positive value (*red shading* >=+0.5 °C) indicates El Niño or positive IOD events, while significant negative value (*blue shading* <=-0.5 °C) represents La Niña or negative IOD events.

The Dipole Mode Index (DMI) was calculated as a gradient of sea surface temperature anomalies (SSTA) averaged over the western (10°S–10°N, 50°E–70°E) and the eastern (10°S–Equator, 90°E–110°E) regions [9]. In addition, the monthly Niño3.4 index used to represent the occurrence of the ENSO events was defined as the average equatorial SSTA over a region of 5°S–5°N and 120°W–170°W. Note that the ENSO event was characterized by a five consecutive 5-month running mean of SSTA exceeding one standard deviation [10]. Based on those criteria, the IOD and the ENSO events are shows in figure 2.

3. Results and Discussion

3.1. Seasonal Variation

The monthly variability of the chl-a in the northern part of Sumatra is presented in figure 3. Generally, the northeastern coast of Sumatra (including the Malacca Strait) has much higher chl-a concentration than the northwestern coast of Sumatra (along the coast of Aceh, Medan and Padang). It is shown that the observed chl-a concentration in the northeastern coast of Sumatra is up to 0.3- 5 mg m^{-3}, while that in the northwestern coast of Sumatra is only 0.1-0.2 mg m^{-3}. The distribution of surface chl-a in the northeastern coast of Sumatra is strongly affected by the nutrient supply from the river run-off product of the land and the mixing process [11].

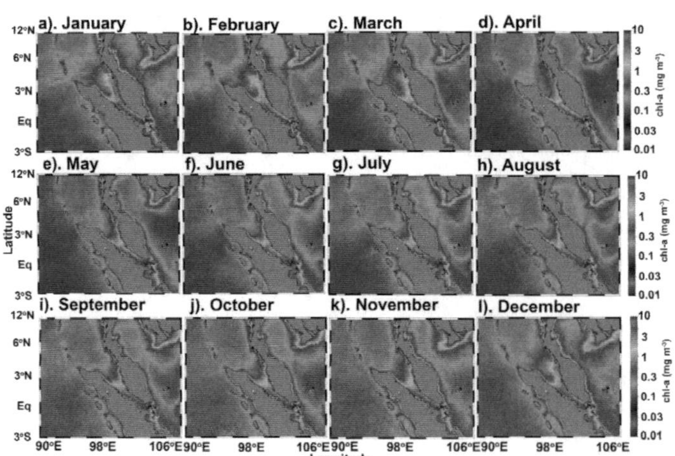

Figure 3. Monthly climatology of chl-a distribution are derived by MODIS satellite-observed.

IOP Conf. Series: Journal of Physics: Conf. Series **1080** (2018) 012045 doi:10.1088/1742-6596/1080/1/012045

The concentration of the surface chl-a in the northwestern coast of Sumatra during boreal winter (DJF) and spring (MAM) season is much lower than it is during boreal summer (JJA) and boreal fall (SON). Noted that the increasing of the chl-a concentration during boreal summer (JJA) is related to intensification of upwelling region because of appearance of the eddies current [1].

Figure 4 shows the monthly climatology of surface wind superimposed on the monthly climatology of SST. It clearly shows that the wind direction reverses seasonally from the northeastward (DJF) to the southwestward (JJA). Weak winds (<10 m s^{-1}) are observed during transition periods (MAM and SON). Meanwhile, the SST during the transition periods (MAM) is warmer (29-30°C) along the northern Sumatra than that observed during the boreal summer in JJA (28-29°C).

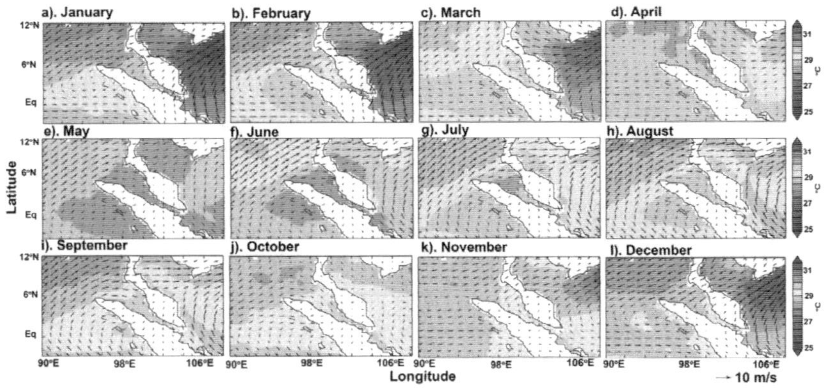

Figure 4. Monthly climatology of surface wind 10 m (vector in m s-1) superimposed on SST (shading °C) in the northern Sumatra.

The evaluation of upwelling region is determined by the wind curl mechanisms. Figure 5 presents the monthly climatology of the Ekman pumping computed from the surface wind data. The positive Ekman pumping indicates upwelling, while negative Ekman pumping indicates downwelling mechanisms. The positive Ekman pumping along the northern Sumatra reached its peak in boreal early winter (DJF) followed by a maximum surface chl-a concentration. On the other hand, the negative Ekman pumping observed during boreal fall (SON) is accompanied by the lower surface chl-a concentration.

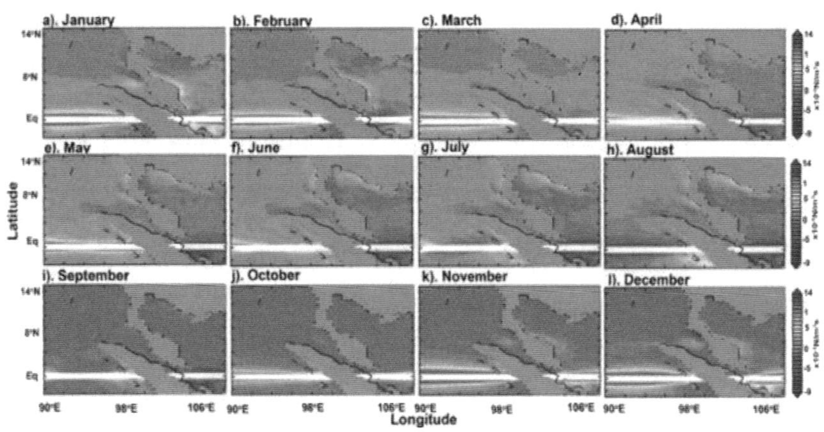

Figure 5. Monthly climatology of the Ekman pumping. Positive (negative) values indicate an upwelling (downwelling) regions.

3.2. *Effect of the IOD and ENSO on the Surface Chl-a Distributions*

The chl-a distribution in the northern Sumatra is more extensive during 2006 and 2015 than that during the normal year (figure 6). This conditions were associated with the El Niño and positive IOD events that were concurrently occurred [12–14]. During the peak of positive IOD event, almost all of the region in the northern coast of Sumatra experienced positive chl-a anomalies. The significant positive chl-a anomalies first appeared over the northern of Sumatra in July 2006. This condition continued and peaked in September-November 2006. This result was confirmed by previous study about the chl-a distribution in Southern Java [13,15]. The positive chl-a anomalies in the region is increase up to ~1 mg m^{-3}. The surface chl-a anomalies observed during 2006 were more extensive than those observed in 2015. Note that the IOD in 2015 was co-occurred with the El Niño event. However, the 2015 IOD is short-lived event compared to the 2006 IOD event.

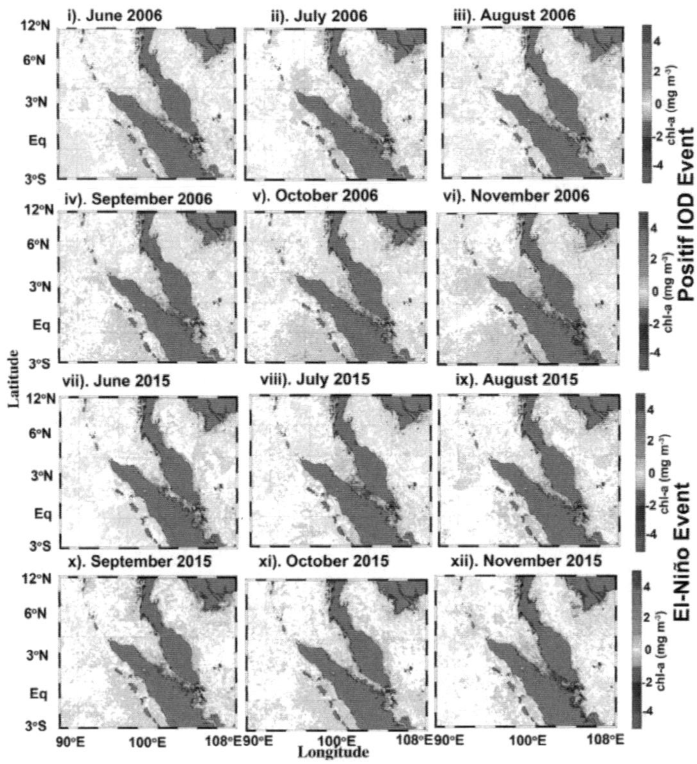

Figure 6. Comparison of the monthly anomalies pattern of chl-a distribution while strong positive IOD event in 2006 (i-vi) and El Niño event (vi-xii) in 2015 over the northern Sumatra (12°S–3°N and 90°S–108°N).

It is clearly shown that a strong pIOD in 2006 and a strong El Niño event in 2015 have caused two different oceanic and atmospheric conditions in the northern Sumatra. During the evolution phase of the El Niño and the pIOD events, the most prominent features are the different in the pattern of SSTA, the surface wind anomaly and the Ekman pumping anomaly. The differences of the SSTA patterns and the surface wind anomaly between two events (e.g. El Niño and pIOD) affect the surface chl-a distribution in the study area. The strong surface winds leaded the cold SSTA in 2006 was more prolonged than those observed in 2015. The results found that the negative SSTA over the northern Sumatra was associated with related changes in the stronger southwesterly surface winds that were shown by figure 7. The positive anomalies of the chl-a concentration pattern agree with the negative SST anomaly. In 2006, the evolution of the SSTA started in the early boreal summer (July) over the northern of Sumatra (figure 7 ii) and reached a minimum of about 0.6°C in boreal fall (October). This condition has been confirmed by Iskandar, *et al* (2009) whose discussed the chl-a blooms in the southern Indian Ocean [13]. Meanwhile, during the El Niño in 2015, the observation did not discover a clear cold SSTA in the northern Sumatra.

It is also interesting to evaluate the impact of the Ekman pumping anomaly on the chl-a distribution in 2006 and in 2015 (figure 8). The results revealed that the positive Ekman pumping coincided with higher surface chl-a concentration. This upward Ekman pumping was associated with upwelling causing negative SST anomaly and enhancing surface chl-a concentration. The northeasterly wind anomalies generated downward Ekman pumping (negative anomaly) leading to warm SST anomaly. This downward Ekman pumping caused downwelling that depressed surface chl-a concentration. During both 2006 pIOD and 2015 El Niño events, strong southeasterly wind anomalies were observed leading to upward Ekman pumping (positive anomaly). However, the southeasterly wind anomalies in 2006 were much stronger than those observed in 2015 (figure 8).

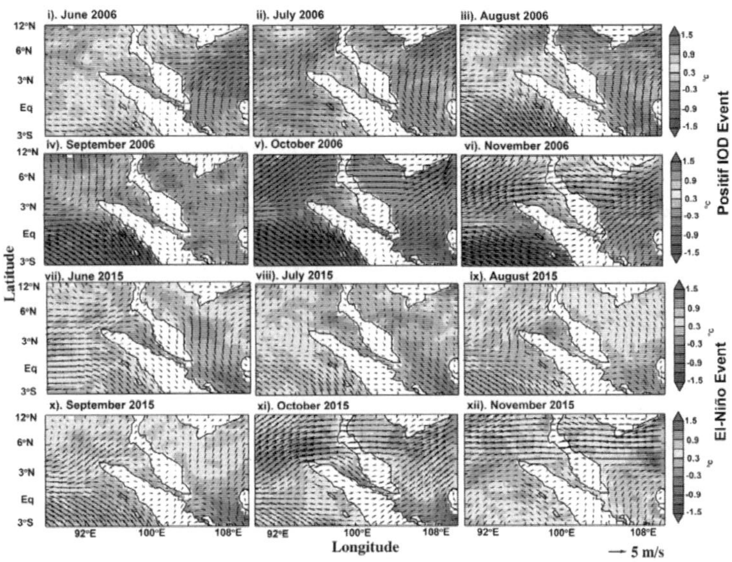

Figure 7. Comparison of the monthly anomalies pattern of SST overlay 10 m surface wind while strong positive IOD event in 2006 (i-vi) and El Niño event in 2015 (vii-xii) over the northern Sumatra (12°S–3°N and 90°S–108°N).

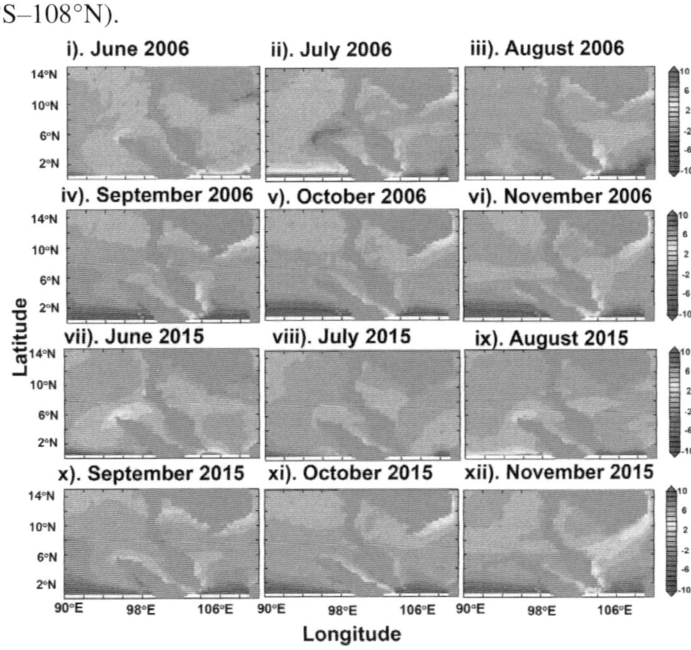

Figure 8. Comparison of the monthly anomalies pattern of Ekman pumping while strong positive IOD event in 2006 (i-vi) and El Niño event in 2015 (vii-xii) over the northern Sumatra (12°S–3°N and 90°S–108°N).

4. Conclusion

In this study, we found that the variability of surface chl-a distribution in the northwestern Sumatra has seasonal variation. The observed chl-a concentrations show higher concentration during the boreal spring and summer compared to those observed during boreal fall and winter. In 2006, the positive IOD event caused higher chl-a bloom almost in the entire northwestern coast of Sumatra. The highest chl-a bloom event occurred in boreal fall (SON) when the northwesterly wind reached its peak. In 2015, the El Niño has lower impact on the chl-a distribution in the northwestern coast of Sumatra in early boreal fall (September-October). The variability of wind direction over the north Sumatra contributed to the chl-a distribution in this region during 2006 positive IOD and 2015 El Niño events.

Acknowledgement

This research was supported by the PMDSU Scholarship funded the Ministry of Research, Technology and Higher Education, Indonesia. The last author is supported by the Ministry of Research, Technology and Higher Education, Indonesia through *Penelitian Berbasis Kompetensi 2017* (No. 102/SP2H/LT/DRPM/IV/2017) and by the University of Sriwijaya through *Hibah Unggulan Profesi 2017* (No. 987/UN9.3.1/PP/2017).

References

[1] Kumar S P, Nuncio M, Narvekar J, Kumar A, Sardesai S, De Souza S N, Gauns M, Ramaiah N and Madhupratap M 2004 *Geophys. Res. Lett.* **31** 2
[2] Susanto R D, Gordon A L and Zheng Q 2001 *Geophys. Res. Lett.* **28** 1599
[3] Kumar S P, Narvekar J, Kumar A, Shaji C, Anand P, Sabu P, Rijomon G, Josia J, Jayaraj K A, Radhika A and Nair K K C 2004 *Geophys. Res. Lett.* **31** 4
[4] Currie J C, Lengaigne M, Vialard J, Kaplan D M, Aumont O, Naqvi S W A and Maury O 2013 *Biogeosciences* **10** 6677
[5] Aparna S G, McCreary J P, Shankar D and Vinayachandran P N 2012 *Geophys. Res.* **117** 1
[6] Chen G, Han W, Li Y, Mcphaden M J, Chen J, Wang W and Wang D 2017 *J. Phys. Oceanogr.* **47** 979
[7] Reynolds R W, Rayner N A, Smith T M, Stokes D C and Wang W 2002 *J. Clim.* **15** 1609
[8] Weisberg R H and Wang C 1997 *J. Clim.* **10** 1998
[9] Saji N, Goswami B N, Vinayachandran P N and Yamagata T 1999 *Nature* **401** 360
[10] Trenberth K E 1997 *Bull. Am. Meteorol. Soc.* **78** 2771
[11] Tan C K, Ishizaka J, Matsumura S, Yusoff F M and Mohamed M I H 2006 *Cont. Shelf Res.* **26** 168
[12] Susanto R D, Moore T S and Marra J 2006 *Geochemistry, Geophys. Geosystems* **7** 1
[13] Iskandar I, Rao S A and Tozuka T 2009 *Int. J. Remote Sens.* **30** 663
[14] Annamalai H, Xie S-P and McCreary J P 2005 *J. Clim.* **18** 302
[15] Iskandar I, Sasaki H, Sasai Y, Masumoto Y and Mizuno K 2010 *Ocean Dyn.* **60** 731

IOP Conf. Series: Journal of Physics: Conf. Series **1080** (2018) 012046 doi:10.1088/1742-6596/1080/1/012046

Optimization of parameters and fabrication of Al-air batteries with economically feasible materials

C Panatarani[1,2*], M Vanitha[1], L Nulhakim[1], Z I Hauna[1,2] and I M Joni[1,2]

[1]Nanotechnology and Graphene Research Centre, Jl. Raya Bandung-Sumedang km 21, Jatinangor, West Java, Indonesia 45363
[2]Department of Physics, Faculty of Mathematics and Natural Sciences, Universitas Padjadjaran, Jl. Raya Bandung-Sumedang km 21, Jatinangor, West Java, Indonesia 45363

[*]c.panatarani@phys.unpad.ac.id

Abstract. Design and fabrication of Al-air battery were done in this present work with economically viable raw materials. Optimization of anode material, catalyst concentration, current density and electrolyte concentration were carried out. Al 5083 as an anode material results in longer discharge time than Al 7075 and Al 6061 anode used in this study. Catalyst loading of 5% $TiCl_3$, current density of 5 mA cm^{-2} and 10% of NaCl electrolyte were optimized. A single battery cell is fabricated using Al 5083 as an anode, air cathode which consists of graphite, paint as a binder with 5% $TiCl_3$ as a catalyst and 10% NaCl as the electrolyte. The single cell exhibits voltage of about 0.8 V with 20 mA h^{-1} current capacity. The lifecycle of the fabricated battery is tested for four consecutive cycles which demonstrates almost similar shape in the charging and discharging curves representing better stability of the fabricated battery.

1. Introduction

In the current scenario, one of the problem in the society that needs to be addressed is the energy crisis. With the depletion of fossil fuel resources and deteriorating renewable energy sources, the quest for an alternative energy has become the spotlight in research [1]. Since early 1990's lithium-ion batteries (LIBs) has been the most widely employed portable electronic devices and also used in hybrid electric vehicles owing to their high energy density and low self-discharge [2]. Metal-air batteries have higher capacity and energy density than other devices. In specific, Al-air battery is considered as the novel energy resource compared to the expensive and highly active Li in LIBs. Al-air battery is a type of metal-air battery with an air cathode, Al or Al alloy as the anode and liquid-based electrolyte. The use of air cathode will enhance the energy density. Air cathode assists in carrying oxygen and interact with the catalyst producing energy density that exceeds other commercial rechargeable batteries [1, 3].

The performance of battery depends on the nature of anode, cathode and electrolytes used. Aluminium as an anode is one of promising material owing to its abundance in nature, low cost and ease in processing with high energy capacity. Another abundance material in nature especially in Indonesia is graphite, one form of carbon [4]. Researcher use graphite as cathode material for metal-air battery by integration with catalyst, to enhance the electronic conductivity. The most widely used electrolytes for aluminium battery are NaCl, NaOH and KOH or a combination between them [5]. In this study, NaCl has been selected as an electrolyte based on the abundance of the salt in the sea and hence battery could be used for marine applications. Therefore, the present work employs aluminium as an anode, air cathode (integrated with graphite, paint and $TiCl_3$) and NaCl as an electrolyte for the fabrication of metal-air battery.

The main objective of this present work is to fabricate the Al-air battery with the low-cost materials and to optimize the parameters to obtain an enhanced battery performance. The catalyst used is graphite along with $TiCl_3$, graphite is obtained from the natural resources from Kalimantan region in Indonesia and was used as received. Graphite and graphite-based materials have been used as an electrode material in LIBs, supercapacitor and fuel cells [6]. The presence of graphite is expected to improve the current capacity of the battery owing to its high electrical conductivity. The design, fabrication, and optimization of parameters such as current density, anode material and electrolyte concentration will be discussed and investigated.

2. Experimental Procedure

2.1. Materials

The materials used for the fabrication of Al-air battery are aluminium 5083 (aluminium alloy with Mg and traces of Mn and Cr), air cathode with naturally available graphite as carbon source, black paint (maxwille) as binder, 5% $TiCl_3$ as catalyst, nickel substrate (in the form of a net) with dimension of 300 as current collectors and 10% NaCl as electrolyte concentration, distilled water as solvent for electrolyte and acrylic for battery fabrication. Different aluminium alloys such as Al 6061, Al 7075 which has a varying percent of metals such as Mg, Mn and Cr are tested as anode materials.

2.2. Fabrication of Al-air battery

The methodology involved in the fabrication of Al-air battery involves three stages: Fabrication of battery cell, battery characterization and battery performance analysis. Parameters such as Al alloy anode, current density, catalyst concentration and electrolyte concentration were optimized. With the optimized results the life cycle of the battery was evaluated. The anode is comprised of Al 5083 with a thickness of about 2.5 mm, length 6 cm, and width 5 cm. The steps involved in cathode fabrication are the preparation of cathode active material, manual deposition on Ni substrate and drying. The cathode active material consisting of air cathode (air from the surrounding) and the cathode active material i.e. graphite and paint in the ratio 3:1, it is mixed manually with 5% $TiCl_3$ catalyst. The cathode active material is manually cast on the Ni substrate i.e. Ni mesh 300 and dried at 80°C for 24 h. The electrolyte 10% NaCl solution was added onto the separator (membrane sheet) using a dropper.

Al 5083 anode is placed at both the end of the cell in between that Ni current collector coated with the cathode active material is placed. The separator i.e. sheet of membrane is placed in between the anode and the current collector (Ni mesh). The arrangement of the battery components anode, separator, cathode, net substrate and electrolyte is illustrated in figure 1.

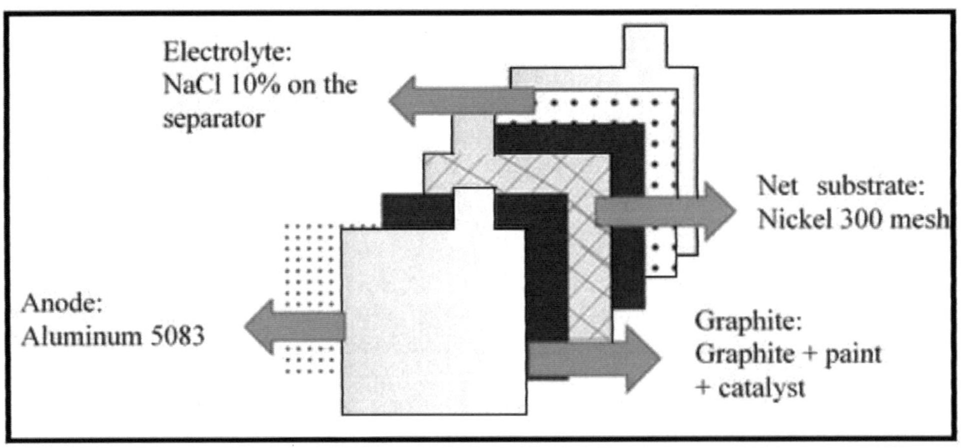

Figure 1. Fabrication of Al-air battery components.

2.3. Characterization

The constant current discharge was characterized by using Mayuno DC electric load with the current load of 5 mA cm^{-2}, 10 mA cm^{-2}, 20 mA cm^{-2} and 30 mA cm^{-2}. The current capacity of the battery was calculated by using the equation $C = I * t$, where C is battery capacity (Ah), I is current (A) and t is time (s) [4]. The life cycle of the battery was obtained by releasing battery cell voltage using a current density of 5 mA cm^{-2}, then after the zero voltage, then releasing current load until the initial battery voltage is achieved. The life cycle of the battery was tested using the optimized parameters for four consecutive cycles to analyse the cyclic stability.

3. Results and Discussion

3.1. Effect of anode material

The electrode materials in Al-air battery has a significant role in determining the performance of battery [5, 7]. To study this three different Al alloys were selected as anode materials such as Al 5083, Al 6061, and Al 7075. The potential selected to optimize the anode material is from 0.0 to 0.8 V and 10% NaCl was used as the electrolyte concentration. The discharge behaviour of different anode materials is shown in figure 2, Al 5083 as anode material exhibit a decrease in voltage initially, then the voltage remains constant up to 1500 s of time. In the case of Al 7075, the voltage dropped abruptly and slowly discharged up to 800 s, the decrease in the voltage may be due to the internal resistance of the battery [3]. The performance of Al 6061 was not uniform and the initial potential was low compared to the other anode materials.

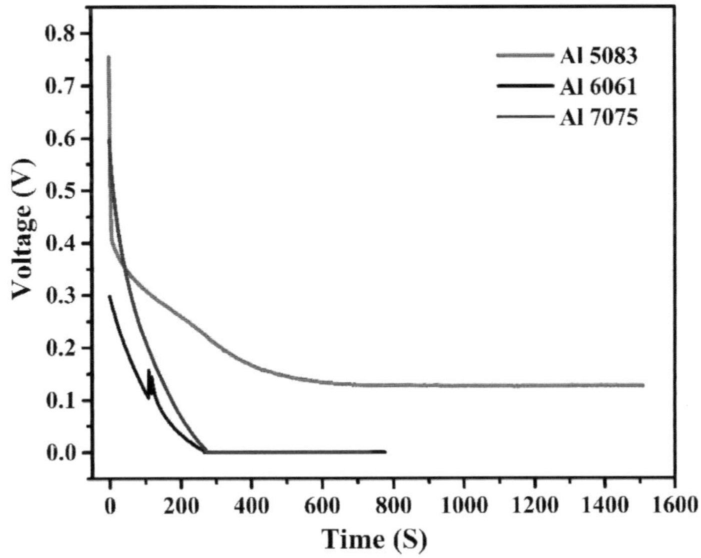

Figure 2. Effect of different anode materials on the discharge performance of the Al-air battery.

3.2. Effect of current density

The discharge behaviour of metal-air battery with Al 5083 as an anode with different current densities such as 5, 10, 20 and 30 mA cm^{-2} was studied. The operating voltage declined in the early stage of discharging and then reached a constant value for 5 and 20 mA cm^{-2} current density as depicted in figure 3. Whereas current density of 30 mA cm^{-2} has the shortest discharge time followed by 10 mA cm^{-2} with a discharge time of 600 s. The current density of 5 mA cm^{-2} is optimized for further studies since it exhibits the highest voltage with a stabilized discharge rate.

3.3. Effect of catalyst loading

Figure 4 shows the effect of the catalyst on the performance of Al-air battery with discharge current density of 5 mA cm^{-2}. The air cathode consists of graphite with paint as a binder in the ratio 3:1 with $TiCl_3$ as a catalyst which is stirred magnetically for 30 min followed by manual casting of the material on the Al anode. The battery was fabricated with Al 5083 as an anode, 10% NaCl as electrolyte and air cathode without $TiCl_3$, with 5% and 10% $TiCl_3$ respectively. The metal catalyst plays a crucial role by providing the necessary active sites for the reaction, whereas graphite improves the electronic conduction pathways and retains it during battery operation. It can be analysed from the figure 4 that 5% $TiCl_3$ addition has a prolonged discharge capacity than 10% $TiCl_3$, without a catalyst, the system exhibits a very low discharge capacity and is unstable.

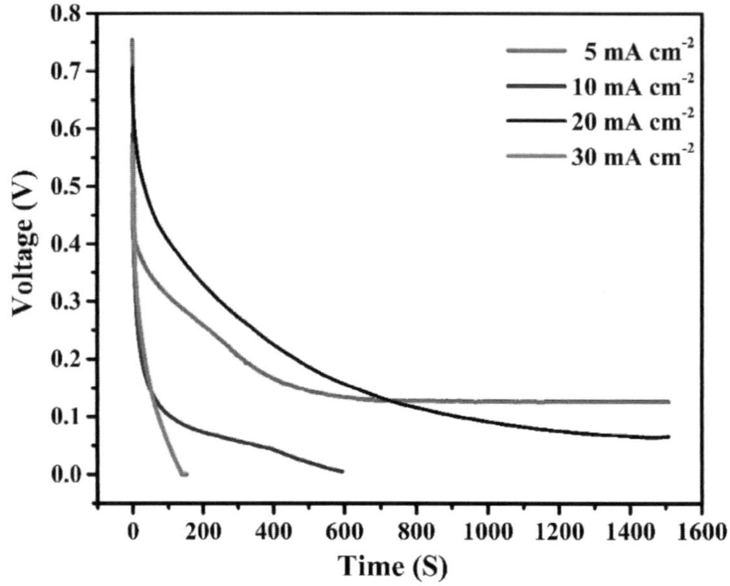

Figure 3. Effect of different current density and its discharge curve for the Al-air battery.

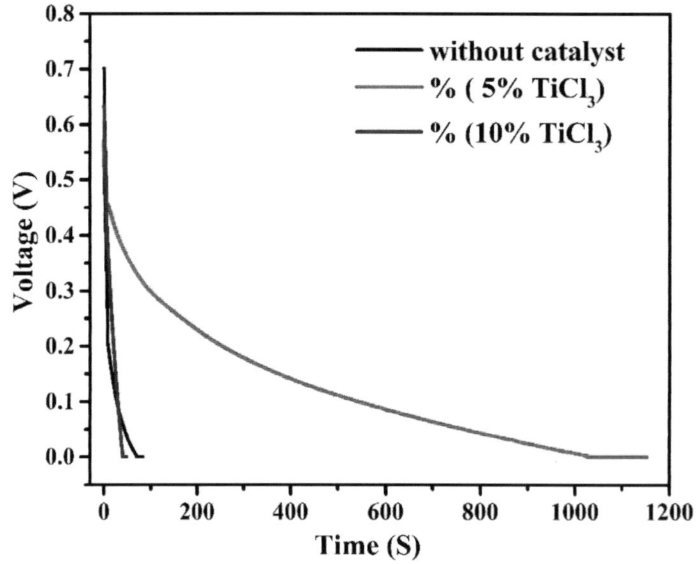

Figure 4. Effect of catalyst loading for the fabricated Al-air battery single cell.

3.4. Effect of electrolyte concentration

The voltage-time curve for the fabricated Al-air battery with different electrolyte concentration is shown in figure 5. Electrolyte concentration was varied in the range of 5-15%. When the concentration of NaCl is 10%, which is close to the saline water concentration, it can be predicted that the conductivity of the electrolyte is better leading to a higher discharge time. When the electrolyte concentration is increased to 15%, the electrolyte marginally inhibits the flow of electrons leading to self-discharging with a gradual decrease in voltage [8, 9].

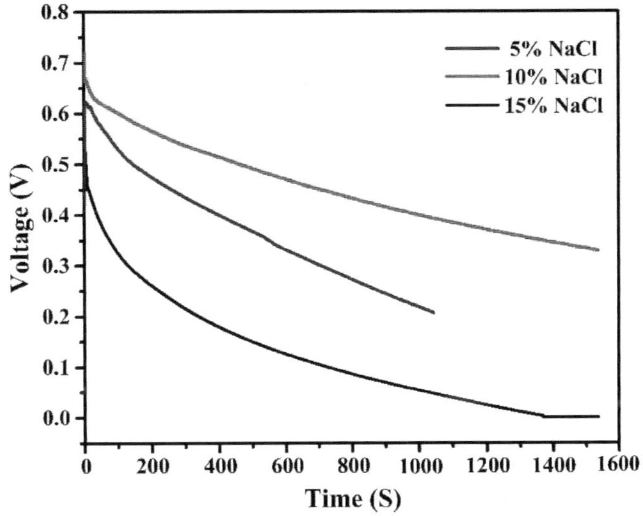

Figure 5. Effect of electrolyte concentration on the performance of the Al-air battery.

3.5. Lifecycle analysis for the fabricated single cell

Lifecycle analysis of battery is vital to demonstrate the stability of the electrode material developed, in the present case the material exhibits a better stability and can be employed for real-time applications. The shape of the charging-discharging curves obtained was uniform and the current capacity was calculated to be 20 mAh^{-1}. Al 5083 was used as an anode with a current density of 5 mA cm^{-2}, 10% NaCl as electrolyte concentration and catalyst loading of about 5% $TiCl_3$.

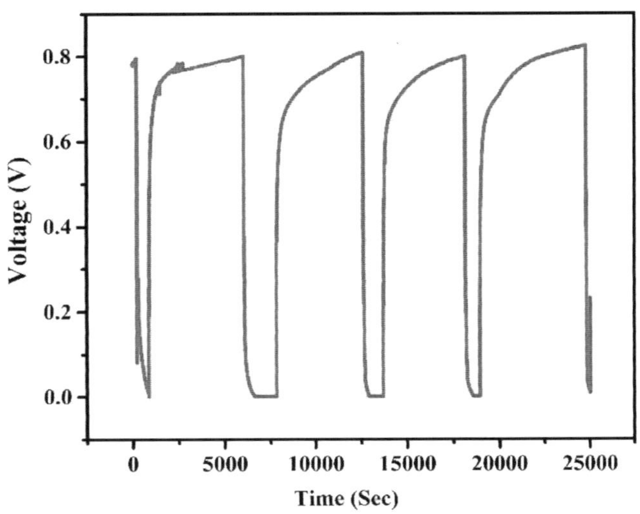

Figure 6. Lifecycle test for the Al-air battery with Al 5083 anode, current density 5 mA cm^{-2}, 10% NaCl as electrolyte and 5% $TiCl_3$ as a catalyst.

This work can be extended by stacking many single cells in series and study its performance analysis. The battery developed with the economic materials can be effectively used in marine applications for power backup.

4. Conclusion

The process parameters such as an electrode, catalyst concentration, electrolyte concentration, current density influences the performance of Al-air battery. Al 5083 as electrode material exhibits higher discharge time for a single cell fabricated compared to Al 6061 and Al 7075. Presence of graphite in the air cathode along with $TiCl_3$ as catalyst improves the discharge performance of Al-air battery. The stability of the battery was tested by performing lifecycle analysis for four consecutive cycles with a discharge time of 7h and current capacity of 20 mAh^{-1}. The increased current capacity is owing to the presence of highly conducting graphite in the cathode active material. It provides the necessary conduction pathway for the redox reaction to occur and enhances the current capacity of the fabricated battery.

Acknowledgement

The authors would like to acknowledge PUPT Research Scheme, Ministry of Research, Technology and Higher Education of the Republic of Indonesia, contract no. 30/E/KPT/2017 for funding this research work.

References

[1] Zu L, Junhong Z, Yanping C and Bin X 2017 *AIP Conf. Proc.* **1794** 04006
[2] Vanitha M and Balasubramanian 2013 *Environ. Technol. Rev.* **2** 101
[3] Jingling M, Jiuba W, Junwei G and Quanan L, 2014 *J. Power Sources* **253** 419
[4] Ghenadi A, Anton R, Ferry F, Hidayat D, Hidayat S, Panatarani C and Joni I M 2016 *AIP Conf. Proc.* **1719** 030055
[5] Egan D R, Ponce de Leon C, Wood R J K, Jones R L, Stokes K R and Walsh F C 2013 *J. Power Sources* **236** 293
[6] Joni I M, Panatarani C, Hidayat D, Wijatmoko B M, Anton R and Husni T 2013 *AIP Conf. Proc.* **1554** 20
[7] Wang H Z, Leung D Y C, Leung M K H and Ni M 2009 *Renew. Sust. Energ*. Rev. **13** 845
[8] Liang F, Huimin L and Jing L, 2015 *Electrochim. Acta,* **165** 22
[9] Liu Z, Junhong Z, Yanping C and Bin X 2017 *AIP Conf. Proc.* **1794** 040006

IOP Conf. Series: Journal of Physics: Conf. Series **1080** (2018) 012047 doi:10.1088/1742-6596/1080/1/012047

The activation of microorganism inoculum from rumen of beef cattle

Y A Hidayati*, T B A Kurnani, E T Marlina, K N Rahmah and E Harlia

Faculty of Animal Husbandry, Universitas Padjadjaran, Jl. Raya Bandung-Sumedang km 21 Jatinangor, Sumedang, West Java, Indonesia 45363

*yuli.astuti@unpad.ac.id

Abstract. The aim of this research was to investigate the activation of microorganism inoculum from rumen of beef cattle on lignite coal against number of anaerobic bacteria, the amount of volatile fatty acid (VFA) and biogas production. The research utilized completely randomized design with five treatments of various dosages of microbial inoculum from beef cattle rumen on lignite coal: P1 = 2%; P2 = 4%; P3 = 6%; P4 = 8%; P5 = 10%, and four repetitions. Parameters observed were the amount of VFA, the number of anaerobic bacteria and biogas production on day 2, 5, 10 and 15. Initial activation of beef cattle rumen microorganism inoculum utilized in-vitro technique with MC Dougall's artificial saliva methods, by using combination of 70% concentrate + 30% forage as beef cattle feed. The rate of biogas production was observed on hour 2, 4, 6, 8, 24, 48 and 72. The highest biogas produced was selected as the optimal hour of inoculum source. The treatment used 250 mL bottle serum, as biogas digester, filled with 100 mL 98-5 media and 12 g lignite coal. Microbial inoculum inoculated in digester using 10 mL syringe and incubated at 39°C. The result showed that the highest rate of biogas production on initial activation was 140 mL at hour 8. The highest average number of anaerobic bacteria was reached in P1, approximately 1068 x 10^{10} cfu/mL; the highest amount of VFA on day 2 and 5 in P3, accounted by 181 mM (acetate acid 29.76 mM/L; propionate acid 6.93 mM/L; butyrate acid 2.48 mM/L); the highest biogas production was reached on P1, accounted by 37.17 mL (CH_4 = 0.68%; CO_2 = 99.32%).

1. Introduction

Microorganisms in the rumen of beef cattle have an important role in degrading organic materials from feed. Rumen microbes live under strictly anaerobic environment. Number and type of rumen microbes depend on type of diet. Organic material in rumen is degraded into a simpler organic compound, is break into organic acids and followed by gas formation, such as methane (CH_4). Thus, rumen microbes, in particular beef cattle rumen microbes, can be utilized as starter in biogas formation.

Coal Bed Methane (CBM) is a natural gas formed through absorption processed of methane in liquid condition which trapped into coal pores. Similar with biogas, CBM contain carbon-dioxide (CO_2), N_2, H_2, hydrogen sulfide (H_2S), and O_2. CBM is formed along with coalification. It can be formed through chemical changes from heat, namely thermogenesis, and microorganism activity under anaerobic condition, namely biogenesis. The quality of CBM relate to coal's calorific rank which depends on coalification process: peat, lignite, bituminous and anthracite. In anthracite the quantity of CBM decreases.

This research study the activation of beef cattle rumen microbial inoculum in lignite coal against number of anaerobic bacteria, VFA and biogas production. This refers to CBM in lignite coal can be form biogenesis by microorganism activity under anaerobic condition. Result of this research is expected that microbial inoculum from rumen beef cattle can be utilized and activated in assisting methane gas formation process in CBM by observing the amount of anaerobic bacteria growing and the amount of VFA produced and the amount of biogas that is formed, so that in the end it can produce alternative energy in the form of methane gas (CH_4).

2. Materials and Methods

2.1 Tools and materials

The tools used in this research were Erlenmeyer, 250 ml serum bottle, 10 ml syringe, roll tube, and GC to analyze gas proportion on biogas produced. The materials used were rumen liquid of beef cattle, lignite coal, artificial saliva (Mac Dougall), concentrate, forage, 98-5 culture media consist of: 15.0 ml mineral liquid I and II, 0.1 ml Resazurin 0.1%, 50.0 ml distilled water, 2 g Bacto-agar Difco powder, 40.0 ml rumen fluid, 0.05 g glucose, 0.05 g Cellobiose, 0.05 g soluble starch, 5.0 ml Cysteine-HCl.H_2O-Na_2S.9H_2O liquid, 5.0 ml Na_2CO_3, 8%) [1].

2.2 Methods

In vitro fermentation: Initial rumen microbe activation utilized in vitro technique by using artificial saliva (Mac Dougall) and diet consists of 70% concentrate + 30% forage. Rate of biogas formation was observed at hour 2; 4; 6; 8; 24; 48; 72. The highest biogas production was selected as inoculum source.

Research design: Completely Randomized Design was utilized with five treatments: P1 = 2% inoculum of beef cattle rumen microbes in lignite, P2 = 4% inoculum of beef cattle rumen microbes in lignite, P3 = 6% inoculum of beef cattle rumen microbes in lignite, P4 = 8% inoculum of beef cattle rumen microbes in lignite, and P5 = 10% inoculum of beef cattle rumen microbes in lignite, with 4 replications.

Parameter observed: Parameter observed were number of anaerobic bacteria, volatile fatty acid (VFA), biogas production Observation period was at day 2, 5, 10 and 15 after inoculation. The data were analyzed statistically with ANOVA test and followed with Duncan test using SPSS 23.

Research procedure: All treatments were utilized 250 ml bottle serum as biogas digester. Each bottle was filled by 100 ml 98-5 media and 12 g lignite. Microbial inoculum was inoculated by using syringe and incubated at 39°C for 7 days.

3. Result and Discussion

3.1 In vitro fermentation

The initial activation of rumen microbes using in vitro technique with artificial saliva (Mac Dougall) and diet combination of 70% concentrate + 30% forage can be seen in table 1.

Table 1. Biogas production in various observation periods.

Hour	Biogas Production (ml)
2	40
4	110
6	140
8	140
24	100
48	132
72	36

Biogas production was an indicator to show the rumen microbial activity. The highest biogas production was reached on hour 8 accounted by 140 ml. It assumed that anaerobic microorganism from beef cattle rumen actively degraded organic material in diet consist of 70% concentrate and 30% forage with artificial saliva as buffer. It has a good agreement with other researcher who stated that growth curve of bacteria consists of adaptation, logarithmic, stationer and death phase [2].

3.2 Number of Anaerobic Bacteria

Average number of anaerobic bacteria growing on various treatments of inoculum dosages can be seen in figure 1.

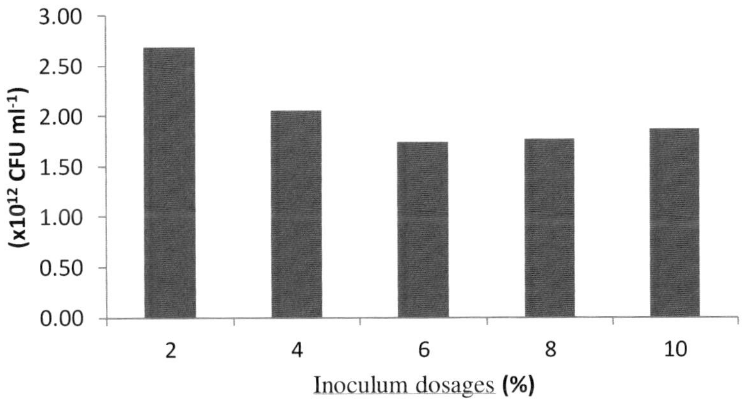

Figure 1. Number of anaerobic bacteria on various treatments at 15 days of incubation period.

Note: treatments were not significantly different ($P>0.05$).

The result showed that the highest number of anaerobic bacteria were found in P1, however based on analysis of variance the average number of anaerobic bacteria growing not significantly different among the treatments ($P>0.05$). This condition occurred because the growth of anaerobic bacteria was affected by the ratio of available medium and number of inoculum as starter. Even though inoculum dosage that inoculated was different, the average of anaerobic bacteria grow was almost similar. It was allegedly because the growth of anaerobic bacteria greatly influenced by the nutrient availability. Other researcher stated that the growth of bacteria was affected by ratio of bacteria population-available nutrient and environment condition (temperature, pH) [3, 4]. Jing Yi *et al.* suggested that the content of total solid materials affected the anaerobic digestion performance and changes on total solid content resulted on changes of microbial morphology in rumen [5].

3.3 VFA Production

The average production of VFA on various treatments can be seen in Figure 4. Based on the result, addition of inoculum dosage until 6% tended to increase VFA production. However, the addition of more than 6% decreased VFA production. The highest VFA production at day 2 and 5 was reached on P3, accounted by 181 mM (acetate acid 29.76 mM/l; propionate acid 6.93 mM/l; butyrate acid 2.48 mM/l). This condition occurred because VFA production was influenced by the ratio of nutrient source (medium and lignite) and inoculum dosage. The optimum ratio of nutrient and inoculum was reached at 6% (P3) which showed the highest VFA production. Too much microbial population resulted competition in obtaining substrate, otherwise small population of microbes resulted substrate degradation was not maximal. This has good agreement with other researcher who argued that methane and VFA concentration had bimodal pattern which related to nutrient cycle [6, 7]. On other reference, it is suggested that although VFA production and organic polymeric hydrolysis increased, community of primary bacteria with different treatments had no substantial difference [8].

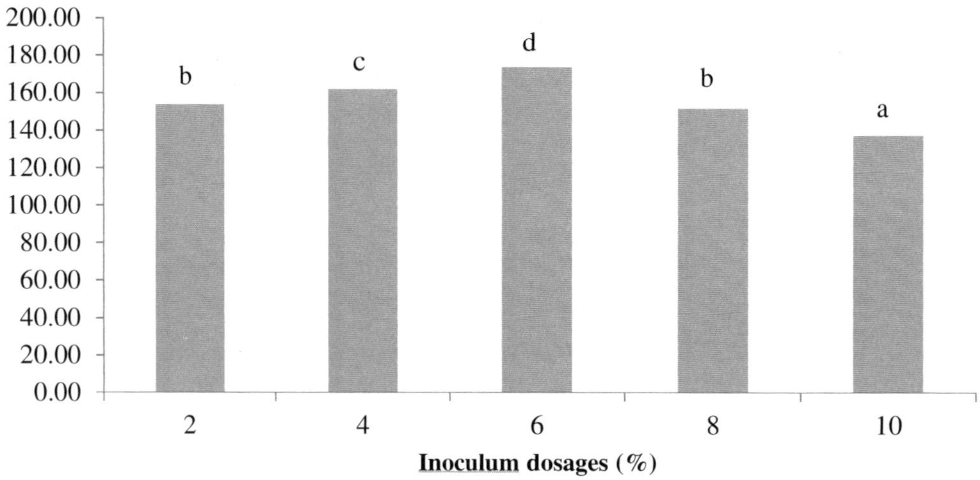

Figure 2. VFA production on various treatments at 5 days of incubation period.

Notes: treatments followed with same letter are not significantly different according to Duncan Multiple Test at 5% significance level.

Biogas, CH_4 and CO_2 Production

The average number of biogas, CH_4 and CO_2 production on various dosages of inoculum can be seen in figure 3 and 4. The highest biogas production was reached in P1. Based on analysis of variance, the biogas production was not significantly different (P>0.05). This related to number of anaerobic bacteria that grow. Biogas production was determined by organic material which was degraded and the activity of anaerobic bacteria. On other reference, it is suggested that ratio of carbon and nitrogen on degraded organic material affected biogas production. In addition, other researcher state that the addition of organic materials will affect the population of bacteria that activity so that will increase the biogas [9, 10].

Biogas production consists of CH_4 and CO_2 with different proportion. The highest CH_4 production obtained on P5 = 10% accounted by 13101.13 ppm, and the lowest CO_2 accounted by 271617.4 ppm. This condition illustrated that biogas formation was still in phase 3 where the proportion of CH_4 was low and CO_2 was high, in the next phase CO_2 could be reduced to CH_4 and H_2O [11].

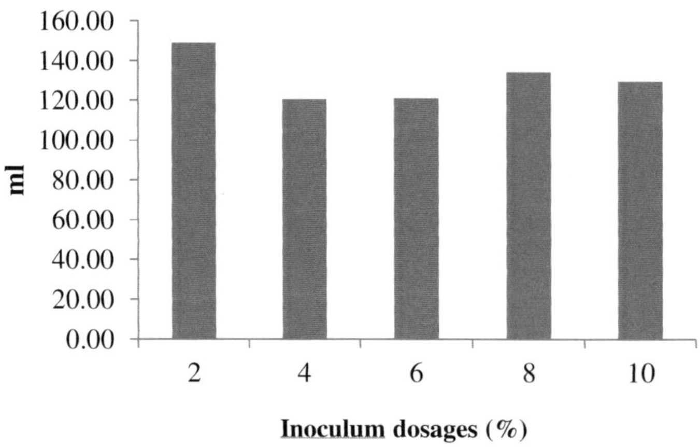

Figure 3. Biogas production on various treatments.

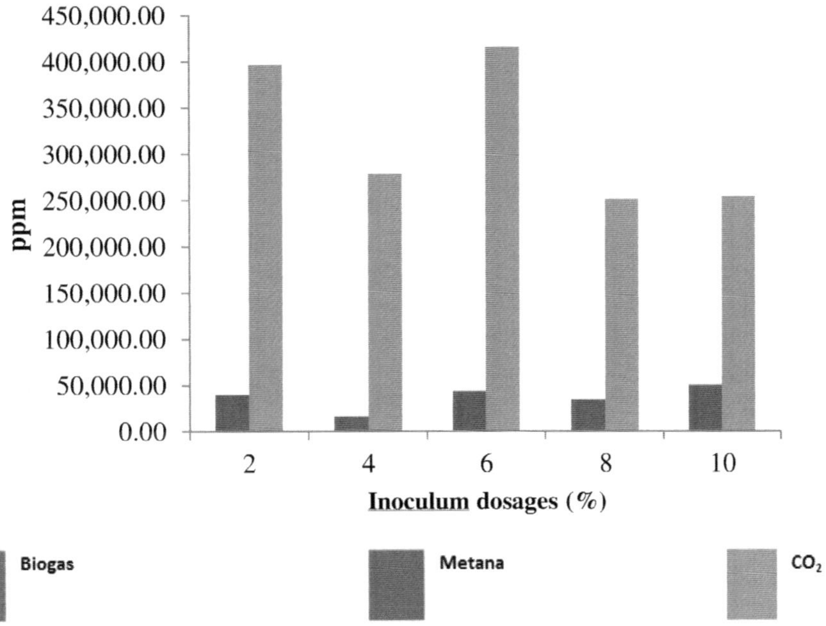

Figure 4. The proportion of methane (CH_4) and CO_2 at 15 days of incubation period.

Note: treatments were not significantly different (P>0.05).

4. Conclusion

The addition of inoculum of beef cattle rumen on lignite at 15 days' incubation period was not affected number of anaerobic bacteria and biogas production, as well as CH_4 and CO_2 produced. However, the treatment of inoculum dosages significantly affected VFA production. Inoculum dose of microorganism from rumen of beef cattle up to 10% added to lignite coal with incubation period for 15 days, anaerobic bacteria growing 2 x 10^{12} cfu / ml and the amount of biogas produced as much as 130 ml and CH_4 and CO_2 produced 50000 ppm and 250000 ppm, but the dose of inoculum affects the amount of Flying Fatty Acids produced on the 5th day P 3 = 181 mM (Acetic Acid 29.76 mM / l; Propionic Acid 6.93 mM / l; Butyric Acid 2.48 mM / l).

Acknowledgement

We are profoundly indebted to Rector of Universitas Padjadjaran through the Academic Leadership Grant (ALG) and DPRM Universitas Padjadjaran. We thank Prof. Dr. Ellin Harlia, MS who led the Program at the Faculty of Animal Husbandry and provided insight and expertise that greatly assisted the research.

References

[1] Bryant M P and Robinson I M 1961 *J. Dairy Sci.* **44** 1446
[2] Rolfe M D, Rice C J, Lucchini S, Pin C, Thompson A, Cameron A D S, Alston M, Stringer M F, Betts R P, Baranyi J, Peck M W and Hinton J C D 2012 *J Bacteriol.* 2012 Feb **194** 686
[3] Leung D Y C and Wang J 2016 *International Journal of Green Energy* Volume **13** 119
[4] Shah F A, Mahmood Q, Shah M M, Perves A and Asad S A 2014 *Scientific World Journal* **2014** 183752
[5] Jing Yi, Bin Dong, Jingwei Jin and Xiaohu Dai 2014 *PLoS ONE* **9** e102548
[6] Robinson D L, Goopy J and Hegarty R S 2010 *Animal Production Science* **50** 630
[7] Dong R and Zhao G 2014 *PLoS ONE* **9** e116290
[8] Yang X, Liu X, Chen S, Liu G, Wu S and Wan C 2016 *Archaea* **2016** 1698163

IOP Conf. Series: Journal of Physics: Conf. Series **1080** (2018) 012047 doi:10.1088/1742-6596/1080/1/012047

[9] Dioha J, Ikeme C H, Nafi'u T, Soba N I and Yusuf M B S 2013 *International Research Journal of Natural Sciences* **1** 1
[10] Cioablăa A E, Djuricb A, Dumitrelc G A, Ăd D C and Pode V 2017 *Studia Ubb Chemia* **LXII** 51
[11] Chen J, Falivene L, Caporaso L, Cavallo L and Chen E Y X 2016 *J. Am. Chem. Soc.* **138** 5321

IOP Conf. Series: Journal of Physics: Conf. Series **1080** (2018) 012048 doi:10.1088/1742-6596/1080/1/012048

Application of convolutional neural network classifier for wireless arrhythmia detection

D E Kusumandari[1], M I Rizqyawan[1], M Yazir[2], M Turnip[2], A Darma[2] and A Turnip[1*]

[1]Technical Implementation Unit for Instrumentation Development, Indonesian Institute of Sciences, Bandung, Indonesia
[2]Faculty of Technology and Computer Science, Universitas Prima Indonesia, Medan, Indonesia

[*]jujhin@gmail.com

Abstract. The heart as the main system of blood circulation in our body can certainly experience abnormalities called arrhythmia. Information of the arrhythmia is very important in the field of health and electrocardiogram is one of the most commonly used tool to detect the heart abnormalities. A wireless monitoring of the heart activities has been developed. An application of convolutional neural network classifier for arrhythmia detection is proposed. Twenty four health subjects were tested in the three experiment methods. The classification accuracy about average 93,97% is obtained.

1. Introduction

Heart disease is the number one killer in the world. Every year, more than 2 million Americans die of heart disease/stroke [1-4]. Heart disease is a disorder that occurs in large blood vessel systems. So that causes the heart and blood circulation is not working properly. Heart disease is often known as Suddent Death. The high number of death caused by heart disease can be prevented and suppressed by increasing the public's knowledge of heart disease symptoms, improving the quality of equipment, medical examinations, and supported by a healthy lifestyle. Due to the large number of inpatients and increasing of the deaths, heart disease has become a special concern for the government. Cardiac arrhythmias or heart rhythm (i.e., abnormal changes in cardiac electrical activity) are a common feature of heart disease. Such abnormalities are identified from the heartbeats of too slow, too fast or even irregular. This condition can cause the heart is not able to pump the blood throughout the body well and can even result in an emergency of the heart failure. Each heartbeat will produce an electrical impulse that flows throughout the body to stimulate muscle contraction according to a particular pattern. The pattern of electrical activity of the heart or caused by the heartbeat can be measured using an electrocardiogram (ECG) which captures the time-series electrical activity of the heart. The information content present in the ECG signals can be used to identify various forms of cardiac arrhythmias. Therefore, the development and application of certain methods to classify ECG data has been of particular concern to the current researchers. With a reliable method, it is expected that built medical instruments can be used for the proper diagnosis, treatment and prevention of arrhythmias or abnormal heart conditions.

Various methods have been proposed in the past for the classification of ECG signals based on signal processing techniques such as frequency analysis [5], wavelet transforms [6-8], and band pass filters [3, 9-10], statistics [11] and heuristic approaches [12] hidden Markov model [13], supporting vector machines [2, 4, 14], artificial neural networks, and other mixed methods [15-20]. But overall each method is still not perfect, especially in the application. The problems often encountered in algorithm development are the variation of ECG signals between patients and limitations of accuracy when used to classify new patient ECG signals. This makes the method not widely used, especially for clinical, and tends to have a high degree of accuracy and efficiency for larger databases [21, 22]. To overcome these weaknesses, several methods with patient-specific designs of late have been developed [22, 23-25]. From the test results indicate that its performance has a significant increase in performance against the results of conventional classification methods. The classification method has a general approach with two main operations: feature extraction and classification of extracted features.

To minimize the deficiencies of previous methods and to improve the classification performance, an adaptive 1D convolutional neural networks (CNNs) classifier are used for ECG signal classification and its anomaly detection. The CNN consist of hierarchical neural networks with each layer of convolution alternating with a subsampling layer. The principle can be analogous to simple and complex cells in the human visual cortex. Basically the ECG signal consists of many waves, each wave representing one heartbeat (one heart electrical activity). The types of waves represent each combined segment form. In one wave the ECG signal consists of several wave points called intervals and segments. The point consists of points P, Q, R, S, and T. whereas te intervals consist of PR intervals, QRS intervals and QT intervals as well as the segments composed of PR segments, and ST segments. The applied algorithm is used to identify the intervals and segments.

2. Method

To evaluate reliability and ensure ease of operation, ECG signals from 24 subjects, aged 24 ± 3 years, were recorded. All subjects are healthy both physically and mentally. Next, each experiment is divided into two session conditions: relax (14 subjects) and type (10 subjects), respectively. Each is equipped with three channel sensors with each location relating to cardiac activity. Once completed, each electrode must be covered with an electrolyte gel to minimize the effect of noise. ECG data were collected from three Ag / AgCl electrodes embedded in a developed wireless ECG system.

The ECG signals processing methods consists of several stages including pre-processing, algorithm processing, and classification. In the preparing pre-processing the data set that used as input for processing is generated. 30 minutes length recording data with 360 Hz frequency at 48 two-channel with resolution of 11 bits perchanel with a distance of 10 mV. Further process of algorithm processing is algorithm design which designed using Python programming and applied to convolutional neural network method for classification. The design of the algorithm is divided into three main parts which are generate data, skew data and run experiment while for data training optimizer used stochastic gradient descent model. The last process is classification which is the result of classification of the input and algorithm that has been designed. The output generated from this program is the accuracy of arrhythmia detection. This research method can be illustrated as in figure 1.

IOP Conf. Series: Journal of Physics: Conf. Series **1080** (2018) 012048 doi:10.1088/1742-6596/1080/1/012048

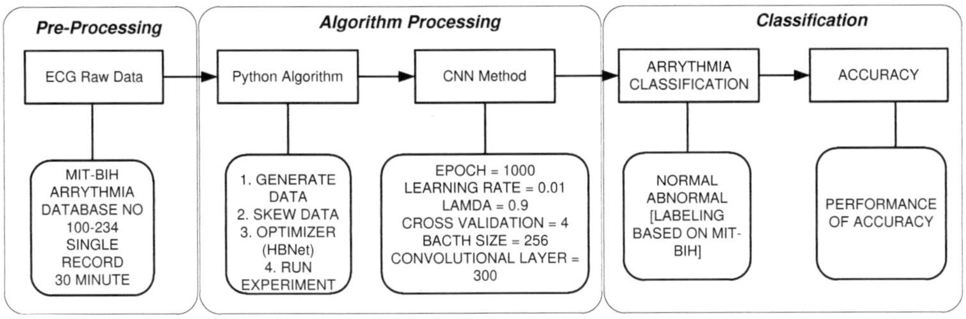

Figure 1. ECG processing Scheme.

Classification is a process of arrhythmia grouping which is output is an average accuracy of the epoch process for each type of arrhythmia. These results will serve as an evaluation of the convolutional neural network method as a classifier. The first processing is the determination of RR-interval using Weavelet and DWT method. Then the results will be adjusted to the arrhythmia category and will be made into 1 bacth file with arff format to further processed by using software WEKA.

3. Signal Processing

Band pass filter is used to reduce the effect of noise due to the muscle contraction. The design with a frequency of 60 Hz, used as an initial damping and T-wave interference. The bandpass is desirable to maximize QRS energy around 5-15 Hz that is quick and recursively applied in real-time with poles are located in z plane of the unit circles to cancel zeros. This approach results in a filter design with integer coefficients, this is because only integer arithmetic is required. The filter class that has the poles and zeroes only on the unit circle allows for limited passband design flexibility.

Convolutional neural networks use relatively less complex pre-processing, especially when performing the data labeling process and inputting input values compared to other classification algorithms. Convolutional neural network is divided into several layers: the input layer and the output layer and some hidden layers. Part of the hidden layer in this algorithm is convolutional, pooling and fully connected layer. Convolutional Layers apply convolution operations to data inputs and emulate individual neuron responses to be forwarded to the next layer. Pooling is a non-linear down-sampling whereas a fully connected layer connects each neuron in one layer to each neuron in another. This is in principle the same as the multi-layer perceptron neural network.

4. Results and Discussions

The problem encountered in the classification process of arrhythmia is feature extraction. Feature extraction is required to obtain R-R intervals from the recorded ECG signal. An applied of incorrect methods in feature extraction can lead to calculation delays in subsequent processes. If the extracted feature is not appropriate then it could be serious, where the result represent someone who has no arrhythmia (normal) detected as an arrhythmia (abnormal). Another problem is over fitting data, where the results will show good accuracy but when the data is applied to the real data then the result become less accurate.

At the time of testing the data is divided into two parts, namely 80% as a training file and 20% as a test file. By converting 2-dimensional convolution to 1-dimensional convolution as a representation of convolutional neural network algorithm, the program runs self leraning to get the processing results that have been designed on the previous algorithm. The main provision in this process is that the epoch process will be done as much as 1500 times. During training, the selected learning rate and lamda were 0.01 and 0.9 with covolutional layers of 300 (using a batch stochastic gradient with batch size 256). The process is done with cross validation 4 times to avoid over fitting data. The output obtained from the algorithm is the final result of the signal processing. Cross entropy is used for loss function testing and regularization, where the regularized coefficient value is λ. Initial test showed that choosing λ with value more than 0,01 gave as significant performance degradation. Therefore, the uniform distance value test in the range of [0,0,01] (0,0, 0,001, 0,002, ..., 0,01) is done. The obtained results with an average accuracy of 99.15% against the λ test with a value of 0.01 as shown in figure 2 are achieved.

The classification result of WEKA software using convolutional neural network method is obtained as in figure 3. From figure 3 it can be explained that the confusion matrix obtained from each class is as follows: For class B arrhythmia detected 10 as class B and 1 as class T so that TP rate of 0.909; for N class arrhythmia detected 254 as N so TP rate of 1,000; for class T detected 1 as class T and 7 as class B and 4 as class N so TP rate of 0.083; and for the VT class is not detected at all. The results are obtained based on the classification process using WEKA program and cconvolution neural network as classifier.

In the analysis results into two experiments sheme that are in relaxed and typed conditions there are some differences in the results of QRS detection by applying the Pan-Tompkins algorithm. The difference occurs due to the activities of each groups such that affecting the detection value of desired QRS waves. The results are represented in the form of tables with outputs categorized based on the variables required for subsequent calculations. The required variables are as follows: True Possitive (TP) ie the exact number of peaks which detected as a peak; True Negative (TN) ie the number of peaks that are not peaks and detected as non peak; False Possitive (FP) ie the number of detected peaks as a peak but not peak; False Negative (FN) is the number of peaks that are not detected as peak but peak. As shown in table 1(relaxed) and table 2 (Typed), the average accuracy, sensitivity, and positive predicticity are 98%, 98%, 100% and 59%, 64% and 86%, respectively.

Figures 4 and 5 shows the results on each of the QRS detection stages using Pan-Tompkins algorithm for each experimental group. In the QRS detection, the length of the used raw data is about 2 minutes with the results as shown in figures 4(a) and 5(a). The upper part is the raw data, the second row is the filtered signals using bandpass and derivative filters, respectively, the third row is the average windowing and QRS detection which is characterized by red round point, respectively. The zooming results of figures 4(a) and 5(a) are shown as in figures 4(b) and 5(b) for each group which are the detected QRS are clearly shown.

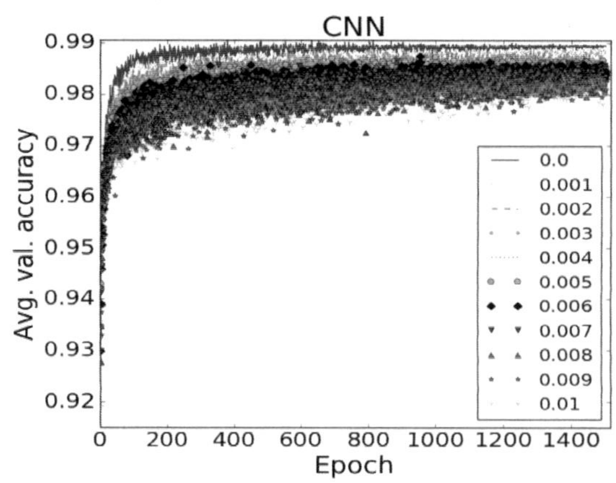

Figure 2. Algorithm training test with Pyhton.

```
Classifier output

Time taken to test model on test split: 0.03 seconds

=== Summary ===

Correctly Classified Instances         265               93.9716 %
Incorrectly Classified Instances        17                6.0284 %
Kappa statistic                          0.6168
Mean absolute error                      0.0489
Root mean squared error                  0.1389
Relative absolute error                 50.7411 %
Root relative squared error             64.53   %
Total Number of Instances              282

=== Detailed Accuracy By Class ===

                 TP Rate  FP Rate  Precision  Recall   F-Measure  MCC      ROC Area  PRC Area  Class
                 0.909    0.026    0.588      0.909    0.714      0.718    0.986     0.665     (B
                 1.000    0.321    0.966      1.000    0.983      0.810    0.998     1.000     (N
                 0.083    0.004    0.500      0.083    0.143      0.192    0.972     0.451     (T
                 0.000    0.000    0.000      0.000    0.000      0.000    0.996     0.891     (VT
Weighted Avg.    0.940    0.291    0.914      0.940    0.919      0.765    0.996     0.961

=== Confusion Matrix ===

   a   b   c   d   <-- classified as
  10   0   1   0 |   a = (B
   0 254   0   0 |   b = (N
   7   4   1   0 |   c = (T
   0   5   0   0 |   d = (VT
```

Figure 3. Classification results with WEKA CNN algorithm.

Table 1. Relax condition: the QRS Detection using Pan-Tompkins algorithm for the average accuracy, sensitivity, and positive predicticity.

NO	File	Original Peak	Detected peak	Undetected Peak	Non Peak	TN	TP	FN	FP	Accuracy (%)	Sensitivity (%)	PP (%)
1	14.07.57	169	169	0	0	0	169	0	0	100%	100%	100%
2	09.51.19	175	167	0	0	0	167	0	0	100%	100%	100%
3	09.57.37	156	156	0	0	0	156	10	0	100%	100%	100%
4	10.04.55	156	150	10	9	0	141	0	9	88%	93%	94%
5	10.09.29	150	149	0	0	0	149	0	0	100%	99%	100%
6	13.20.22	189	189	0	0	0	189	0	0	100%	100%	100%
7	13.23.41	192	190	2	0	0	190	2	0	99%	99%	100%
8	13.27.21	176	176	0	0	0	176	0	0	100%	100%	100%
9	13.31.31	166	166	0	0	0	166	0	0	100%	100%	100%
10	13.49.42	223	220	3	0	0	220	3	3	99%	99%	100%
11	13.56.54	169	169	0	0	0	169	0	0	100%	100%	100%
12	14.01.00	174	163	14	0	0	163	14	0	92%	942%	100%
13	14.03.47	221	211	10	1	0	211	10	1	95%	95%	100%
14	16.06.17	151	151	0	0	0	151	0	0	100%	100%	100%
15	Total	2467	2426	39	10	0	2417	39	10	98%	98%	100%

Table 2. Typing condition: the QRS Detection using Pan-Tompkins algorithm for the average accuracy, sensitivity, and positive predicticity.

NO	File	Original Peak	Detected peak	Undetected Peak	Non Peak	TN	TP	FN	FP	Accuracy (%)	Sensitivity (%)	PP (%)
1	Data 1	168	28	140	0	0	28	148	0	100%	16%	100%
2	Data 2	176	117	50	9	0	117	50	9	100%	66%	93%
3	Data 3	206	200	6	0	0	200	6	0	100%	97%	100%
4	Data 4	178	174	110	106	0	68	110	106	88%	38%	39%
5	Data 5	198	186	22	10	0	176	22	10	100%	89%	95%
6	Data 6	187	172	29	14	0	158	29	14	100%	84%	92%
7	Data 7	214	191	67	44	0	147	67	44	99%	69%	77%
8	Data 8	165	157	14	6	0	151	14	6	100%	92%	96%
9	Data 9	202	113	110	21	0	92	110	21	100%	46%	81%
10	Data 10	172	90	107	8	0	82	107	8	99%	43%	91%
	Total	1866	1428	655	218	0	1219	663	218	98%	64%	86%

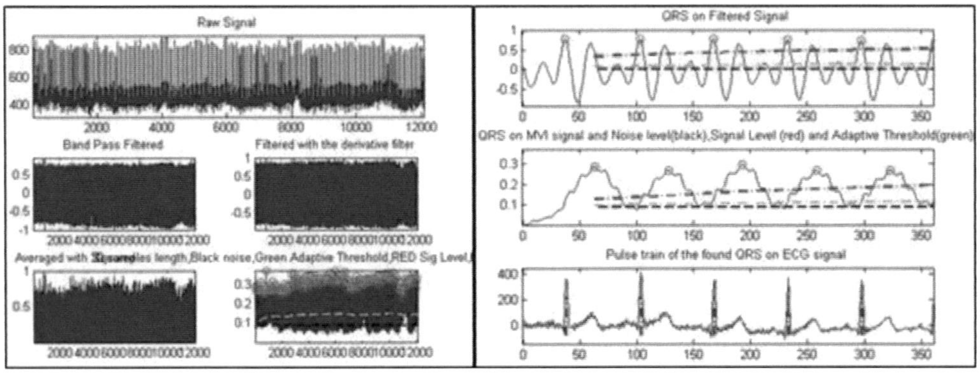

(a) (b)

Figure 4. The detected QRS for Relaxed subjects: (a) QRS process, (b) Zooming Detected QRS.

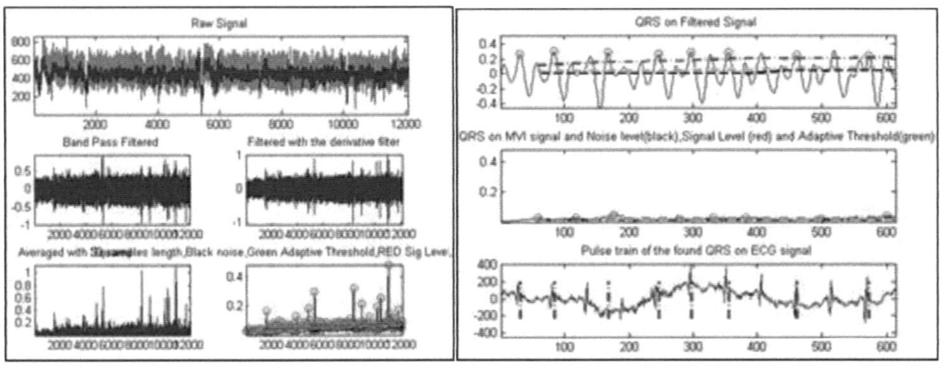

(a) (b)

Figure 5. The detected QRS for Typing subjects: (a) QRS process, (b) Zooming Detected QRS.

5. Conclusions

A wireless monitoring of the heart activities is developed. An application of convolutional neural network classifier for arrhythmia detection is proposed. Twenty four health subjects were tested in the two experiment methods. The average accuracy, sensitivity, and positive predictivity are 98%, 98%, 100% (for relax subjects) and 59%, 64% and 86% (for typing subjects), respectively. The arrhythmia classification accuracy of 93,97% with misclassification of 6,028 using weka algorithm based convolutional neural network classifier is obtained.

Acknowledgment

This research was supported by the Technical Implementation Unit for Instrumentation Development, Indonesian Institute of Sciences and, Insentif Riset Sistem Inovasi Nasional (INSINAS), Kementerian Riset, Teknologi, dan Pendidikan Tinggi (Ristekdikti), Indonesia.

References

[1] Freiden T R and Berwick D 2011 N Engl J Med **365** E27
[2] Turnip A, Rizgyawan M I, Kusumandari D E and Hermida I D P 2016 System *International Journal of Pharma Medicine and Biological Sciences* **5** 189
[3] Turnip M, Saragih R I E, Dharma A, Kusumandari D E, Turnip A, Sitanggang D and Aisyah S 2018 *Journal of Physics: Conference Series* **1007** 012019
[4] Turnip A, Rizqywan M I, Kusumandari D E, Turnip M and Sihombing P 2018 *Journal of*

Physics: Conference Series **970** 012012
[5] Minami K, Nakajima H and Toyoshima T 1999 *network IEEE Trans. Biomed. Eng.* **46** 179
[6] Shyu L Y, Wu Y H and Hu W C 2004 *IEEE Trans. Biomed. Eng.* **51** 1269
[7] Inan *et al.* 2006 *IEEE Trans.Biomed. Eng.* **53** 2507
[8] Amri M F, Rizqyawan M I and Turnip A 2016 *IEEE Information Technology, Computer, and Electrical Engineering* (ICITACEE) p 1
[9] Turnip A, Suhendra M A, Kusumandari D E and Sihombing P 2017 *IEEE Instrumentation, Control, and Automation* (ICA) p 121
[10] Alfonso X and Nguyen TQ 1999 *IEEE Trans. Biomed. Eng.* **46** 192
[11] Willems J L and Lesaffre E 1987 *J. Electrocardiol.* **20** 83
[12] Talmon J L Pattern Recognition of the ECG. Berlin, Germany: Akademisch Proefscrift, 1983
[13] Coast D A, Stern R M, Cano G G and Briller S A 1990 *IEEE Trans. Biomed. Eng.* **37** 826
[14] Osowski S, Hoai L T and Markiewicz T 2004 *IEEE Trans.Biomed. Eng.* **51** 582
[15] Hu Y, Palreddy S and Tompkins W J 1997 *IEEE Trans. Biomed. Eng.* **44** 891
[16] Garda Y R, Caesarendra W, Tjahjowidodo T, Turnip A, Wahyudati S, Nurhasanah L and Sutopo D 2018 *Journal of Physics: Conference Series* **1007** 012069
[17] Sihombing P, Siregar Y M, Tarigan J T, Jaya I and Turnip A 2018 *Journal of Physics: Conference Series* **978** 012105
[18] Turnip A and Hong K S *Int. J. Innov. Comp. Inf. Control* **8** 5839
[19] Turnip A, Hutagalung S S, Pardede J and Soetraprawata D 2013 *International Journal of Brain and Cognitive Sciences* **2** 63
[20] Turnip A, Simbolon A I, Amri M F, Sihombing P, Setiadi R H and Mulyana E 2017 Internetworking Indonesia Journal **9** 53
[21] Lee S C 1989 *IEEE Proc. Conf. Neural Information Processing Systems*: Seatle
[22] de Chazal P and Reilly R B 2006 *IEEE Trans. Biomed. Eng.* **53** 2535
[23] Jiang W and Kong SG 2007 *IEEE Trans. Neural Networks* **18** 1750
[24] Ince T, Kiranyaz S and Gabbouj M 2009 *IEEE Transactions on Biomedical Engineering* **56** 1415
[25] Kiranyaz S, Ince T and Gabbouj M 2013 *Book: Springer* 383

Subsurface structure of Tangkuban Parahu area derived from CSAMT and gravity investigation

E J Mustopa[*], D Sutarno, Nurhasan, A R Naulia, M L Baskoro and N S Asih

Earth Physics and Complex System Research Divison, Faculty of Mathematics and Natural Sciences, Bandung Institute of Technology, Indonesia

[*]enjang@fi.itb.ac.id

Abstract. An integrated Controlled-source audio frequency magnetotelluric (CSAMT) and gravity investigations across the Tangkuban Parahu area has been carried out. The Tangkuban Parahu is located in Bandung, West Java, one of a volcano area. The purpose of the investigations was to locate the subsurface structures of the area that may correlate to a geothermal reservoir in unknown geothermal potential area. The CSAMT survey has been conducted in the area using transmitter locatedin 3-5 km from the survey area. The objective of the survey was to determine a subsurface resistivity structures. On the other hand, the gravity survey was carried out in the same sites with the CSAMT sites. The results of the integrated interpretation of the CSAMT and gravity data show that the subsurface of the area has a correlation with a geothermal reservoir structure. It is also confirmed by several surface manifestations in the Ciater area, the north of the Tangkuban Parahu volcano.

1. Introduction

Tangkuban Parahu is located in Bandung, West Java, Indonesia. The topographic elevation of the field is ranging from 800 to 2200 m above sea level. An integrated Controlled-source audio frequency magnetotelluric (CSAMT) and gravity investigations across the Tangkuban Parahu area has been carried out. The aim of the CSAMT method is to determine a subsurface resistivity structure, while the aim of gravity method is to determine a subsurface density structure according to a different gravitational value in a measurement point.

The CSAMT measurement, in the operating frequency range of 0.25 Hz to 8192 Hz, has been carried out in the area using transmitter that located about 3 to 7.5 km in the South from the sounding site area to acquire 42 soundings with the irregular space stations, varied from 200 to 500 m between sounding, covering the Tangkuban Parahu and its environs. On the other hand, 42 sites of the Gravity survey using gravity meter was carried out in the same sites with the CSAMT survey. The location survey Parahu is located in a volcanic region in Bandung, West Java, Indonesia. The field in UTM (Universal Transverse Mercator) zone 48 is geographically situated between 790,000 – 795,000 East and in the range of 9,200,000 – 9,257,000 North. The topographic elevation of the field is ranging from 800 to 2200 m above sea level (figure 1).

The objective of the CSAMT and Gravity Surveys were to determine subsurface structures of the area based on resistivity and density characteristics. These characteristics can be used to define whether there is a geothermal prospect in unknown geothermal potential area or there is no prospect. Furthermore, if it is possible in the future, the area can be exploited to produce a geothermal energy for generating the electricity after doing several geophysical, geological and geochemical exploration. For the reason, the geothermal energy is one of the alternatives in Indonesia.

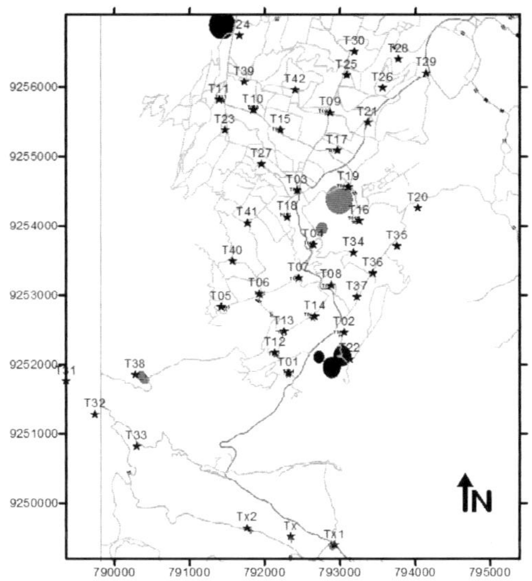

Figure 1. Location of CSAMT and Gravity survey[1)].

2. Basic Theory

2.1. Basic Theory of CSAMT method

The basic concept used in the CSAMT method can be derived from the solution of Maxwell's equations, these are

$$\nabla \mathrm{x} \mathbf{E} = -\mu \frac{\partial \mathbf{H}}{\partial t} \tag{1}$$

$$\nabla \mathrm{x} \boldsymbol{H} = \mathbf{J} + \frac{\partial \boldsymbol{D}}{\partial t} \tag{2}$$

$$\nabla . \mathbf{D} = 0 \tag{3}$$

$$\nabla . \mathbf{B} = 0 \tag{4}$$

where **E** is the electric field intensity (V/m), **B** is the magnetic induction (tesla), **H** is the magnetic field intensity (A/m), ω is the angular frequency, μo is the magnetic permeability (H/m).

The apparent resistivity on the earth surface derived from Maxwell's equation in a far-field zone is

$$\rho_{axy} = \frac{1}{\mu\omega} \left| \frac{E_x}{H_y} \right|^2 \tag{5}$$

In this paper, a 2D inversion technique based on a non-linier conjugate gradient method [3] was applied to invert the selected CSAMT data.

2.2. Basic Theory of Gravity method

Gravity method is one of the geophysics methods to determine a subsurface density structure based on a different gravity value on measurement point. The method is one of passive methods because there is nothing sources inducting to the earth surface when the measuring is occurred. Gravity method utilized difference between gravity values at each measurement points. There are factors that affected these values such as difference in rocks' density beneath the surface, earth's tides, latitude angle, sea surface height, topography, and others. Therefore, several corrections are needed to obtain the real gravity value where it can only interpret the value of subsurface density. The Newton's Gravity Law can be expressed as follows:

$$\mathbf{F}(r) = G(\frac{m_1 m_2}{r^2})\hat{\mathbf{r}} \tag{6}$$

where **F** is the force between $m1$and $m2$, $m1$ is the mass of object 1, $m2$is the mass of object 2, r is the distance between objects centre, G is the gravitational constant and $\hat{r}$ as unit vector [4].

3. Interpretation

3.1 Interpretation of CSAMT data

Employing a 2D inversion scheme to the CSAMT data in Tangkuban Parahu, we constructed the 2D resistivity section for several profiles. The resistivity section obtained from 2D inversion of Line-3 indicates that the distribution of the low resistivity anomaly (<10 ohm.m) in the depth of 700-1000 m is distributed in the northeast from t-19 to t-28 (figure 2). This anomaly is may considered as hydrothermal alteration zone in a geothermal reservoir structure. On the contrary, the low resistivity anomaly is not found in the southeast. From the figure also shows that the electrical discontinuity is found below the sounding site t-07, it may correspond to the existing fault in the area [1].

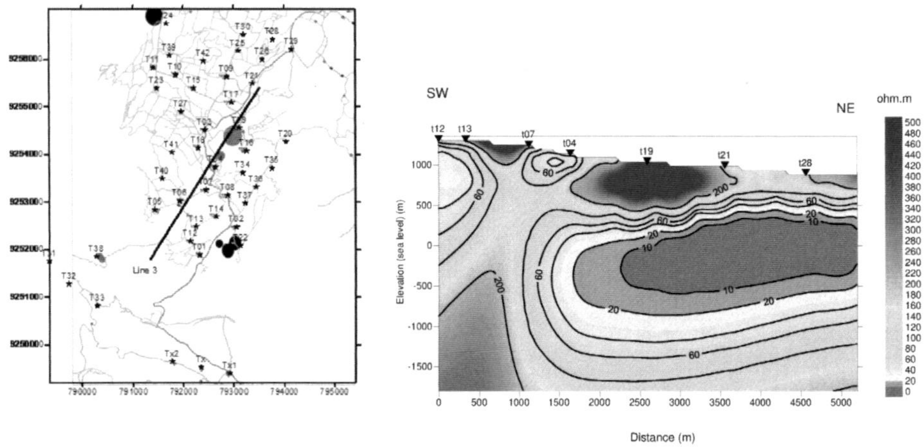

Figure 2. Resistivity section derived from 2D inversion result at Line-3.

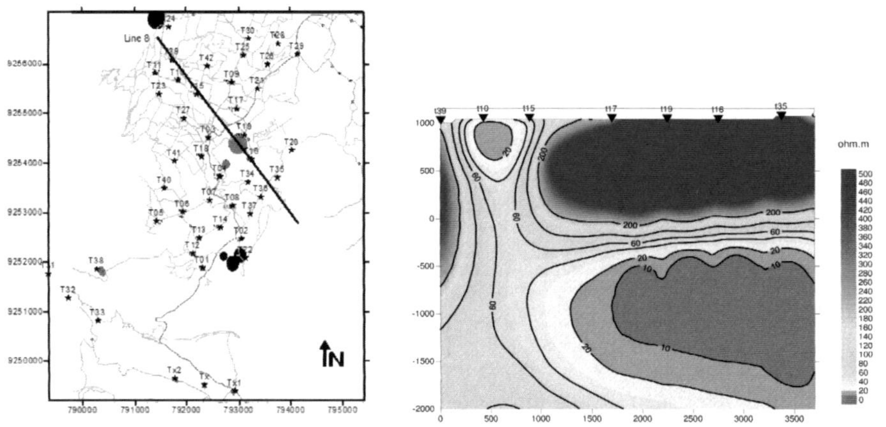

Figure 3. Resistivity section derived from 2D inversion result at Line-8.

Figure 3 shows 2D resistivity section of Line-8 from Northwest to Southeast direction, the distribution of the low resistivity anomaly is also found below sounding sites t-17 to t-35. It is also considered as the hydrothermal alteration zone. On the other hand, figure 4 shows 2D resistivity section of Line-9 from Northwest to Southeast direction, it indicates that the subsurface resistivity of this line is mainly composed of 3 layers. The first layer (overburden) is high resistivity (50 to 500 ohm-m) that is between 800 to 1000 meters thick. These anomalous features may be due to faulting, fracturing, and hydrothermal alteration along the significant fault.

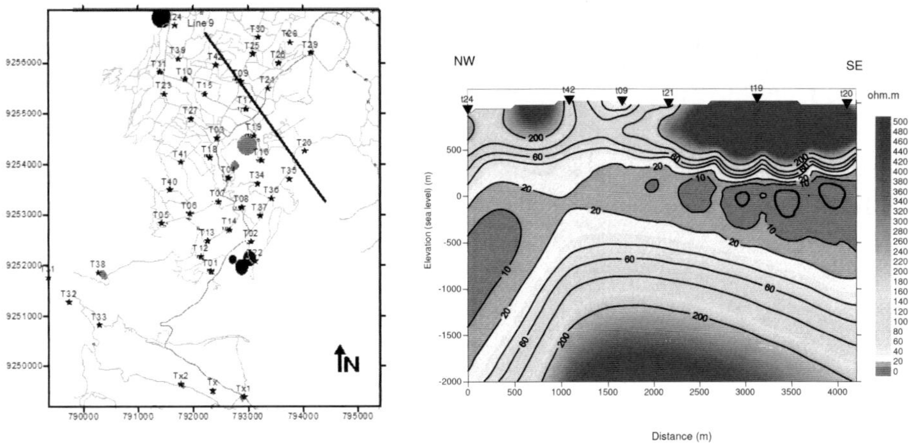

Figure 4. Resistivity section derived from 2D inversion result at Line-9.

3.2. Interpretation of Gravity data

Measurement of gravity method in this survey used a Lacoste Romberg gravimeter that measuring adifferent gravitational value at each measurement points caused by differences in rock density beneath the earth's surface. The value of gravity is also affected by factors such as earth's tidal, latitude angle, and others. Therefore, corrections such as drift correction, tide correction, latitude correction, free-air correction, Bouger correction, terrain correction and complete Bouger correction are required to obtain the real gravity value where it can only interpret the value of subsurface density. Gravity datacorrection results are presented in the complete Bouger anomalies contour with values varying between 369-395 miligal shown in figure 5. The high anomalously values are conformed to high rock densities while the low anomalously values are indicated with the low rock densities, this area is predicted as a source of geothermal potential.

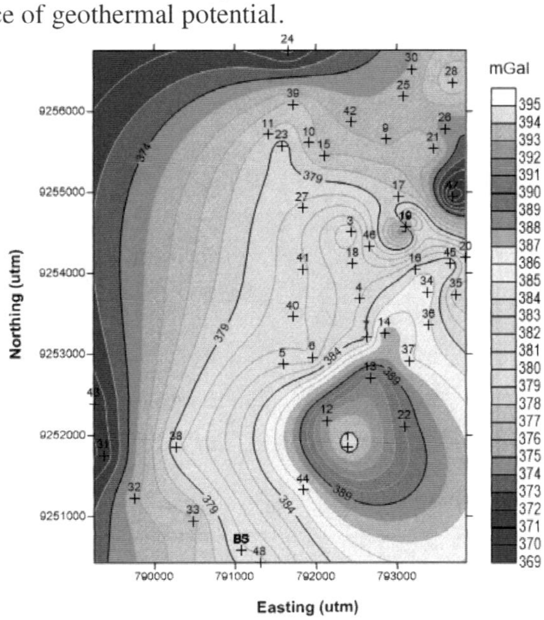

Figure 5. Contour map of Compelete Bouger Anomaly.

The anomalies will be separated into two categories such as the residual and regional anomalies. Both anomalies will be modeled by slicing the residual contour maps shown in figure 6 (in miligal). From the residual anomaly contourshows that the average density value of 2.44 g/cm^3.

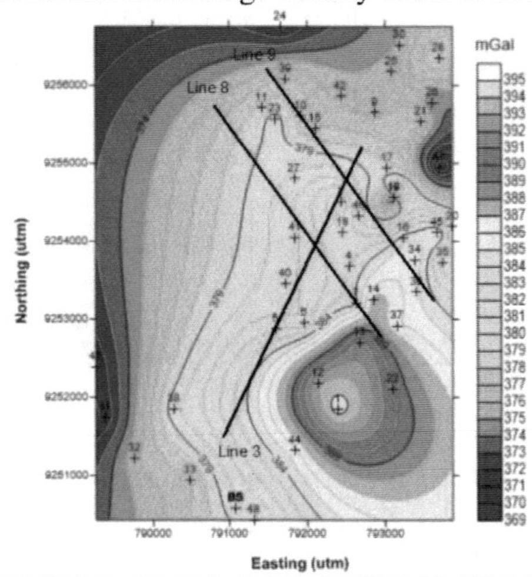

Figure 6. Contour map of residual anomaly with cross-section.

Using software using GRAV2DC modelling [5] for cross-section of Line 3 with the direction from Southwest to Northeast, it is confirmed that the density contrast varies between -0.1 gr/cm^3, 0 gr/cm^{3}and 0.065 gr/cm^3 (figure 7). The created model has a depth of 0-3 km with an error of 2.73% generating a variety density as shown in table 1.

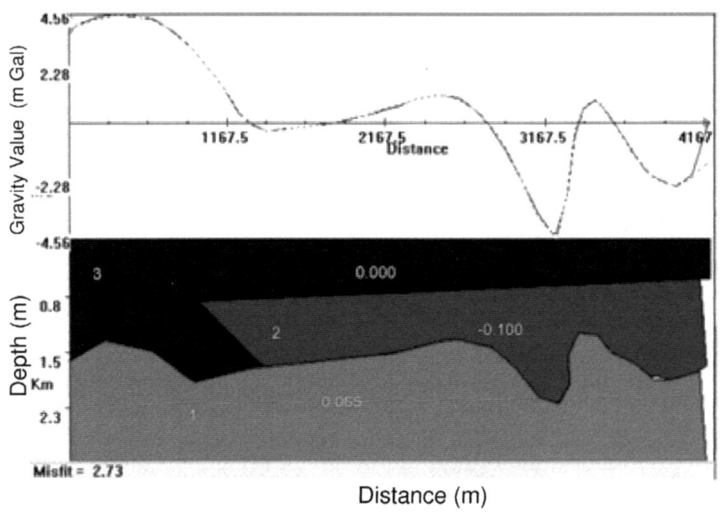

Figure 7. Result of subsurface modelling from Line 3.

Table 1. Interpretation from line 3.

Number	Depth (km)	Contrast Density (gr/cm^3)	Density (gr/cm^3)	Interpretation
1	0.52 - 2.3	0.065	2.505	Andesite
2	1.2 - 3	-0.100	2.34	Rhyolite
3	0 – 1.5	0	2.44	Andesite

Cross section of Line 8 is a direction from Northwest to Southeast resulting a variety density contrast between 0 gr/cm^3, -0.1 gr/cm^3, -0.06 gr/cm^3 and 0.065 gr/cm^3 (figure 8). The created model has depth of 0-3 km with an error of 1.52% generating a variety density is shown in table 2.

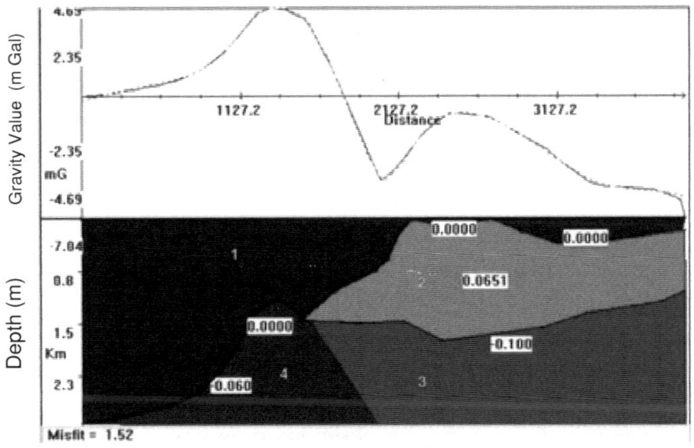

Figure 8. Result of subsurface modelling from Line 8.

Table 2. Interpretation from line 8.

Number	Depth (km)	Contrast Density (gr/cm^3)	Density (gr/cm^3)	Interpretation
1	0 – 3	0	2.44	Andesite
2	0.01 – 1.6	0.065	2.505	Andesite
3	1 - 3	-0.100	2.34	Rhyolite
4	1.15 - 3	-0.060	2.38	Rhyolite

Finally, the Line 9 is sliced from Northwest to Southeast, producing variety densities between 0 gr/cm^3, -0.1 gr/cm^3, -0.06 gr/cm^3, 0.1 gr/cm^{3}and 0.0651 gr/cm^3 (figure 9). The created model has depth of 0-3 km with an error of 1.52% generating variety densities as shown in table 3.

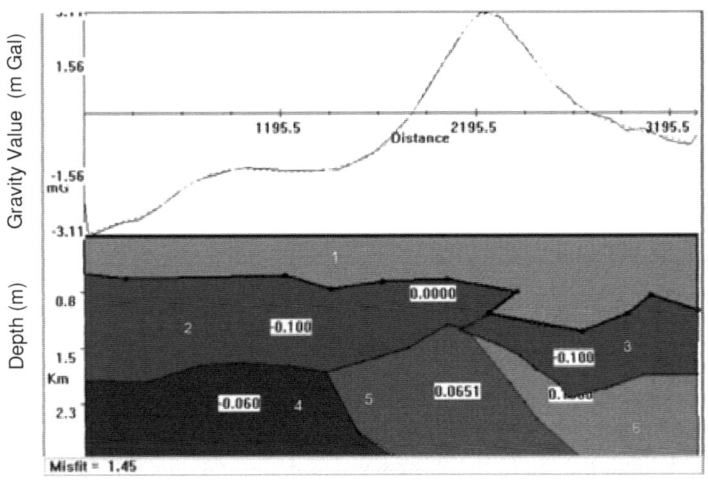

Figure 9. Result of subsurface modelling from Line 9.

Table 3. Interpretation from line 9.

Number	Depth (km)	Contrast Density (gr/cm^3)	Density (gr/cm^3)	Interpretation
1	0 – 1	0	2.44	Andesite
2	0.5 – 1.9	-0.100	2.34	Rhyolite
3	0.7 – 2.2	-0.100	2.34	Rhyolite
4	1.6 - 3	-0.060	2.38	Rhyolite
5	1.2 - 3	0.065	2.505	Andesite
6	1.2 - 3	0.100	2.54	Andesite

Generally, the modelling results for each section confirmed that in the subsurface survey area has a different type of rock based on its density. Commonly, the subsurface layer is composed by andesite and rhyolite. Rhyolite layer has a density value lower than andesite density, it means that the rhyolite has a higher porosity than andesite. The geothermal reservoirs probably present in a low porosity rock that is in rhyolite structure in the depth in variety of depth between 1.5–3 km.

4. Conclusion

The CSAMT and Gravity methods applied at Tangkuban Parahu area was successful to locate an anomalously low resistivity zone that may indicate the existing of a potential geothermal reservoir. From the CSAMT interpretation, results there are three layers of subsurface in the measurement area, that is the first layer a resistivity of 30 – 150 ohm-m and thickness of 800 to 1000 m, second layer has an extremely low resistivity of 3 – 15 ohm.m with 1000 – 1500 m thick. Finally, the basement (third) layer that is relatively more resistive than the second layer with resistivity of 30 – 100 ohm-m.

From the gravity interpretation results geological structure beneath the surface of Mount Tangkuban Parahu consist of andesite with density value of 2.4 – 2.6 gr/cm^3 and rhyolite with density value of 2.3 – 2.4 gr/cm^3. From the interpretation result, the geothermal reservoir is estimated to be in rhyolite.

From the model of CSAMT and Gravity in Line-3, both models show that there is anomaly distributed in the Northeast area. This may consider as hydrothermal alteration zone in geothermal reservoir structure. From Line-8 both models show anomaly in the Southeast area. And from Line-9, both model show the low resistivity in the depth around 800 – 1000 meters.

Furthermore, if it is possible in the future, the area can be exploited to produce a geothermal energy for generating the electricity after doing several geophysical, geological and geochemical exploration. For the reason, the geothermal energy is one of the energy alternatives in Indonesia.

Acknowledgments

Authors wishing to acknowledge to Program Penelitian, Pengabdian kepada Masyarakat, dan Inovasi (*P3MI*) *ITB* for funding of the research in the year of 2017.

References

[1] Mustopa E J, Nurhasan, Srigutomo W and Sutarno D 2017 *J. Phys Conf. Ser.* **877** 012055
[2] Kaufman AA and Keller G V 1983 *Frequency and transient sounding method* (Amsterdam: Elsevier Science Publ. Co. Inc.)
[3] Rodi W and Mackie R L 2001 *Geophysics* **66** 174
[4] Telford W G 1990 *Applied Geophysics 2nd edition* (Melbourne: University of Cambridge)
[5] Talwani M, Worzel J L and Landisman M 1959 *J. of Geophysical Research* **64** 49

IOP Conf. Series: Journal of Physics: Conf. Series **1080** (2018) 012050 doi:10.1088/1742-6596/1080/1/012050

Comparison of EEG sleep characteristic with music and aromatherapy stimuli

D E Kusumandari[1*], M A Suhendra[1], M F Amri [1], A I Simbolon[1], M I Rizqyawan[1], P Wardono[2], A D Fauzan[3] and A Turnip[1]

[1]Technical Implementation Unit for Instrumentation Development, Indonesian Institute of Sciences, Bandung, Indonesia.
[2]Faculty of Art and Design, Bandung Institute of Tecnology, Bandung, Indonesia.
[3]School of Electrical Engineering, Telkom University, Bandung, Indonesia.

[*]esti_dek@yahoo.com

Abstract. The quality of human's sleep can be determined by reading the bio-potentials signals in the brain, commonly called as electroencephalograph (EEG). In order to detect this signal, we placed an EEG sensor at some points around the head. In this research, we investigated the effect of Sundanese music and aromatherapy bio-composite from Myristica Fragrans on day sleep quality. We compared the EEG sleep characteristic with music and aromatherapy stimuli. The feature extraction uses wavelet transform. Finally, the experimental result shows that EEG signal frequency with music stimuli is higher than aromatherapy stimuli statistically.

1. Introduction

Sleep is a natural activity in human life. In general, adults need to sleep for 7-8 hours every night to keep the body on top form. In addition to adequate sleep, good sleep quality is also an important factor. The better quality of someone's sleep will have a positive impact in the morning such as feeling calm and not easily drowsy. While poor sleep quality will have a negative impact such as feeling tired, easily agitated, lethargic, difficulty concentrating, etc. Poor sleep quality is usually caused by several factors, such as insomnia, sleep apnea, snoring, circadian rhythm abnormalities, shift work system, parasomnia, etc.

Medical researches reveal that the smells have a significant impact on feelings. Smells directly affect the brain [1, 2]. In previous studies, lavender and sandalwood aromatherapy has proven to help depression, anxiety, insomnia and headaches [1 - 4].

When sleeping, the human brain will keep working, especially when dreaming, the brain will work as it is doing activities during the day. From this fact, we knew the condition of the body and the quality of one's sleep through brain signals. The brain signals commonly referred to as electroencephalograph (EEG) are wave signals that have low voltage amplitude, which is in the orde of microvolts (100 μV-1 mV). The humans produce five EEG signals, namely Gamma (30 Hz-45 Hz), Beta (15 Hz-30 Hz), Alpha (8 Hz-15 Hz), Theta (4 Hz-8 Hz) and Delta (0.1 Hz-4 Hz). Each brain signal has a frequency that varies according to the activity being done, whether it is during sleep or awake. The shape and activity of each brain signal are shown in figure 1.

In this research, system testing will be done with aromatherapy and music stimulation. Aromatherapy is the use of a variety of essential oils derived from plants in helping to treat various physical and emotional health problems. Each type of aromatherapy has different benefits to the body, one of them is to make the body relax and easy to fall asleep. Music or sound stimulation is one of the variables that affect the electrical waves in the brain. Therefore, the purposes of emotional and psychiatric therapy is often used voice stimulation to achieve the increase of certain waves in the brain. Although a person's perception of the sound received is different, but in general, a sound stimulus will form a certain pattern on the EEG signal.

Frequency Band Name	Frequency Bandwidth	State Associated with Bandwidth	Example of Filtered Bandwidth
Raw EEG	0–45 Hz	Awake	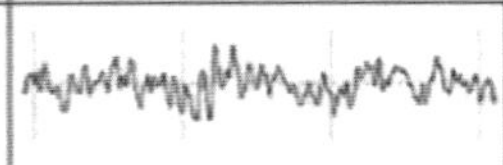
Delta	0.5–3.5 Hz	Deep Sleep	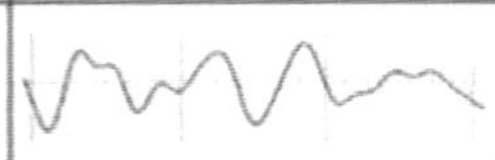
Theta	4–7.5 Hz	Drowsy	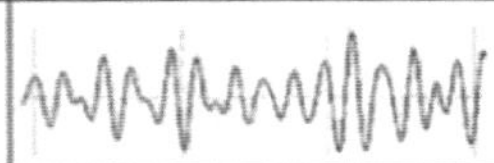
Alpha	8–12 Hz	Relaxed	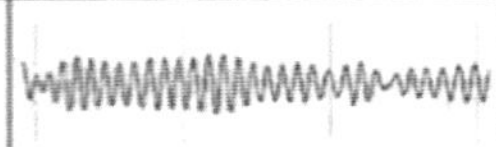
Beta	13–35 Hz	Engaged	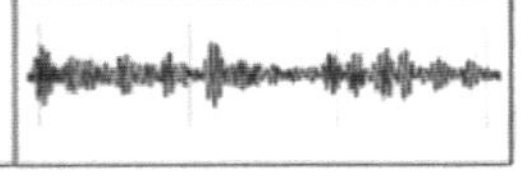

Figure 1. Shape and activity of EEG signal. Source: Conorrus Somanno.

2. Physiology of Smell and Auditory

Aromatherapy and music stimuli are related to smell and auditory in which the physiology field has a specially region in the brain as presented in figure 2. An olfactory bulb explains how the nose to receive the odor information in which it has a glomerular cell as a receiver and data storage of them to send to a special area on frontal lobe [5]. Figure 3 illustrates the principle of ear operation when it received sound. Theoretically, there are two types of auditory receptor, i.e inner and outer auditory hair cell. The hair cells form synapses with dendrites of bipolar neurons whose axons bring auditory information to the brain and their location at Brodmann Area (BA) 41 and 42 or temporal lobe [5, 6].

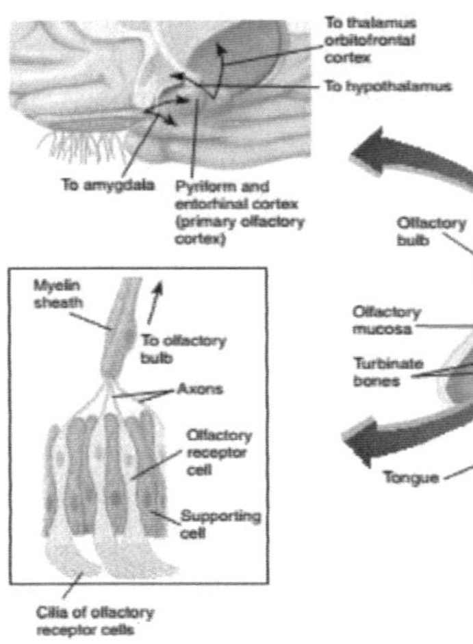

Figure 2. Physiology of smell [2].

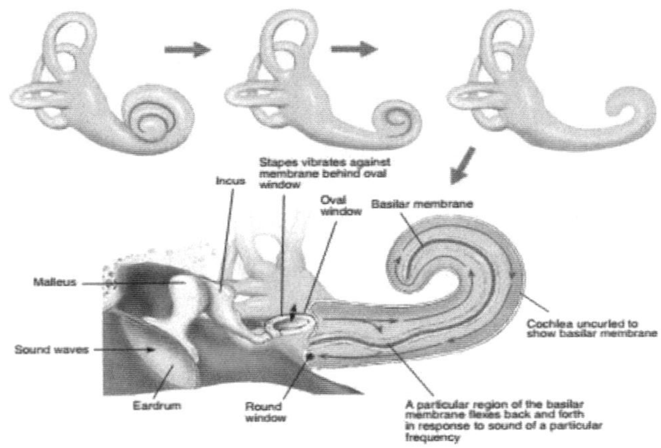

Figure 3. Physiology of auditory [5].

3. Material and Method

3.1. Data Acquisition

Ten subjects were selected for this experiment which consisted five male and five female. Characteristics of subjects are students with average of age is 20.6 years. The EEG signal of ten subjects were recorded on daytime by NicoletOne EEG with 23-channels according to 10/20 principle (Fp1-F7, F7 - T3, T3 - T5, T5 - O1, Fp2 - F8, F8 - T4, T4 - T6, T6 - O2, Fp1 - F3, F3 - C3, C3 - P3, P3 - O1, Fp2 - F4, F4 - C4, F4 - P4, P4 - O2, Fz - Cz, Cz – Pz) as shown in figure 4. We divided the experiment into three steps, sleep without treatment, aromatherapy treatment and music treatment. We recorded all the treatments for 20 minutes. Properties of aromatherapy based on essential oil from bio-composite Myristica Fragrans (MF) and properties of music based on Sundanese music (SM) from Bandung, West Java, Indonesia. The experimental setup given in figure 5.

The ideas of experiment is to provide chairs especially for relax, so that subject would give several information of acquisition. When the information of acquisition take places, the subjects do not allowed to moving their limb because it could produce a noise and artifact of EEG signal. To obtain information of acquisition, subjects sleep without treatment (non-aromatic and music) for 20 minutes during the day. After 20 minutes, they have to be awakened and then we put the MF bio-composite as stimuli in a few minutes to let the subject sleep again for the next 20 minutes. We repeated the same treatment using sound stimuli SM for 20 minutes.

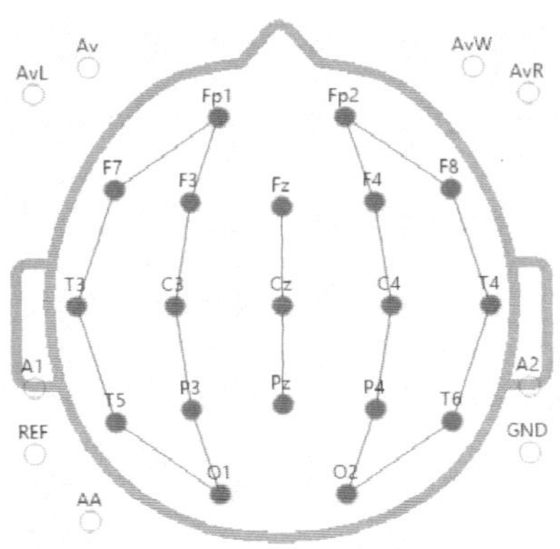

Figure 4. EEG-Bipolar Montage.

Figure 5. Experimental Setup.

3.2. Feature Extraction

Figure 6 is scheme of signal processing in which wavelet transform (WT) become feature extraction. Theoretically, WT is a mathematical basic for signal processing to several applications included EEG data. In general, it provides to signal analyze by time-scale signal analysis, signal decomposition and signal compression [7]. Type of analyze commonly used to signal processing is Discrete Wavelet Transform (DWT) which generate a signal into two type of signals. Equation (1) describes the approximation and detail coefficient of DWT.

$$\psi_{(a,b)}(t) = 2^{\frac{a}{b}} \psi(2^{-\frac{a}{b}}(t-b)) \quad (1)$$

Where (a) is scales and (b) positions of the wavelet mother (ψ) t [8]. The original EEG signal is time domain signal, the signal energy distribution is scattered and non-linear.

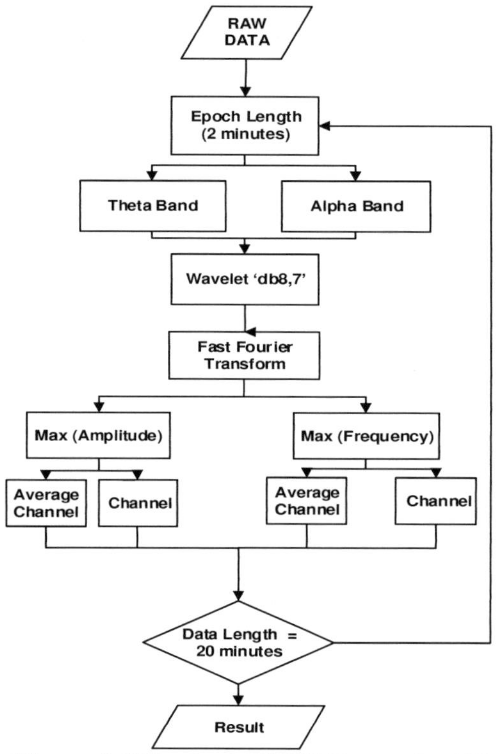

Figure 6. Flowchart of EEG Processing [5].

The EEG signal characteristic was analyzed using function wavelets ($\psi_{a,b}$, mother wavelet) described in equation (2).

$$\psi_{a,b} = \frac{1}{\sqrt{a}} \psi(\frac{t-b}{a}) \quad (2)$$

where a, b $\in$ R, $a > 0$, and R is the wavelet space [6].

The signal features are buried away in the noise. In order to extract the features, the EEG signal is analyzed to give a description of the signal energy as a function of time and frequency. Based on previous studies, the feature extracted in the frequency domain is one of the best to recognize the mental task based on EEG signals [9]. EEG signal is nonstationary that means its spectrum changes with time, such a signal can be approximated as piecewise stationary, a sequence of independent stationary signal segments [9].

The Fourier transform converts time domain signals into frequency domain representations defined in equation (3).

$$X(f) = F\{x(t)\} = \int_{-\infty}^{\infty} x(t)e^{-j2ft}\,dt \tag{3}$$

And similarly, the DFT converts discrete-time sequences into discrete-frequency versions derived by equation (4).

$$X_k = \sum_{i=o}^{n-1} X_i e^{-2j\pi\frac{ik}{n}} \quad for\ k = 0,1,\ldots,n-1 \tag{4}$$

In equation (3), *x(t) i*s the time domain signal, and *X(f)* is its Fourier Transform. In equation (4), *x* is the input sequence, *X* is DFT, and *n* is the number of samples.

The Fast Fourier Transform (FFT) is an optimized implementation of a DFT that takes less computation to perform. In 1965, J.W.Cooley and J.W.Tuckey reinvented the FFT for fast computation of the DFT. We apply FFT to those signals with N=2*n* number of samples. Benefit of FFT algorithm is reducing the computation time by a factor of the orde N/log2N, i.e. more than 100 times for a sample of 1024 samples.

4. Results and Analysis

In general, the raw data resulted from NicoletOne EEG is shown in figure 7 which has three signals, i.e. non-treatment, aromatherapy treatment and music treatment and each signal separated by line signal for each treatment. We obtain raw data in figure 7 because we recorded EEG directly and one-time for all treatments. Typical of acquisition is one of disadvantages in our experiment.

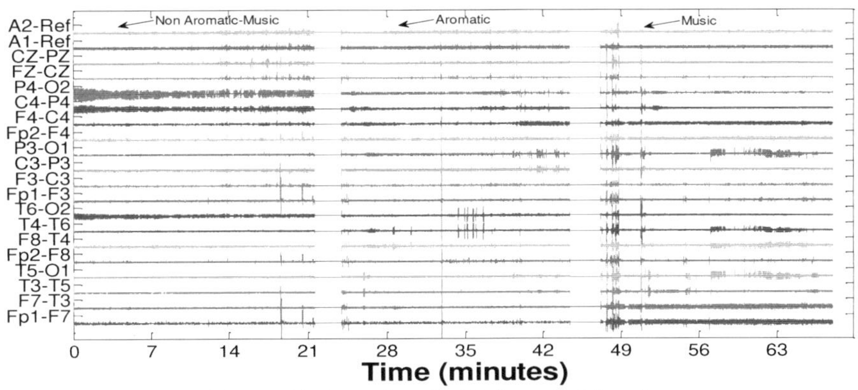

Figure 7. Raw data EEG from Subject 1.

The raw data in figure 7 has been discovered noise and artifact so it requires signal processing to remove them. The goal from signal processing is to yield the real signal and exactly to obtain EEG sleep characteristic. We used Band Pass Filter (BPF) with cut-off frequencies alpha band (8-15 Hz) and theta band (4–8 Hz) because those frequencies involved significant in sleep stages. Furthermore, analyze EEG data by WT with function of mother wavelet "Daubechies D8" levels 7 resulted detail coefficient and sleep characteristic. The results from WT is shown in figure 8.

Figure 9 shows brain activity before and after giving treatment. It shows that giving treatment decrease brain activity until ten minutes, while the before treatment phase show a random brain activity. We assumed that by using Sundanese music as a sleep treatment could make someone feel calm and relax.

Figure 10 shows the chart of data of signal processing for amplitude and frequency of alpha and theta wave with music stimuli, aromatherapy stimuli, and non-stimuli. The green color show the results with aromatherapy stimuli, while the red and blue color show results with music stimuli and non-stimuli, respectively. The processes signal was taken at 2, 4, 6, 8, 10, 12, 14, 16, 18, and 20 minutes. The result of this signal processing is then tested by T test to know its significance.

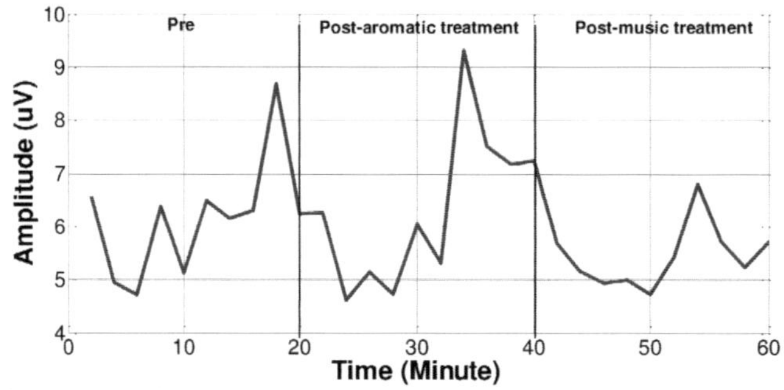

Figure 8. Brain activity for average subjects (Pre, post aromatic and music treatment).

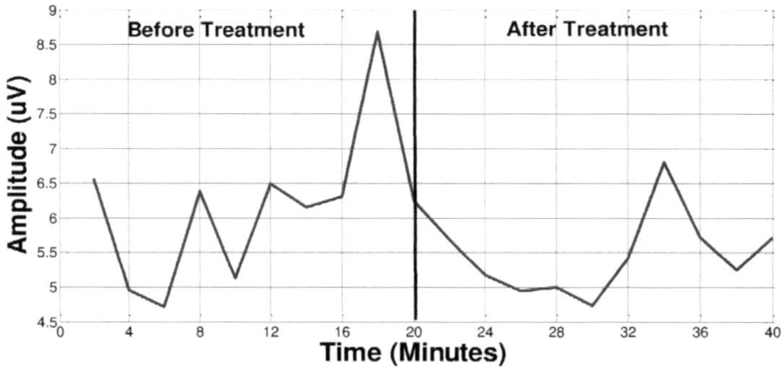

Figure 9. EEG Signal pattern before and after giving treatment.

The result of the T test is shown in table 1. The T test is to analyze whether giving stimulus music and aromatherapy affect the quality of one's sleep. Significant value is indicated when p value less than or equal to 0.05. From the results of T test shows that the existence of music stimuli have greater significance compared with aromatherapy stimuli. The effect of music stimulation for sleep quality is seen after 12 minutes.

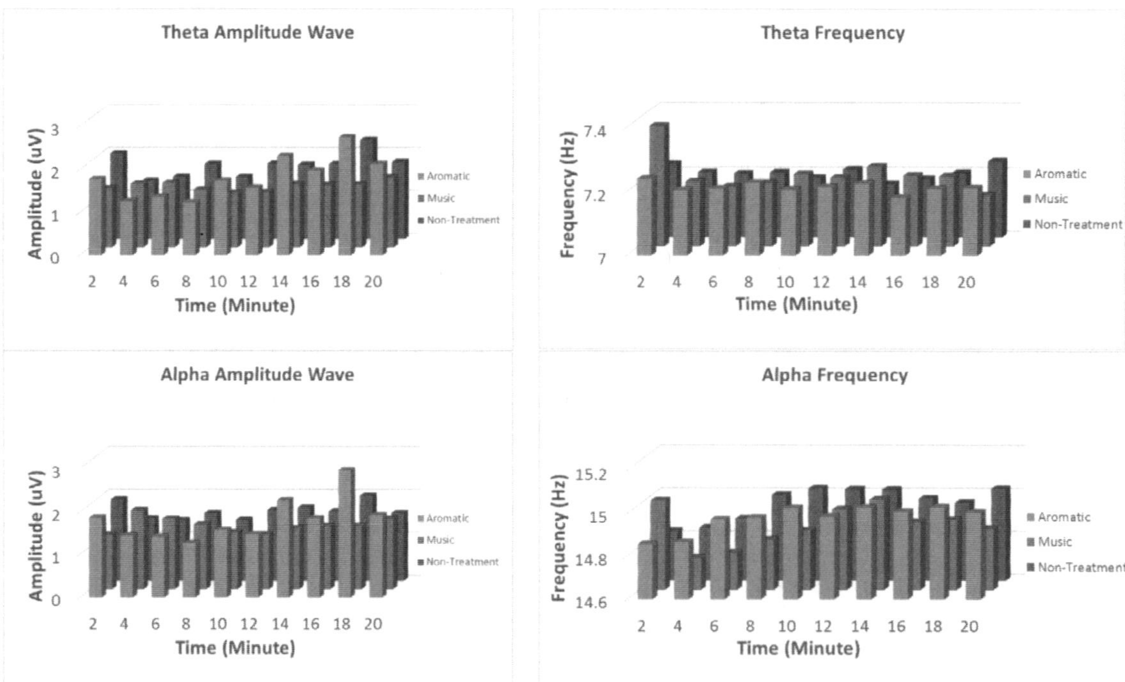

Figure 10. The Result of Average EEG Signal according to band range (alpha and theta).

Table 1. The Result of T test of music and aromatherapy stimuli to alpha and theta wave.

t	Alpha Wave		Theta Wave		Alpha Frequency		Theta Frequency	
	Aromatic	Music	Aromatic	Music	Aromatic	Music	Aromatic	Music
2	0.937	0.223	0.710	0.304	0.616	0.222	0.791	0.079
4	0.872	0.527	0.580	0.756	0.824	0.338	0.955	0.991
6	0.952	0.519	0.729	0.882	0.420	0.108	0.760	0.771
8	0.182	0.815	0.148	0.247	0.590	0.228	0.356	0.909
10	0.277	0.575	0.128	0.431	0.891	0.168	0.238	0.154
12	0.208	0.022	0.396	0.004	0.327	0.156	0.892	0.882
14	0.065	0.034	0.268	0.122	0.915	0.957	0.041	0.050
16	0.083	0.078	0.178	0.061	0.594	0.558	0.988	0.300
18	0.403	0.091	0.548	0.051	0.357	0.744	0.777	0.529
20	0.129	0.784	0.328	0.551	0.559	0.170	0.446	0.030

5. Conclusions

The Experiments to test the effect of music stimulation and aromatherapy on sleep quality have been carried out. The T test gives the result that the existence of music stimuli is significantly greater than aromatherapy.

Acknowleadments

This research was supported by the thematic program through the Bandung Technical Management Unit for Instrumentation Development (Deputy for Scientific Services) funded by Indonesian Institute of Sciences, Indonesia.

References

[1] Kurnia D K, Wardhani V and Rusca K T 2009 *Journal of Medical* Brawijaya **XXI** 83
[2] Sowndhararajan K and Kim S 2016 *Scientia Pharmaceutica* **84** 724
[3] Karadag E, Samancioglu S, Ozden D and Bakir E 2017 *Nursing in Critical Care* **22** 105
[4] Li X, Chen X, Yan Y, Wei W and Wang Z J 2014 *Sensors* **14** 12784
[5] Carlson N R 2013 *Physiology of Behavior* (USA: Pearson Education Inc. University of Massachusetts Amherst)
[6] Pickles Cf, James O 2012 *An Introduction to the Physiology of Hearing (4th ed.)* (Bingley UK: Emerald Group Publishing Limited p 215)
[7] Prochazka A, Kukal J and Vysata O 2008 *Proceeding of IEEE ISCCSP* p 719
[8] Maan M Shaker 2007 *International Journal of Medical, Health, Biomedical, Bioengineering and Pharmaceutical Engineering* **1**
[9] Murugappan M, Nagarajan R and Yaacob S 2010 *Journal of Medical and Biological Engineering* **31** 45

Observing micro-meter object using photonic jet

A Abdurrochman*, D Naufal, D S Sofiati and A Bahtiar

Department of Physics, Faculty of Mathematics and Natural Sciences, Universitas Padjadjaran, Jl. Raya Bandung-Sumedang km 21 Jatinangor, Sumedang 45363, Indonesia

* a.andri@phys.unpad.ac.id

Abstract. Applying photonic jet for daily basis is getting near. Many potential applications have been proposed. Previously, we reported photonic jet etching using near-infrared lasers. Now we are reporting our works on applying photonic jet in observing micro-meter object. Our experiments laying a microsphere on micro-meter object confirm that the observing object is magnified by photonic jet. We also propose a simple practical method on how to apply photonic jet on an objective lens of microscope so it is enable to observe smaller object.

1. Introduction

The photonic jet (PJ) is non-evanescent beam generated at the shadow side of a dielectric micro-cylinder or microsphere that is larger than the wavelength (λ_0) of incidence light [1, 2]. The beam is highly concentrated with the full-width at half-maximum (FWHM) that can go beyond the optical diffraction limit. These two main features are mostly related to the PJ applications. For 35 μm diameter glass microsphere in free-space, the highest intensity is 729 times of the initial intensity with the FWHM around λ_0 [3]. There are many potential application of the PJ, including: nanoparticle detector [4-7], data storage [8, 9], waveguide [10-12], laser surgery [13-15], fluorescence enhancement [16-20] and microscopy [1, 21-23].Our previous works on PJ are concerning the PJ micro-etching by nanosecond near-infrared lasers. We started with PJ micro-etching using microspheres [3, 24, 25] then continued with PJ micro-etching using shaped tip optical fibers [26-30]; therefore we have a patent in device of laser processing and work station comprising the device (*Dispositif de traitement laser et station de travail un tel dispositif*) published in Jan 5, 2017 [29].

In fact that photonic nano-jet was coined in proposing a new microscopy technique [1] encourages us engaging research of photonic jet microscopy. In this paper we are reporting the upgrading of a common optical microscope to have higher magnification by applying photonic jet. Some attemps have been done [23, 31], but we designed the system in free-space (figure 1). The objective lens, microsphere and object are not immerse in any liquid medium; so it will be operationally simpler, and it will not kill or ruined the object if it is a living biological object. The objective lens we used are 10×, 20× and 40× with numerical aperture NA= 0.75. The glass microspheres we used are made of glass-lime (refractive index, n=1.5) with three groups of diameter, i.e.: 38-45μm, 75-90μm and 111-125μm. And, the observed object we used is a stage micrometer with the smallest division of 0.01 mm. The data we obtained are the ratios of stage micrometer sizes with and without the microsphere.

IOP Conf. Series: Journal of Physics: Conf. Series **1080** (2018) 012051 doi:10.1088/1742-6596/1080/1/012051

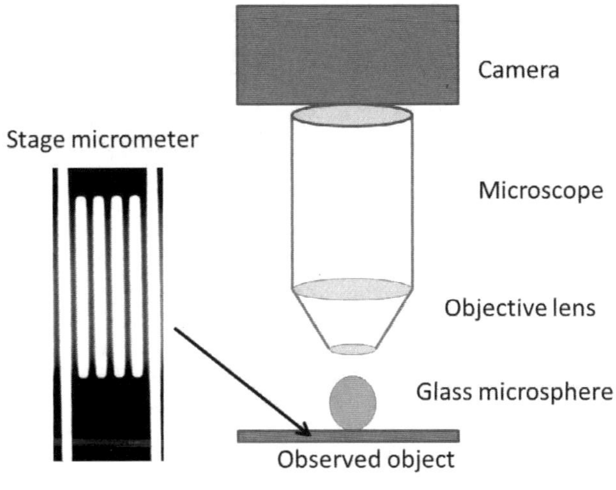

Figure 1. The experiment setup: observing micro-meter object under optical microscope applying photonic jet.

2. Experiment observation and discussions

And so, we laid some glass microspheres on a stage micrometer which then put on microscope stage to be observed using different objective lenses. Figure 2(a), (b) and (c) show some glass microspheres on micrometers observed using three different objective lenses: 10×, 20× and 40×. Here we measured the diameters of glass microspheres (*D*), the widths of stage micrometer lines (*h*) and the divisions of these lines (*h*); also the positions of microscope stage to the objective lens (*s*). Afterward, the tuning button was turned to acquire the images of stage micrometer lines under the glass microspheres. Figure 2(d), (e) and (f) show some images of stage micrometer lines under the glass microspheres with the diameter range of 111-125 μm using three different objective lenses: 10×, 20× and 40×. Here we measured the widths of stage micrometer lines (*h'*) and the divisions of these lines (*h'*); also the positions of microscope stage to the objective lens (*s'*). The whole data we have acquired from these observations are presented in table 1.

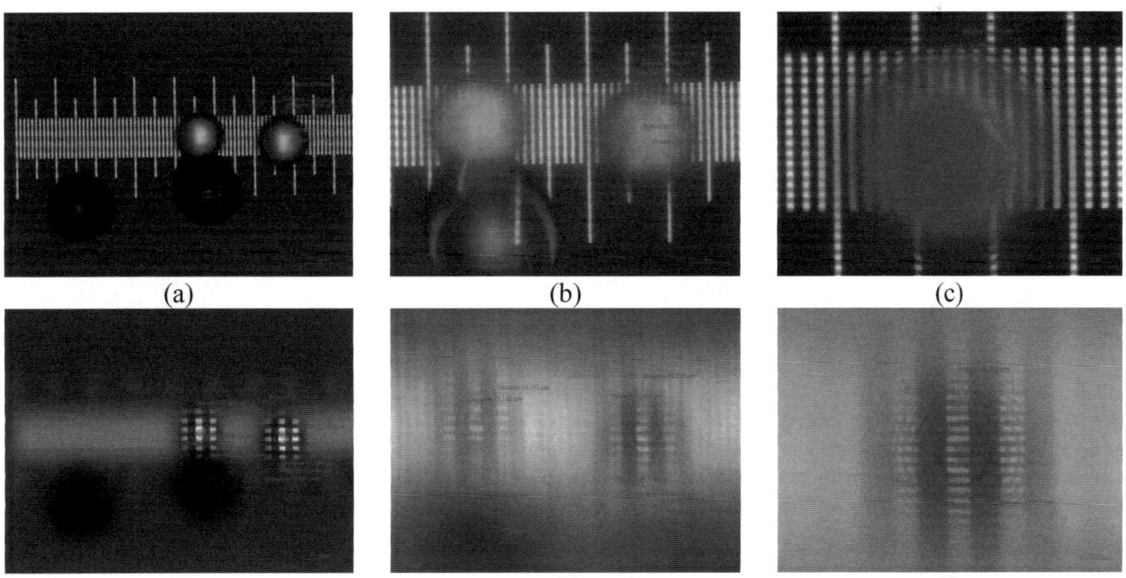

(a) (b) (c)

(d) (e) (f)

Figure 2. The original width of stage micrometer observed using objective lens: (a) 10×, (b) 20× and (c) 40×. The image of stage micrometer observed under microspheres (diameter range of 111-125μm) using objective lens: (d) 10×, (e) 20× and (f) 40×.

Table 1. Magnification of microsphere as the ratio of image width obtained due to microsphere and stage micrometer width.

No	Microsphere diameter (μm)		Width of stage micrometer line (μm)		Position of microscope stage (mm)	
	Range	$\bar{D}$	$\bar{h}$	$\bar{h}'$	$\bar{z}$	$\bar{z}'$
1	111-125	120	4.84	15.03	23.96	24.10
2	75-90	88		15.97	23.93	23.05
3	38-45	40		16.96	23.92	23.97

We can see clearly from table 1 that the microsphere increases the image width of stage micrometer, i.e. at the first row: the image width of stage micrometer is increasing from $\bar{h}$ = 4.84 μm to $\bar{h}'$ = 15.03 μm; which is magnified 3.1 times. This magnification is known as longitudinal magnification. Nevertheless, it is also increased the working distance of microscope from z to z'. These further working distances due to additional microspheres is known as lateral magnification, which makes the microsphere behave as a positive lens. In addition, the shifting of microscope stage (z'- z) is about in the range of microsphere size. It indicates that the focal point of PJ is at a sub-micron away from the shadow-side of microsphere which ought to be smaller than the microsphere size. So, the microsphere must be laid directly on the stage micrometer. Prior to this experiment above, we put a glass cover-slide (thickness of 130 – 160μm) between stage micrometer and microsphere. But, we did not obtain any image as large as figure 2. Figure 3 shows a submicron focal distance of microsphere that is at the order of wavelength, so that our experiment is of the mesoscopic region.

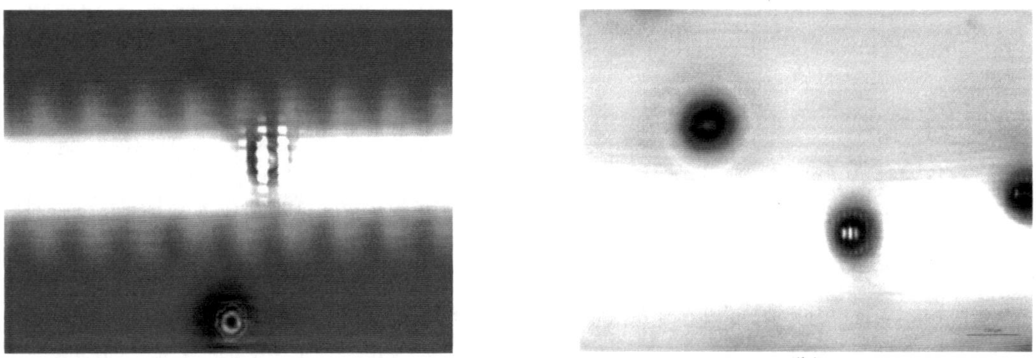

(a) (b)

Figure 3. The image of stage micrometer is observed under a microsphere (a) with and (b) without a glass cover-slide between them.

Also, figure 3 gives us a referential notice if there must not be any dense media (such as a glass slide), so the object can be propagated closer to the microsphere since it has a submicron focal distance. This must be the reason why most researcher hovering the microsphere over the object [23, 31].

Our experiment is not of macroscopic regime; since the visible light wavelength we used is not much smaller than the size of the object. While, in optical geometric analysis the objects are commonly much smaller than the visible light wavelength. This makes a simply optical geometric theory is not suitable assessing our experiments. However, we may use some models in geometrical optics, i.e. by considering the microsphere as a lens or a ball or a microsphere. And we find that the optical model proposed by Li [31] is the most satisfied model in accordance to our experiment results (the last column of table 2) but not for the smallest diameters *D*.

Table 2. Image magnifications results from experiments is compared to geometrical optics model of Li [31].

Microsphere diameter $\bar{D}$ (μm)	**Image magnification from experiments, $\bar{M}_{\mu s}= \|h'/h\|$**	**Image magnification from geometrical optics model, $\bar{M}$**
120	3.1	3.11
88	3.3	3.30
40	3.5	5.84

There is no mathematical models for near field magnification with field singularities [32]. Since our experiment is of mesoscopic system, the assessment must be considered as electromagnetic. Therefore the magnification of microsphere can be written based on maximal field enhancement for the constant of $\beta \approx 0.34$ [32] as:

$$M \approx (I_{max}/I_0)^{\beta} \quad (1)$$

Therefore, it is compulsory to conduct electromagnetic calculation in order to acquire the best satisfying agreement of our experiment results.

3. Conclusion and perspectives

Our experiment observing micrometer object using photonic jet has shown the potential application of photonic jet in optical microscope. Nevertheless, it also gives a notified if optical geometry cannot elaborate our results, so that electromagnetic theories may have the capability.

Acknowledgment
This research is funded as a scheme of Riset Fundamental Universitas Padjadjaran in 2017.

References

[1] Chen Z G, Taflove A and Backman V 2004 *Optics Express* **12** 1214
[2] Lecler V, Takakura Y and Meyrueis P 2005 *Optics Letters* **30** 2641
[3] Abdurrochman A *et al.* 2014 *Laser Sources and Applications II* **9135** 913523
[4] Li X, Chen Z, Taflove A and Backman V 2005 *Optics Express* **13** 526
[5] Li X, Chen Z G, Siegel M P, Taflove A and Backman V 2005 *Nanobiophotonics and Biomedical Applications II* **5705** 14
[6] Heifetz A, Huang K, Sahakian A V, Li X, Taflove A and Backman V 2006 *Applied Physics Letters* **89**
[7] Heifetz A, Simpson J J, Kong S- C, Taflove A and Backman V 2007 *Optics Express* **15** 17334
[8] Kong S C, Sahakian A, Taflove A and Backman V 2008 *Optics Express* **16** 13713
[9] Kong S C, Sahakian A V, Heifetz A, Taflove A and Backman V 2008 *Applied Physics Letters* **92** 211102
[10] Kapitonov A and Astratov V N 2007 *Optics Letters* **32** 409
[11] Halaq H, Lecler S, Takakura Y, Robert S, and Sauviac B 2010 *Waveguide driven photonic jet: FEM, issues and perspectives* (Presented at the Workshop EM IEEE 2010, Saint-Malo, France)
[12] Astratov V N, Darafsheh A, Kerr M D, Allen K W and Fried N M 2010 *PIERS Online* **6** 793
[13] Allen K W *et al.* 2013 *Transparent Optical Networks (ICTON), 15th International Conference*
[14] V. N. Astratov *et al.* 2010 Photonic nanojets for laser surgery. SPIE-Newsroom.
[15] Darafsheh A, Fardad A, Fried N M, Antoszyk A N , Ying H S and Astratov V N 2011 *Optics Express* **19** 3440

[16] Aouani H, Deiss F, Wenger J, Ferrand P, Sojic N and Rigneault H 2009 *Optics Express* **17** 19085
[17] Lecler S, Haacke S, Le C N, Cregut O, Rehspringer J L and Hirlimann C 2007 *Ultrafast Phenomena XV: Springer Series in Chemical Physics* **88** 181
[18] Gérard D *et al.* 2008 Optics Express **16** 15297
[19] Lecler S, Haacke S, Lecong N, Cregut O, Rehspringer J L and Hirlimann C 2007 *Optics Express* **15** 4935
[20] Lecler S *et al.* 2006 *Solid State Lasers and Amplifiers II: Proceedings of the Society of Photo-Optical Instrumentation Engineers (Spie)* **6190** R1900
[21] Chen Z, Taflove A, Li X and Backman V 2006 *Optics Letters* **31** 196
[22] Hoang T X, Duan Y, Chen X and Barbastathis G 2015 *Optics Express* **23** 12337
[23] Lee S, Li L, Wang Z, Guo W, Yan Y and Wang T 2013 *Applied Optics* **52** 7265
[24] Abdurrochman A, Lecler S, Mermet F, Tumbelaka B Y, Serio B and Fontaine J 2014 *Applied Optics* **53** 7202
[25] Abdurrochman A, Lecler S, Mermet F, Fontaine J and Tumbelaka B Y 2014 *Photonic jets for micro-etching* (International Conference on Physics (ICP 2014))
[26] Zelgowski J *et al.* 2016 presented at the *SPIE-Photonic Europe*, Brussels
[27] Abdurrochman A, Zelgowski J, Lecler S, Mermet F, Tumbelaka B and Fontaine J 2016 *AIP Conference Proceedings* **1712** 030001
[28] Zelgowski J, Abdurrochman A, Mermet F, Pfeiffer P, Fontaine J and Lecler S 2016 *Optics Letters* **41** 2073
[29] Lecler S, Abdurrochman A, Mermet F and Fontaine J 2017 *Dispositif de traitement laser et station de travail comportant un tel dispositif* (ed: Google Patents)
[30] Abdurrochman A, Lecler S, Zelgowski J, Mermet F, Fontaine J and Tumbelaka B Y 2017 *IOP Conference Series: Materials Science and Engineering* **196** 012043
[31] Li L, Guo W, Yan Y, Lee S and Wang T 2013 *Light: Science &Amp; Applications* **2** e104,
[32] Wang Z *et al.* 2011 *Nature Communications* **2** 1211

Institute of Physics
Dirac House, Temple Back
Bristol BS1 6BE UK

ISSN: 1742-6588
ISBN 978-1-5108-7202-8

9 781510 872028